Advances in Phytomedicine Series, Volume One

Editor: Maurice M Iwu

Ethnomedicine and Drug Discovery

Advances in Phytomedicine 1

Ethnomedicine and Drug Discovery

Maurice M Iwu
Executive Director, Bioresources Development and Conservation Programme,
Silver Spring, MD 20902, USA; InterCEDD, Nsukka, Nigeria

Jacqueline C Wootton
Director, Alternative Medicine Foundation, Bethesda, MD 20814, USA

2002
ELSEVIER
Amsterdam – London – New York – Oxford – Paris – Shannon – Tokyo

ELSEVIER SCIENCE B.V.
Sara Burgerhartstraat 25
PO Box 211, 1000 AE Amsterdam, The Netherlands

First edition 2002

Library of Congress Cataloging in Publication Data.

Ethnomedicine and drug discovery / [edited by] Maurice M. Iwu, Jackie Wootton. – 1st ed.
p. ; cm. – (Advances in phytomedicine ; 1)
ISBN: 0–444–50852 X (HC : alk. paper)
1. Pharmacognosy. 2. Ethnobotany. 3. Medicinal plants. 4. Materia medica, Vegetable.
I. Iwu, Maurice M. II. Wootton, Jackie. III. Series.
[DNLM: 1. Ethnobotany. 2. Medicine, Traditional. 3. Medicine, Herbal. 4. Plants,
Medicinal. QV 766 E84 2002]
RS160.E847 2002
615'.321 – dc21   2001055558

http://www.elsevier.com/locate/isbn/044450852X

⊗   The paper used in this publication meets the requirements of ANSI/NISO Z39.48-1992 (Permanence of Paper).
Printed in Great Britain.

Contents

Preface to the Series

The systematic study of herbal medicinal products and the investigation of the biologically active principles of phytomedicines, including their clinical applications, standardization, quality control, mode of action and potential drug interactions have emerged as one of the most exciting developments in modern therapeutics and medicine. Studies in phytomedicine have moved from purely descriptive analytical studies to conceptual inquiries on the pharmacodynamic advantages and limitations of plant medicines for the treatment of moderate or moderately severe diseases and prevention. Healthcare practitioners and medical scientists have come to accept herbal medicinal products as drugs that are different from the pharmacologically active molecules that they may contain. Several comparative clinical studies have been published to show that these plant medicines could have full therapeutic equivalence with chemotherapeutic agents, while retaining the simultaneous advantage of being devoid of serious adverse effects. Developments in molecular biology and information technology have now made it possible for us to begin to understand the mechanism of action of many herbal drugs and the associated phytomedicines, which differ in many respects from that of synthetic drugs or single chemical entities. Herbal medicinal products are now generally available in both industrialized countries and traditional societies. With the current lack of standardization and regulation of herbal products, it is important to develop common criteria for judging safety and efficacy of phytotherapeutic agents.

This 'new' science demands different approaches to the classical methods of drug analysis, dosage formulations, manufacturing and claims substantiation. The therapeutic response observed with most herbs and phytomedicines are often not fully explainable using the currently available methods. Their activity usually characterized as polyvalent and interpreted as an aggregate or additive outcome of several constituents in the plant medicines are subject of intense pharmacological studies. In most cases, a rationale does not even exist for the observed pharmacodynamic effects of very low doses of phytomedicines after prolonged or long-term application.

The public press is replete with lay information and claims about the use of herbal remedies, however, there is scarcity of scientifically accurate reviews and guides on the efficacy and safety of plant medicines. The time therefore seemed ripe to broaden the communication on the use and benefits of phytomedicines as safe and useful natural health care products aimed at the health professionals and scientists. It is also important to provide a broader dissemination of the extant scientific literature on phytomedicines, to enable the conventional medical community to fully appreciate the fact that plant extracts specifically, and natural products generally, offer valuable and needed benefits in the treatment and prevention of diseases, especially for conditions where there is no effective or generally acceptable drugs.

Although there are many papers published yearly on the use and analysis of plants as sources of biologically active molecules and some published materials on the use of plants as medicinal substances, volumes specifically addressing the needs of scientists and clinicians in the use of herbal products and standardized extracts as medicinal agents are far-in-between. We considered it therefore appropriate to fill this gap by producing an entirely new multi-volume series on the sourcing, selection, standardization, safety and clinical application of herbal medicinal products. As the name of the series implies, emphasis will be on those herbal medicinal products that are well characterized, standardized and substantiated as phytomedicines. The series will also provide timely reviews on the industrial production, regulatory and policy issues related to the use of phytomedicines. Although the literature in this field is evolving so rapidly that books on the subject become obsolete soon after they leave the press, the volumes in this series will aim at capturing the fundamental framework of each topic while remaining thoroughly up-to-date and comprehensive in scope.

The aim of this series is to present to the scientists and clinicians the state of current knowledge in various fields of phytomedicine research, development and use. The approach is to provide the historical background to each topic, discuss methodological issues and illustrate current trends with case studies and critical examples, and when ever possible, the authors will indicate those plant medicines that are available for immediate use in clinical settings. The series is not intended to serve or replace the many excellent journals in this field but it will rather attempt to distill information from primary references and introduce elements of medicinal plants research and development that are in transition from speculative knowledge to standard practice. *Advances in Phytomedicine* is also not meant to be a textbook of the various topics covered in the series. It will, however, provide a guide to specialized articles and books on the topics that are relevant to scientific research and development of plant medicines, as well as information on the regulatory issues, clinical trials and application of phytomedicines for healthcare.

Advances in Phytomedicine will therefore serve as a platform for reviewing recent developments in the use of herbal medicinal products. The coverage will include reviews of studies and use of all plant medicines, phytotherapeutic agents, nutraceuticals, plant cosmetics and therapeutically important molecules derived from these plant medicines. The first volume in this series has been devoted to exploring the ethnomedical approaches to drug discovery. It is indeed a very important starting point in addressing the relationship between plants and human health. Subsequent volumes will deal with other aspects of the use of herbal medicinal products. Selection of subjects will be through consultations with experts in the various fields of interest. We shall be guided by the principles of the so-called 6S, that is, herb selection, sourcing, structure, standardization, safety and substantiation. Acknowledged experts and authorities in the various aspects of phytomedicine will be invited to assemble and edit specific volumes in the series.

The series is the outcome of extensive consultations among several biomedical scientists and clinicians who participated in the workshops and conferences organized by the Bioresources Development and Conservation program (BDCP) on related subjects. I am immensely grateful to these colleagues for their support in developing the original concept. Many thanks go also to Ms Kim Briggs and Ms Joke Zwetsloot of Elsevier Science for their suggestions and help in producing this series. I am indebted to my colleagues at BDCP and the International Centre for Ethnomedicine and Drug Development for their contribution; and to my wife, Kate for her love and support. I acknowledge the International Cooperative Biodiversity Group (ICBG) of the Fogarty International Centre, United States National Institutes of Health for providing financial support to my research group at the Walter Reed Army Institute of Research and BDCP.

Maurice M. Iwu
M.Pharm., Ph.D.
Séries Editor

Preface

The first book in the series, Advances in Phytomedicine, provides an interface between ethnomedical approaches to new drug discovery and advances in biotechnology, molecular and clinical sciences that have made it increasingly feasible to transform traditional medicines into modern drugs. The contributions are drawn primarily from an international conference with the same title, Ethnomedicine and Drug Discovery, held in Silver Spring, Maryland, USA, November 3–5, 1999. Speakers and delegates attended from 17 different countries, spanning 5 continents. Different systems of traditional medicine were represented: African, North American Indian, South American, Ayurvedic, Chinese, and the European Eclectics. This book, however, is not a conference proceedings. Not all speakers from the conference contributed papers, and some additional papers were specifically commissioned to ensure full coverage of all aspects of the topic. This volume represents a rare collection of contributions from eminent scholars and experts in this field.

Professor Iwu's introduction, Therapeutic Agents from Ethnomedicine, unpacks the conceptual ambiguities underlying ethnomedical and ethnobotanical research. Herbs contain pharmacodynamic compounds that can be standardized and incorporated into modern Western phytomedicines or nutraceuticals, based on the drug model of isolating a single active ingredient. This is a standard approach to new drug design, but quite different from the whole herb or multi-component herbal traditions from which the natural products derive, where the components are known to act synergistically and often non-specifically. Such herbal traditions are neither unscientific nor static but are based on systematic experimentation and, particularly in the Asian systems, careful documentation. Iwu also draws an important distinction between an eclectic approach to integrative medicine that allows pragmatic borrowing between conceptually distinct systems and recognition of the parallel development of both traditional systems of medicine and modern western medicine as different and largely incompatible. Most of the following papers can be seen to fall into one of the two approaches.

Cragg (Chapter 1) sets the scene with a rich history of drugs developed from nature by successfully isolating the main active ingredient. While so much of the world's biodiversity remains unexplored the potential for future drug development is enormous. Harvey's paper (Chapter 2) discusses a way to make the process more efficient with high-throughput screening, while Carlson (Chapter 3) suggests multidisciplinary teams of researchers. Chapter 17 on *Garcinia kola* by Iwu et al. provides a case study for discussing many of the sensitive issues involved in developing a modern ethnobotanical product.

Screening of plants for development into pharmaceuticals is usually based on local community knowledge. This raises a host of legal and ethical issues surrounding intellectual property rights. Moran (Chapter 16) provides an overview of the

Convention on Biological Diversity. The underlying issues are discussed by Lettington, Gollin, and Guerin-MacManus et al. (Chapter 7, 18–19). The situation is compounded by problems of sustainability and how to restore populations of medicinal plants, Cech (Chapter 8). Gericke (Chapter 13) gives a Southern African perspective on the failure of authorities to validate local knowledge. Schuster in Chapter 14 outlines the development of antimalarial agents based on leads from traditional medicine. The paper by Balick et al. (Chapter 23) describes the ongoing Belize ethnobotany project that was started in 1988 and encompasses all the interwoven issues of conservation, declining local knowledge, and benefit sharing. A further case study of the collaboration between the Buganda Traditional Healers Association and Shaman Pharmaceuticals, is provided by Nelson-Harrison et al. (Chapter 24).

Some researchers from developing countries call for raising the profile of their contribution to drug development by increased training and expertise in medicinal chemistry, Efange (Chapter 5), or by using methods of bioprospecting for economic development, Onugu (Chapter 8). One example of an herbal drug developed in Nigeria used for sickle cell anemia and derived from ethnomedicine is discussed by Okogun. In contrast, Noé (Chapter 10) celebrates the traditional healing systems that incorporate spiritual and cosmological belief systems. Such whole systems cannot be incorporated into the framework of modern western medicine. The conference incorporated a Festival of Living Culture with a celebration of plants, music, dance, and incantations reminding delegates that science, spirit, art, dance and ritual are integral to the world's main healing traditions.

Borrowing of traditions cuts both ways. Obijiofor (Chapter 6) develops a model for integrating the best of western biomedicine and traditional ethnomedicine in South-eastern Nigeria. Integrative approaches to modern medicine usually involve a distinction between pharmaceutical drugs and dietary supplements and how to regulate them. Gruenwald (Chapter 20) provides an explanation of the regulatory situation in Europe; In Chapter 21, Osuide discusses the regulatory system in Nigeria; and Srinivasan (Chapter 22) provides the US perspective. Wootton (Chapter 4) presents the HerbMed database that gives access to neutral categorized scientific data on which regulatory and drug development decisions are based.

The ultimate goal should be to obtain therapeutic agents from ethnomedicines that are safe, consistent, and effective remedies based on whole herbs, and standardized methods of cultivation and preparation. Meserole (Chapter 15) elaborates on the concept using the example of health foods in anti-aging therapy while Elisabetsky (Chapter 11) proposes ways to develop truly innovative paradigms in drug usage and development using psychotropic drug actions as a model. Finally, Iwu in the concluding chapter, weighs up the strengths and limitations of an ethnobotanical approach to drug discovery.

We were privileged to have Representative, John Edward Porter, give the opening keynote address at a meeting on Capitol Hill. Mr Porter, since retired from office, has been a great champion of science funding and was a key driving force behind the International Cooperative Biodiversity Group grant program in the United States. Special thanks are due to the staff of the Bioresources Development and

Conservation program (BDCP), in particular Dr Magnus Azuine, the conference assistant, who worked indefatigably to ensure the conference ran smoothly. The team were able to combine a cheerful, relaxed atmosphere with sharp academic rigor and efficient organization. It was a great pleasure to be involved in all aspects of the conference organization, the conference sessions and exhibits, and as joint editors of this volume.

We would like to pay tribute to the many individuals and groups who have made this volume and the preceding event possible: Chris Okunji, Angela Duncan Diop, Chioma Obijiofor, Tony Onugu, Sam Iwu, Baljit Wadhwa and to other members of the organizing committee: Alice Clark, Ph.D.; University of Mississippi; James Miller, Ph.D., Missouri Botanical Garden; Katy Moran, the Healing Forest Conservancy; and to Mendy Marsh, the editorial assistant who collected, collated and copy-edited all the papers for this volume. We are grateful to Kim Briggs, Joke Zwetsloot and Anthony Prukar of Elsevier Science for their invaluable assistance and hard work in producing the book. We are indebted to the Ford Foundation, whose generous financial support of the International Conference on Ethnomedicine and Drug Discovery made this volume possible.

Jacqueline Wootton
Maurice M. Iwu

Iwu and Wootton (eds.), Ethnomedicine and Drug Discovery

Introduction: therapeutic agents from ethnomedicine

MAURICE M IWU

Abstract

Ethnomedicine covers healthcare systems that include beliefs and practices relating to diseases and health, which are products of indigenous cultural development and are not explicitly derived from a conceptual framework of modern medicine. Ingredients used in the preparation of ethnomedical remedies provide attractive templates for the development of new pharmaceutical products. They contain compounds with pharmacological activities, which could be transformed into modern therapeutic agents and personal care products. Herbal medicinal plants used in ethnomedical practice are much more than vessels that contain lead compounds for drug development. They are medicinal agents in their own right, even as crude unprocessed materials. They are also considered in some settings as possessing vital healing energy that exceeds the measurable effects of their physical properties. Furthermore, they can be developed into dietary supplements, nutraceuticals and phytomedicines with precisely defined characteristics and consistent quality but with pharmacodynamic properties that are different from those of pharmaceuticals containing isolated single chemical compounds. The ultimate objective in drug discovery and development should be the production of safe and effective remedies, not the introduction of elegant molecular entities into medicine often without discernible therapeutic advantages over the traditional formulations.

Keywords: *ethnomedicine, healing remedies, drug discovery, phytomedicines*

I. Healing and traditional knowledge

Ethnomedicine encompasses the use of several health-promoting cultural practices and/or the use of minimally processed naturally occurring products for the prevention and treatment of diseases, as well as for the maintenance of optimal physical and emotional health. These indigenous or culturally based forms of medicine have their origin in antiquity, but they are not ancient medicine, so the use of the term 'traditional' to describe ethnomedicine may be misleading. As has been noted by Barsh,[1] the term 'traditional' may erroneously imply the repetition, from generation to generation, of a fixed body of data, or the gradual, unsystematic accumulation of new data. This is hardly the situation in ethnomedicine. On the contrary, ethnomedicine is based on careful observation by healers in a given generation of indigenous people. It is the healers' duty to compare their personal experiences with what they have been told by their teachers and neighbors, conduct experiments to test the reliability of their knowledge, and exchange their findings. As Barsh[1] remarked, 'All "tradition" in actuality is continually undergoing revision. What is "traditional" about traditional knowledge is not its antiquity but the way in which it is acquired and used, which in turn is unique to each indigenous culture. Much of the knowledge is actually quite new, but it has a social meaning and legal character, entirely unlike the knowledge indigenous peoples acquire from settlers and industrialized societies.' The use of the term 'traditional medicine' to describe ethnomedicine in this volume should be understood to denote merely the historical and cultural context of the origin of the healing systems.

Ethnomedicine covers a very wide spectrum, which may be classified into two types, the *personalistic* systems, where supernatural causes ascribed to angry deities, ghosts, ancestors and witches predominate, and the *naturalistic* systems, where illness is explained in impersonal, systemic terms.[2] The personalistic system appears to predominate (although not to the exclusion of the naturalistic explanations) in the traditional medical systems of native America, parts of China, South Asia, Latin America and most of the communities in Africa. Naturalistic explanations (also not to the exclusion of personalistic causation) predominate in *Ayurveda*, *Unani*, *Khampo* (Japan) and traditional Chinese medicine. In the later types, part of the belief is that intrusion of heat or cold into, or their loss from, the body upsets its basic equilibrium; that is, the balance of humors, or of the *dosha* of *Ayurveda*, or the *yin* and *yang* of Chinese medicine, and these must be restored if the patient is to recover.

In the personalistic system, as exemplified by the traditional African system of medicine, healing is concerned with the utilization of human energy, the environment, and the cosmic balance of natural forces as tools in healing. In the African world, the natural environment is a living entity, whose components, the land, sea, atmosphere, and the faunas and floras, are bound to man in an intrinsic manner. Plants therefore play a participatory role in healing. A healer's power is not determined by the number of efficacious herbs they know, but by the magnitude of their understanding of the natural laws. The ability of the healer to utilize these natural laws for the benefit of his patient and the whole community is the ultimate measure of his success. Treatment therefore is not limited to the sterile use of different leaves, roots, fruits, barks, grasses and various objects like minerals, dead insects bones, feathers, shells, eggs, powders, and the smoke from different burning objects for the cure and prevention of diseases. If a sick person is given a leaf infusion to drink, they drink it believing not only in the organic properties of the plant but also in the magical or spiritual force imbibed by nature in all living things and also the role of his ancestors, spirits and gods in the healing processes. The patient also believes in the powers of the incantation recited by the healer and assists them in the designation of the ingredients of the remedies given, from mere objects into healing tools. The patient is also an active participant in the art of healing, not a passive subject of therapy.

According to the World Health Organization, a large segment of human population still depends on traditional medicine, or the so-called alternative medicine, as the preferred form of healthcare, even with improved access to modern hospital medicine and the spectacular advances in molecular biology and physiological chemistry that have greatly enhanced our understanding and treatment of diseases. Traditional medical systems such as meditation, acupuncture, relaxation, training, hypnosis, and energy healing are becoming increasingly popular as alternative or complementary techniques to Western scientific medicine. Herbal-based therapies, once used only in traditional medical systems are now recommended for the treatment of several degenerative disorders and chronic conditions where modern pharmaceutical agents have proved inadequate. The acceptance of these techniques as standard healthcare options has posed serious conceptual problems in the development of public-health programs that are responsive to the real needs of the population, with a tremendous impact in the cost of healthcare interventions, preventive medicine and self-healing.

With globalization, major changes are occurring in the practice of traditional medicine and the role of related therapeutic agents in healthcare delivery. It is changing from a highly personalized form of healing to generic health intervention systems often administered outside its cultural base and for conditions that were neither intended or envisaged by the healers. There is also a shift from dealing with a 'patient' to catering for faceless 'consumers' – often through layers or chains of commercial intermediaries. With an ever-increasing clientele, practitioners of traditional medicine and interested scholars must format a strategy that will make traditional medicine available to a wider population while preserving the cultural bases of the healing systems. This challenge is being met by two approaches: one is the development of integrated approaches, in which elements of the two systems are adopted in an eclectic manner, and the other is the parallel recognition of both traditional medicine and modern Western medicine as legitimate but different types of healthcare. The ingredients used in traditional medicine, therefore, must be recognized and used in modern medicine, not only as therapeutic agents with verifiable pharmacodynamic properties, but as agents of healing with beneficial effects, even when the precise mode of activity has not been properly understood.

II. Traditional medicine and spiritual healing

One aspect of traditional medicine that has remained difficult to understand by conventional Western medicine is spiritual healing. This is due to the differences in the fundamental concepts of health, diseases and healing between Western societies and traditional cultures. This aspect of traditional medicine deals with healing in which intangible energy is used to restore health. The healing ritual, whether pharmacological or spiritual, only reflects a recognition or identification of the part of the multi-layered matrix of human life and existence that is diseased and where the treatment is being targeted.[3] Proponents of ethnomedicine argue that while pharmacologically active herbs do possess beneficial effects in the treatment of several physical components of existence, they are hardly useful in disorders that involve the core of a person's innermost being and reality, or when the non-tangible self is in disequilibrium, or when a person loses the rapture of being alive and merely exists with their spirit long receded. Traditional healers insist that unless this aspect of healing is acknowledged (even if not accepted), our understanding of traditional medicine will remain flawed because of the inherent limitation in our very method of inquiry. In a limited Western sense, spiritual healing could be described as a method of treating patients by positive mental intentions only.[4] This form of treatment is not limited by time and space, therefore, personal contact is no prerequisite for therapy, and healing sessions are known to have occurred over large distances. Although spiritual healing is practised in nearly all human communities, the subject still remains a complicated undertaking because of the very nature of spirituality and healing. As has been noted,[5] the very process of describing it will inevitably be reductionist, since healing and the response to healing are often 'state-of-mind'-dependent and are not readily accessible to outside observers.

It is generally accepted that evaluation of spiritual healing does not fit readily into the usual methodologies used in either laboratory medicine or clinical research.[6] There is a growing body of evidence to suggest that although it is difficult to know when healing is effective (beyond the determination of obvious physiological differences), it is possible to determine clinical outcomes and even undertake randomized clinical studies to establish the effect of spiritual healing practices.[4,7–9] A recent controlled study of patients treated by spiritual healing concluded that chronically ill patients who want to be treated by distant healing and know that they are treated show a statistical improvement in their quality of life.[4] The study reported by German investigators, Wesendanger, Reuter, and Walach has been described as providing an example of a realistic balance between scientific rigor and clinical relevance, by retaining the essential features of proper experimental comparison (random allocation to treatment and good inferential analysis) and, at the same time, staying close to the actual practice situation.[7]

III. Integration and complementarity

Many practitioners of evidence-based medicine view the increasing recognition for traditional health systems as a failure by modern medicine to satisfy the healthcare needs of the society. Some practitioners of biomedicine even feel threatened by a system, which they view as unscientific and beyond rational categorization. Traditional medicine, they argue, is steeped in spiritual and magical practices based on ritual, whereas biomedicine is derived from the belief in materialism and mechanism anchored on experiments and verifiable theories. By this logic, the two forms of healthcare are seen as direct opposites, systems that cannot collaborate in any meaningful way for the greater interest of the population. The issues are often reduced into two intractable questions: should healthcare providers permit the introduction of healing ceremonies, charms, music, incantations into modern healthcare clinics to free the sick persons from demons and ghosts who have possessed them? Is it ethical to allow costumed medicinemen and medicinewomen with their elaborate rituals, feathers, pipes, drums, beads and egg-shells to further scare and confuse already distraught patients with their mumbo-jumbo, when one should be looking for the mutant gene that may be responsible for the illness and use a laser-directed magic bullet to fix it? The answer in both cases is an unqualified 'yes'. Because it is now obvious that no medical intervention will be complete if it does not recognize and accommodate the sociological context in which the patient, healer and remedy exist. It has been observed that traditional medicine is not an alternative to modern medicine but a complementary healthcare that is effective, beneficial and demonstrable. Although the precise mechanism of the observed effects may still not be properly understood in some cases, therapeutic effectiveness of the ethnomedical approach can only be fully realized through parallel practice and research.[10]

Regardless of the preferred form of treatment, a common underlying tenet is that the selection of the therapeutic agent should be based on solid evidence of *efficacy* and *safety*. The establishment of common parameters of safety and efficacy is not always an easy undertaking given the huge differences in both the fundamental framework of the two systems and the methods of practice.

III.A. Plants as medicinal agents

Plants constitute a major ingredient in the practice of ethnomedicine. The study of the use of such plants for healing by indigenous cultures is part of the new discipline called ethnobotany. Ethnobotany involves the study of human interaction with plants, including those used as medicinal plants. The laboratory study of medicinal plants, however, has proceeded along three main disciplines, viz. botany, chemistry and pharmacology. The botanical investigations have been addressed to problems related to medicinal plant anatomy and physiology and sometimes genetic improvement of the existing plants by breeding. The phytochemical study of medicinal plant appears to be the most prolific of the areas of investigation, as evidenced from the number of new compounds isolated and characterized from plants every day. It is the hope that some of these isolates may possess hitherto unknown biological activity or lead to the recognition of new therapeutic agents; such compounds could also be improved upon by synthesis or provide a model for complete synthesis of novel compounds with biological activity. Researches in medicinal plant chemistry have indeed contributed immensely to science by providing challenges and stimuli for molecular structure elucidation that have enriched and considerably advanced our knowledge of organic chemistry. However, unfortunately, such studies have not contributed much either to the understanding of the activity attributed to various medicinal plants or to the discovery of new therapeutic agents. The pharmacological investigation of plant products has not received the same degree of attention as that shown to plant chemistry and this has led to a situation where valuable leads are not usually followed to their logical conclusion, that is, to the identification of new therapeutic agents, or greater insight into the mechanism of action of known compounds. Furthermore, the discovery of the active constituent of a plant used as a folk remedy is not always in the best interest of the patient. More often than not, such a discovery restricts the availability of the herb as a drug to the general public.

With the current legislation in most countries, a plant used in folk medicine is considered a health food and can be cultivated, sold and consumed without any medical or pharmaceutical supervision, but as soon as the active constituent is established, it has to be subjected to a battery of animal studies before it can be accepted as a drug, and at a higher cost. Courses and books on medicinal plants are accordingly structured along these lines. The few existing publications on the utilitarian aspect of medicinal plants have mainly been from proceedings of international conferences. Such books, while excellent for the expert researchers, are not usually very useful to students and scientists interested in medicinal plants as therapeutic agents. Standard textbooks in pharmacognosy and phytochemistry are concerned with medicinal plants as vehicles or containers of active principles, and not necessarily as medicines.

The emphasis has been on plant products whose efficacy, specificity and mode of action are congruent with the conventions of modern pharmacognosy and pharmacology, and less attention to those plants that do not lend themselves to simple tests to establish their efficacy. It is therefore understandable that available literature on medicinal plants concentrates on plants used for treatment of diseases that display simple and discrete symptoms, and not much is known about plants used in folk medicine for the treatment of relatively more complicated disorders that are characterized by multifactorial, inter-related and sometimes externally indiscernible symptom complexes. It is easy to rationalize the use of certain plants as antimicrobial, anti-inflammatory, analgesic, sedative, purgative and anti-ulcerative agents since appropriate animal models allow for a direct correlation of the laboratory experiments with the clinical situation, such direct association may not be easily formulated for complicated diseases with multiple pathological factors, and medicines used in their treatment have not been investigated extensively.

Contribution from non-Western indigenous therapeutics agents in international medicinal plants trade has been rather minimal because of a poor understanding of the socio-cultural features of the indigenous treatments in the assessment of medicinal plants for therapeutic efficacy. It would be useful to consider folk attitudes on disease etiology and treatment, and how they affect the description of the disorder, methods of treatment and selection of healing remedies. A positive clinical profile of a plant drug is the ultimate criterion for its continued use, not whether or not such a therapy is accommodated by the present state of knowledge in pharmacology and medicinal chemistry. Yet, clinical evaluation is seldom included as a criterion for the selection of plants for drug discovery.

IV. Phytomedicines and nutraceuticals

A noteworthy feature in many ethnomedical systems is the use of food plants as ingredients for drug preparation. It has been established that certain types of food substances not only help the body in providing energy sources and assisting in the body's plastic repairs but are also capable of mediating or interfering in the complicated chemical reactions of life processes. These 'non-nutritional', exogenous biochemicals (alkaloids, flavonoids, terpenes, glycosides) have therapeutic potentials comparable to isolates from non-edible or poisonous plants. It is therefore surprising that in the search for potentially effective medicinal plants, much emphasis is placed on harsh poisonous compounds that fit into modern conception of drugs as necessarily 'Poisons, which in low doses could cure diseases'. In many developing countries, plant-based drugs are a common feature in their therapeutic lists, partly because modern synthetic drugs are beyond the reach of a greater proportion of the people inhabiting most Third World countries. It is most unlikely that this situation will change significantly in the immediate future for two notable reasons: the increased awareness among the leaders of developing countries that adherence to herbal medicines would reduce the heavy financial burden of expensive synthetic drugs, and the world-wide resurgence of interest in herbal remedies for their possible incorporation or adoption into modern medicine.

For people living in industrialized countries, phytotherapy provides a complementary form of medicinal agents and, in certain cases, an alternative to modern orthodox medicine. Microbial fermentation and chemical synthesis of antibiotics and improvement in public health care have led to a decline in human mortality due to infectious diseases caused by bacteria. These infections do not constitute a major challenge to modern therapeutics. The main focus of experimental medicine is presently on metabolic disorders and chronic diseases, as well as those due to viral infections and drug-resistant fungi. Phytotherapy is viewed by many medical scientists as a natural approach for the treatment of such diseases. Curtailment of bacterial infections and improved standards of living have also led to longer longevity: a greater number of people are living to very old age with the problems associated with senescence, which require a different kind of 'drug'. The objective is not to provide medication for the treatment of a diseased body but to maintain optimum health, to restore cell and tissue vitality. The main preoccupation of modern medicine has hitherto been the study of diseased states and the restoration of health. Not much has been done to examine this complex problem of being 'healthy' as a whole. Not even pharmacology, which sees its chief task in creating medicines principally for the treatment of illnesses, has examined this obscure province – of maintaining health or what Brekhman called 'pharmacology of health'.[11] Some herbal preparations such as ginseng, *Garcinia kola, Eleutherococcus senticosus, Uapaca, Sphenocentrum* and Yohimbe are dispensed for the enhancement of life, and to help the body adapt to unfavorable physical conditions. There are aphrodisiacs, somantensic drugs, and substances that produce ecstasy to obliterate the pains and the dull sides of life. With the current desire by most people for a 'return to nature', self-medication would likely be on the increase with the inherent danger of drug abuse and toxicity. Scientific validation and pharmaceutical supervision would guarantee good, safe and consistent medicinal preparations from plants, thus minimizing some of the criticism against herbal treatment. When medicinal plant products are subjected to the rigors of applied science, the experimentation reveals and clarifies the conditions and the pharmacodynamic basis for the use of such plants in medical practice. Such re-evaluations would provide a rationale for the continued use of these medicinal plants. The pharmacological evaluation of some of the phytotherapeutic agents may not be possible using presently available techniques, since the process of extraction, standardization and isolation of active constituents creates an artificial situation not originally present in the intact fresh plant material. Most evaluation of ethnomedical remedies records the adaptive physiological responses of the body to newly created medicinal agents.

Furthermore, it has been argued that conducting animal experiments on preparations of plants such as *Coriandrum sativum, Centella asiatica*, or *Momordica charantia*, which are routinely used by people (and have been for several millennia) as drugs and even as food items, would sound as absurd to people in Africa, Asia and Latin America as conducting similar experiments on carrots, lettuce and cabbage would to people in Europe.[12] The advocates of phytotherapy as a complementary form of medication, particularly in chemotherapeutics, have advanced the following arguments in support of the continued use of plant drugs:

- They provide a desirable therapeutic effect with reduced risk of iatrogenous diseases associated with allopathic medication.
- The dosage form of most herbal preparations, the infusion, possesses an obvious bio-availability advantage over conventional dosage forms; the enhanced solubility and dispersion of active molecules at gastro-intestinal absorption sites reduces significantly the problems and uncertainty encountered in the pharmacodynamic phase of drug therapy.
- They are most suitable for diseases requiring prolonged or long-term medication, e.g. in arthritis, hepato-biliary deficiency, spleen diseases, etc. This is because most plant extracts are commonly used not in massive doses for quick results but in repeated low doses that yield the desired results over long periods of administration.
- There is a reduced chance of secondary drug effects and acute toxicity associated with more active synthetic compounds; combined treatment with synthetic and herbal remedies is said to ameliorate some of the side-effects of the potent synthetic compounds.
- The multi-component feature of phytotherapeutic agents is considered a positive attribute in their use as biodynamic agents since the same extract often includes compounds that have synergistic activity, and in some instances even have components that antagonize the main activity of the plant product, thereby maintaining equilibrium.[13]

It is therefore important to study these phytotherapeutic agents since they play such vital roles in clinical medicine. The study of medicinal plants is also important because of the adverse effect sometimes observed with plant drugs; it is of immense clinical importance to determine the acute and chronic toxicity of plant products used as therapeutic agents whether or not they exhibit the expected pharmacological action.

Relatively little effort is spent on investigation of medicinal plants as sources of drugs when compared to the enormous expenditure on the design of synthetic drugs. It has been argued that apart from the considerable space and labor required to cultivate and harvest sufficient quantities of a plant, one has to deal at the same time with seasonal and climatic variations of the biological activities. Each individual batch has to be controled separately. Natural hazards and political restrictions, e.g. war may disrupt the supply of a natural occurring drug as happened with quinine during the Second World War. The same arguments, however, could be made for food crops, yet man still relies on natural food sources.

In nutraceuticals, the source of plant material used as a health food is as important as the activity claimed. The naturally derived vitamin E, for example, is more valued in the trade than the synthetic equivalent prepared from petroleum. The total sales of this vitamin in the US totaled $33 million in 1995, which represents a doubling of the 1991 value of $16.2 million. The product, which is prepared from soybean, corn, peanut, rapeseed, sunflower and cottonseed, remains in constant short supply world-wide.

The availability of modern scientific methods for the cultivation, selection, manufacture and clinical evaluation of herbal remedies has made it increasingly feasible to transform traditional medicine from an almost invisible trade into a modern industrial enterprise capable of making significant contribution to both healthcare delivery and the economic growth of developing countries. In order to exploit fully the enormous potential that medicinal plants hold, it is necessary to understand the major factors that drive medicinal plants trade. The first is the proper management of genetic resources so that they can be used in a sustainable manner and for the benefit of the greater majority of the population. A key starting point is to conduct a reliable survey of the genetic resources available in the area and to document the ethnomedical information. If properly executed, data-gathering and analysis will assist in the economic mapping of medicinal flora, determining the yield and growth density with a view to possible industrial application. Such studies will also indicate the species under pressure and in need of preservation and/or conservation. The second factor is the appreciation that an adequate technological infrastructure is necessary for the conversion of crude plant materials to standardized, effective and safe dosage forms. It is not enough to acquire the scientific know-how without building a supporting technological infrastructure. The third factor is an intimate knowledge of the market place and the ability to create commercial value from what may appear as purely academic or scientific results. The fourth factor is establishment of effective and appropriate regulation that will govern the production and marketing of medicinal plants.

It is estimated that less than 10% of the world's genetic resources have been studied seriously as sources of medicines. Yet, from this small fraction, humanity has reaped enormous benefits. We are grateful to the plant kingdom for such useful drugs like vinblastine from the African periwinkle, *Catharanthus roseus*, used for the treatment of leukemia. The cholinergic drug, physostigmine, used in the treatment of glaucoma, comes from the Calabar bean (*Phyosostigma venenosum*), which was used in southeastern Nigeria as an 'ordeal poison'. Curare for surgery, reserpine for high blood pressure and taxol for cancer are other examples of very important medicines derived from plants. The transformation of these drugs from noxious vegetable preparations, often without precise dosage, took several decades and involved many scientists and at immense cost. Modern technology can now speed up the process, but it still takes about 15 years to develop a single pharmaceutical entity from discovery to market and at an estimated cost of between $150 million and $300 million. Both the cost and time can be drastically reduced by developing medicinal plants as phytomedicines or standardized herbs. This presentation gives an overview of how this can be achieved at a reasonable cost.

This approach to the discovery and development of plant products as medicines and cosmetics holds a lot of promise for scientists and institutions in developing countries. The market is huge, and while still at its growth phase, it has been estimated at over $85 billion per annum. In 1995, China alone generated about $5 billion from the sale of herbal products. The increase in the acceptance of the medical profession of the 'holistic' philosophy of therapy, which the traditional systems exemplify, is a further indication that there will continue to be growing demand for natural products drugs derived from traditional medicine. The global herbal trade is at an early stage in its evolution (or revival), which places the producers from developing countries in a strategically favorable position to compete with the major players, unlike in other biological resources-based industries where new entry into the market is frustrated by already saturated fields with long-established alliances and cooperatives. There are over 2000 herbal medical companies in Europe and over 250 in the USA. India has 46,000 pharmacies manufacturing traditional remedies.[14]

While it takes upwards of 10 years to move a single drug lead from concept to an approved dosage form, it is possible to convert crude drugs into a standardized modern phytomedicine or personal care product in less than two years. This rapid turnover represents enormous savings in R&D costs, but it also poses potential risks for the consumer since products can be rushed to the market place without adequate and appropriate tests to ensure that what is being produced is safe and efficacious. In other words, because of the relatively shorter time required to develop a phytomedical product, extra care must be taken to ensure that rigorous standards are applied at all stages in the research and development of the product.

The problem of standardization is made even more complex by the fact that many traditional remedies usually contain a combination of several plants. It is difficult enough to try to determine the active ingredient in a single plant when it is known that in most cases, the various components of a mixture contributes to the observed pharmacodynamic effect, and in some cases, they may contain substances that help to minimize the side-effects. It is also claimed that in compound prescription, the therapeutic effect can be obtained only if all the different species are used.

The World Health Organization has also emphasized the fact that safety should be the overriding criterion in the selection of herbal medicines for use in healthcare.[15] Proof of safety should therefore take precedence over establishing efficacy. There is no longer any doubt regarding the value of traditional remedies. What is difficult is the conversion of these remedies into modern therapeutic agents in a sustainable and safe manner. Although it is generally recognized that accuracy of labeling the constituents of medicinal plant remedies is critical for their proper use, safety evaluation and drug control. Yet, many people are hesitant to disclose the ingredients in their preparations. This reluctance to divulge what amounts to a trade secret is understandable when one considers the fact that the formulations are unprotected by any legal instrument. This cloak of secrecy leads to a situation where genuine medicines are sold alongside 'snake-oil' remedies with fantastic claims. It is part of the duty of the medicinal plant expert to devise methods for protecting the rights of the medicinal plant inventor and at the same time ensuring availability of quality medicines from plants. This is an area of study, which is grouped under the omnibus term 'intellectual property rights' (IPR). Given the fact that large quantities of plant materials are often required for the production of phytomedicine and nutraceuticals, it is very important that such herbs are collected in sustainable manner.[16]

V. Pharmacological evaluation of ethnomedical remedies

One major factor on which the decision to select one or more medicinal plant candidates for development into phytomedicines on commercial scale will rest, is the wealth of ethnomedical information on its efficacious use in the long history of traditional medicine. The scientific literature is now replete with reproducible experimental results on the pharmacological properties of herbal extracts. These results were obtained using the same laboratory animal models and techniques that have been accepted scientifically for the testing of pharmacological activity in pharmaceutical products.[17,18] A major problem encountered in the pharmacological evaluation of plant drugs is that of solubility. The active constituents occur naturally as soluble salts or organic complexes with solubility-enhancing matrixes, but the extraction process sometimes leads to the disassociation of the organic compounds from the water-soluble component with the resultant production of insoluble extractives. Problems arise from attempts to re-solubilize the material in aqueous media for pharmacological evaluation, and the use of organic solvents is often precluded because of probable interference with the bioassay methods.

V.A. Toxicity

Herbal products used in traditional medicines are generally non-toxic due to a time-tested selection process in favor of non-toxic herbal ingredients. The traditional medical system advocates a liquid dosage form, which encourages the use of extremely low concentrations of the active ingredients in the finished products. The problem of toxicity of pharmaceutical drugs arises mainly due to human manipulation to increase the accumulation of the desired active substances and hence increase the biological activity. The therapeutically active substances in plants co-exist in conjunction, and bond with other substances like tannins, carbohydrate, amino acids, proteins, vitamins, trace metals, etc. When consumed as decoctions and extracts, there is not much change in the molecular integrity of the active constituents. Moreover, the human body is used to these extracts, most of which, in other forms, are consumed as food. When used in the right amount and right form, these herbs do not upset or cause toxicity in the body. Rather it goes to maintain the physiological balance in the body. This explains why herbal extracts produce little or no toxicity, even when large amounts are consumed. Over-processing of food has been shown not to promote health. Single isolated compound drugs have all other substances with which they co-existed in the plant cleaved off leaving free and highly chemical reactive radicals hanging on the compounds. It has been suggested that the reactivity of the single compound drug with the body physiological medium, in order to re-establish bonding, leads to the chelation of substances from the physiological medium. The resultant upset of the balance in the physiological medium caused by chelation leads to the manifestation of the phenomenon described as drug toxicity.

Phytomedicines, with a little amount of processing, can be classified as health foods. They will promote the healthy development of the body since they contain, apart from the active drug molecule, other substances that the body is used to and which it does use in the maintenance of the overall body physiological functions. This is why 'bitter leaf' (*Vernonia amygdalina*) can be used as a food and as a drug in the treatment of diabetes, for example, without any manifestation of toxicity. However, when purified extracts or concentrated isolates are used as phytomedicines, they should be considered as medicines and subjected to the same rigorous standards used to test other medicinal agents. The various national pharmacopoeias provide appropriate guidelines for the relevant toxicity testing methods.

V.B. Phytochemical analysis

Phytochemical studies should be tailored to match the biological activity. The chemical studies should provide information that will be used in the standardization and quality control of the finished product. Even when the product is to be used as whole herb, it is imperative that the chemistry of the plant material be thoroughly studied so that storage conditions, stability and ingredient integrity can be determined. For example, it will be important to determine if the active constituent in a crude drug is a glycoside, which will likely hydrolyze when stored with a high moisture content. Methods for the analysis of herbs used in the preparation of ethnomedical remedies are available in textbooks of pharmacognosy or phytochemistry.[19,20]

VI. Clinical evaluation of ethnomedical remedies

Prior to any detailed laboratory studies of any plant used in traditional medicine, it is important to establish whether the remedy does in fact possess the claimed therapeutic properties. It is important to emphasize the fact that, because a given plant does not show pharmacological activity in laboratory animals, it does not necessarily mean that the plant is devoid of therapeutic application. There are so many unknown elements in our understanding of drug action and mechanism of action of several drugs. A carefully designed clinical evaluation will help determine if the remedy is active within measurable parameters. Such studies should not be confused with a clinical trial of new drug entity. This form of evaluation merely evaluates the *traditional* use of the drug in a *clinical* setting. The usual ethical considerations are followed, and the dosage form is standardized in terms of posology. It is not supposed to address any of the issues meant for a controled clinical study. It is neither randomized nor double-blinded. The number of patients for this type of studies is not usually as large as that expected in randomized clinical trials (RCT). Good reliable results can be obtained for most diseases with a small study population of 8–12. It should, however, provide clear go/no-go decision points. It is estimated that discovery and clinical evaluation accounted for 30% of the $24 billion spent on drug development by pharmaceutical companies in 1990.[21] Conducting a clinical evaluation at the beginning of the study will give the candidate drugs a better chance of making it as a new drug entity and minimize the amount of money 'wasted' on working on a compound that may eventually fail in experimental clinical trials. It does not, however, follow that a drug that makes it in the clinical evaluation will in all cases contain substances that will be feasible pharmaceutical products. Nevertheless, it is a reliable indicator of substances that can be used in the production of effective and safe phytomedicines and natural personal care products.

Perhaps, clinical evaluation of traditional medicine is even more useful not as a tool for pharmaceutical drug discovery but as a method to provide a better understanding of the spectrum of activity or limitations of ethnomedical remedies and practices. In this case, what is required is the establishment of clear and unambiguous parameters for measuring expected outcomes. Three fundamental features determine the scientific rigor and validity of the experimental approach to assessing herbal efficacy; these are randomization, blindness, and measurement of predetermined outcomes. Randomization, which is the hallmark of the randomized controlled trial (RCT), ensures that the various treatment groups are indeed comparable and that the only significant difference is the treatment they receive. Blindness minimizes bias by both patient and investigator and could be considered crucial in a system of medicine that is relatively more subjective than modern medicine. The third leg of the RCT tripod, the measurement of predetermined outcomes, should be complete, appropriate, and accurate. As has been noted by Kravitz, measurement of predetermined outcomes is a property that is recommended not only for RCT but also for good observational studies used in the evaluation of herbal medicines.[22] The overriding consideration in determining the outcomes should be the interest of the patient, since ultimately the objective is to improve their health status. Both generic and disease-specific outcomes are now combined in the determination of the usefulness of therapeutic interventions in complex clinical situations. In certain situations, measurement of various aspects of functional status and well-being and other quality-of-life assessments are considered more relevant to the patient than disease-specific outcome measure-ments.[23] It has been observed that although RCT is a valuable method for the evaluation of ethnomedical remedies and the so-called complementary and alternative medicines, the method may have some important scientific drawbacks that could limit their application in ethnomedical evaluation.[22]

Some of the problems with the RCT method include generalizability of the results produced by this method since in order to enhance internal validity and to keep sample sizes reasonable, the study populations are often limited to a fairly narrow spectrum of patients that may not be representative of either the population or the situation in actual practice.[22] The results obtained from RCTs are also rendered less relevant to real-life situations by the mere fact that the experiments are usually conducted by experienced clinical investigators, and the careful monitoring of the patients in a clinical trial setting may produce results that are better than real life situations. In order to overcome the limitation of the RCT approach to clinical evaluation of ethnomedical remedies, alternative approaches have been suggested that attempt to narrow the differences between *efficacy* (usefulness or activity under clinical trials setting) and *effectiveness* (value in the real world). These alternative approaches include:[22,23]

- *Quasi-experiments*: Patients are assigned to treatment condition not as individuals but as members of a group. This method dispenses with randomization and blindness but solves the problem of selection bias that is common with evaluation of traditional medicines.
- *Regression-discontinuity*: The design is not randomized but provides for full experimental control without randomization. It is based on the assumption that there are predictable relationships between pre-test and post-test scores for all subjects. By using the same patients as controls, it is possible to determine the effect of a given therapy by assessing the observed difference between the regression line obtained by plotting the pre-test scores and the post-treatment effects.
- *Cohort studies*: This involves following a group of patients in time. Both prospective (forward in time) and retrospective cohort studies are valid approaches to assess the effect of a given therapy.
- *n = 1 trials*: These are conducted with just one individual patient, with all the rigor of a true experimental approach. The experiment can be blinded, and definitive decisions can be made regarding the efficacy of a given treatment on the individual patient.

Case-control studies: Patients are selected based on the presence or absence of the disease or outcome of interest at time $t = 0$. Assessment is then made on the exposure status of both cases at an earlier time ($t–1$). This method is particularly useful in epidemiological research.

VI.A. Current trends in cosmetics and personal care products

The key objective in the use of ethnomedical remedy in modern personal care, cosmetics and topical hygiene is the selection of substances with defined constituents and specific biological effect. Whereas, previously, the term cosmetics was used to denote substances that are believed to have no physiological effect but to create an illusion of well-being or beauty, modern cosmetics are expected to possess measurable biological effect. Some of the desirable activity includes moisturizing, soothing, smoothing, protecting against environment, and repairing damaged tissues. These new trends have pushed cosmetics closer and closer to therapeutics, and the line between a drug claim and cosmetics is getting thinner. Some regulatory agencies are insisting that materials intended for personal care must at least be subjected to the same level of control as over-the-counter (OTC) products. For most small manufacturers, the mere thought of submitting a formulation to regulatory agencies conjures fears of endless waiting for approval. And in Third World countries, where governmental agencies are seen at best as centers for illegal collection of fees to enrich the pockets of corrupt officials, any form of control can be used to create artificial bottlenecks, which can be tightened at will to extort money from manufacturers. This situation can be prevented or at least minimized by taking time to study the regulations governing the production and sale of such products in the given country. It is also useful to retain the original form of the natural products as much as possible so that the possibility of adverse effect will be minimized.

Traditional uses of plant extracts in personal care include:

- wound healing
- antiseptic
- anti-irritant
- anti-inflammatory
- anti-infective
- body decorative
- toning (mud packs)
- mouth and teeth-cleaning.

VII. Production of herbal medicinal products[24]

The production of herbal medicinal products based on ethnomedical remedy follows the same basic principle observed in the manufacture of pharmaceuticals. The nature of the active medicinal substance for herbal products, however, requires some peculiar treatment and processes. A major problem that is not encountered in the development or production of single chemical pharmaceuticals is the sourcing of raw materials with consistent quality and in a sustainable manner. Medicinal plants can either be obtained from wild species or cultivated in farms.

VII.A. Collection of medicinal plants gathered by collection from the wild

According to WHO, 70–90% of the more than 21,000 plant species listed as being medicinally used as plant drugs are commercially obtained by collecting the herbs from the natural habitat. Although wild-crafted species present many problems for drug production, herb manufacturers still insist on collecting many of their plant materials from the wild. The reasons for continuing this practice of collection include the fact that some species are difficult to propagate and, in some cases, take a long time to reach a harvestable stage. *Pygeum africanum*, for example, takes up to 25 years to reach a harvestable maturity. Other difficult species include: *Hippocastenum*, *Arctostaphylos uvae ursi* and *Crataegus*.

Others that are not amenable for domestication or large-scale agriculture for a variety of reasons, e.g. symbiotic relationships with other plants, include *Viscum* and *Lischen islandicus*. Some species have also developed highly specialized survival strategies (such as irregular flowering and seed formation, irregular germination parameters, etc.) to adapt to their peculiar ecosystem and may be difficult to grow under agronomic conditions. The biomass produced from agricultural plants may also be chemically different from the wild varieties.

Problems of using wild species for phytomedicine production include:

- overharvesting of endemic species
- reduction and/or elimination of local populations with the result of a decrease of genetic variety
- unnecessary destruction of plants during harvest
- high incidence of misidentification of species, with possible toxicity issues.

VII.B. Plant material from cultivated species

It is estimated that only about 50–100 species of the more than 1000 species used in the preparation of herbal medicinal products are presently undergoing cultivation. This situation has serious environmental and quality-control consequences if it is not reversed. An operational guideline should be that all medicinal plants with a demand for more than 100 tons/annum should be cultivated, with the exception of the recalcitrant species that meet the criteria listed above.

Conditions that require cultivation of desired plant species include:

- when abundance studies or species distribution indicate low availability of the plants in the wild;
- when the wild source is sparsely distributed;
- when the wild plants are inaccessible, e.g. mountainous plants or very tall trees from which leaves have to be collected;
- when there is need to improve the yield of active principles produced by the plants growing wild;
- when there is government control over the plants; for example, plants such as cannabis or plants yielding dangerous drugs (or addictive drugs) are best cultivated under license;
- when only a desired species or variety of a particular plant is to be used for preparing galenicals because of its high yield of active constituents;
- cultivation can also produce more plant growth, and hence a better yield, by introduction of good agricultural practices, good soil, pest control, etc.

VII.C. Sterilization of crude herbal materials

Micro-organisms associated with the environment usually contaminate plant materials. The two methods employed for the sterilization of these plant materials are treatment with ethylene oxide and exposure to γ-rays. Both pose some risks: ethylene oxide may react with some constituents of the plant to generate toxic or carcinogenic compounds, and γ-rays may affect the chemical integrity of the plant materials in a manner not fully understood. For example, γ-irradiation has been reported as reducing the morphine content of opium.[25] Organic farmers use less effective, but more wholesome, methods like hot-water treatment and a high-vacuum autoclave.

VII.D. Formulation of ethnomedical remedies

Formulation and trial production of the dosage forms are usually structured to mimic the traditional use of the herb. The stability of the finished product should be given careful attention when formulating the final dosage form. The dominance of fresh materials in the preparation of traditional remedies sometimes makes it difficult to relate chemical constituents with therapeutic activity. The most common dosage form used in traditional medicine is liquid decoction, in which herbal drugs are extracted in hot or cold water or in alcohol. For example, powders or teas prepared from leaves, roots or barks are macerated in either the cold or hot water, or an alcoholic tincture is taken orally. The ethnomedical information should provide guidance in the selection of a suitable dosage formulation.

VII.D.1. Teas and powders
The simplest form is to process the plant material by powdering or mincing and packaging into individual tea sachet dosage forms with instructions and directions on preparation, e.g. 'add to one teacup of boiling water, leave to infuse for 10 minutes, and drink'. Powder dosage forms can be packed for addition to food preparations like soft drinks, porridge or soups.

VII.D.2. Syrups and extracts
This is the most convenient dosage-form used in the preparation of herbal medicinal products. It is preferable when extracts obtained from hot or cold water or from ethanol extraction processes are formulated into liquid suspensions or syrups and packaged with information on the amount of active material contained in a stated dose.

VII.D.3. Tablets and capsules
Both dried extracts and whole-plant powder preparations can be formulated by the granulation and addition of excipients and preservatives and compressed into tablet dosage forms or encapsulated in hard gelatin capsules.

VII.D.4. Parenteral preparations

Though the formulation of phytomedicines into injectable forms is not recommended at this stage, with a wealth of clinical and pharmacological information on the long-term use, some of the phytomedicines can be developed in the future into parenteral dosage forms.

VII.D.5. Suppositories, pessaries, enemas, snuffs, creams, lotions and ointments

Topical antibacterial and antifungal preparations of plant extracts, such as suppositories incorporating herbal extracts against piles, and vaginal pessaries incorporating plant extracts to treat bacterial and fungal infections can be formulated. Some herbal extracts can be absorbed through the nasal mucosal membrane. Snuffs incorporating plant extracts can be formulated for this purpose. It is very important to stress that ethnomedical information on the mode of drug administration should form the basis for deciding the formulation method to be used for the phytomedicines. Finally, in the formulation of phytomedicines, complex chemical extraction processes are inadvisable.

VII.E. Preservation

There are two main factors responsible for the spoilage of formulated herbal extracts. The first is microbial spoilage, which includes fungus and yeast fermentation. The other is oxidative spoilage. There are literally thousands of tropical plants that have antimicrobial and antifungal properties that can be used as preservatives of phytomedicines. For example, the fruits of *Xylopia* have antimicrobial properties: the oil is an antioxidant. All the *Afromomum* species, which are used as spices, have broad spectrum antimicrobial, antifungal and antioxidant properties. *Ocimum gratissimum* and a host of other aromatic plants that are becoming the basis of another medical fashion in the West, i.e. aromatherapy, contain essential oils that have good antimicrobial, antiseptic and other preservative properties. In most galenical formulations, essential oils are used both as flavoring agents and as preservatives. One reason for the use of two or more different plants in a traditional herbal formulation is that while one may be the active medicine, the other acts as flavoring agent or preservative, to enhance activity or to reduce toxicity. The main area in phytomedicine that will need the use of preservatives is in aqueous decoctions, syrups and parenteral preparations.

VII.F. Shelf life and standardization

Most traditional medicines are extemporaneous preparations, and therefore, the traditional information in this area is not going to be very reliable. The best practice is to prepare the phytomedicines in batches. A sizeable number of samples from each batch are kept on the shelf and labeled with batch numbers and dates of preparation. Samples are retrieved at, say, 6-monthly to 1-yearly intervals for organoleptic testing and for assessment of pharmacological efficacy. In this way, records will be maintained to show that samples of phytomedicines prepared in any given period are continuously monitored. Continuous monitoring of samples of all batches of products is the simplest way to ensure that the standard and quality of the products are maintained. Apart from the above methods, spectrophotometric analysis of batches based on the spectrometric characteristic of a known chemical constituent in the product can be used to ascertain the quantity and quality of the active ingredients in the product.

VII.G. Scale-up production

It is easy to produce from a kitchen-based factory a few thousand dosage forms of phytomedicine daily. In this small-scale processing method, quality is easily monitored. When it comes to large-scale industrial production, however, more sophisticated engineering designs and measures need to be put in place in order to monitor quality. This engineering know-how is already available and can be easily applied once we know what we want to monitor.

VII.H. Dosage packing

Three factors are important considerations in determining the appropriate dosage form of phytomedicines:

1. the targeted consumer
2. the most convenient dosage form for administration
3. handling properties.

Since dosage-form and packaging can significantly affect the cost of a given herbal medicinal agent, care should be taken to select the most cost-effective but suitable dosage form and packaging. The primary objective for the production of phytomedicine is to make healthcare available with acceptable scientific standards to wide segments of the population. It should cater to the estimated 80% of the human population to whom pharmaceutical preparations are not available and affordable, as well as to the more affluent people who elect to use ethnomedical remedies instead of pharmaceutical products. Two types of dosage packaging are generally available:

- cheap packaging in polythene bags or sachets with clearly labeled directions on dosage, etc. should only be dispensed from herbal clinics to low-income patients as an available source of clinical information on the use of the drugs
- a second type of packaging should conform to acceptable national standards of product packaging for commercial scale distribution.

Table 1
Definition of terms

Phytopharmaceutical: A pure isolated chemically defined plant substance (e.g. digitoxin, atropine, escin, etc.) or group of substances used for the diagnoses, prevention or treatment of diseases. Most national authorities regulate them like their synthetic equivalents. The term does not include plant materials used as formulation aides.

Phytomedicine: A medicinal product that contains, as active ingredients, exclusively plant materials and/or standardized herbal medicinal products or preparations thereof containing identified chemically defined substances or groups of substances of plant origin that are known to contribute to the therapeutic activity of the preparation and presented in a suitable dosage form. A phytomedicine for the purposes of Statute is a drug substance but differs from other pharmaceutical products only in terms of its origin and composition (adapted from EU Commission E).

Dietary supplement: A product (other than tobacco) intended to supplement the diet that bears or contains a vitamin, a mineral, an herb or other botanical, an amino acid, a dietary substance for use by human to supplement the diet by increasing the total dietary intake, or a concentrate, metabolite, constituent, extract, or combination of any of the above ingredients. Dietary supplements for the purposes of Statute are substances intended only for ingestion in a form approved by the appropriate regulatory agency and are not represented as conventional food or as a sole item of a meal or the diet, and are labeled as a dietary supplement (taken from US DSHEA, with minor modifications).

Herbal medicine: A product that consists of comminuted or powdered vegetable material, whole herb, plant part or parts, prepared as extracts, tinctures, fatty or essential oils, expressed plant juices, etc. or botanic products described in the relevant official compendia as approved herbs.

Traditional remedy: A natural product (plant, animal or mineral) used in the practice of traditional medicine. They usually contain comminuted or powdered vegetable material, whole herb, plant part or parts, prepared as extracts, tinctures, fatty or essential oils, expressed plant juices, etc. Many regulatory agencies stipulate that a product labeled as 'traditional remedy' must be prepared according to original traditional recipe and must be free of any chemical additives.

Nutraceutical: A product (concentrate, metabolite, constituent, extract, or combination of any of the above ingredients) derived from an herb or other botanicals that are generally used as a dietary substance, which has a beneficial effect for the promotion of optimum health or the prevention of diseases. Although usually applied to botanical materials, the term can also be applied to minerals and isolated dietary ingredients with functional health-promoting properties.

Health food: A dietary substance for use by humans to supplement the diet by increasing the total dietary intake, or a concentrate, metabolite, constituent, extract, or combination of any of the above ingredients taken as part of the diet that promotes general well-being. A food substance presented as a 'heath food' must be free of chemical additives, pesticides or other extraneous material.

Herbal tea: Tea-bagged plant parts, often with natural flavors used as beverages or medicine. The term is correctly used as a dosage formulation.

Essential oils: Volatile oils for fragrance or medicinal uses. The method of production often determines the properties and potential medicinal application.

Oleoresins: Fat-soluble components extracted with a suitable solvent used as a fragrance or medicine.

Liquid extracts: Extracts prepared by maceration and/or percolation of the comminuted plant material with suitable solvents to separate the active ingredients from the raw plant parts. Common extraction solvents include water, alcohol, a combination of the solvents or other permitted solvents such as hexane. Extracts are often described as aqueous, alcoholic or dry to denote the nature of the product, and these terms have a specific technical meaning.

Tinctures and fluid extracts: A form of galenical preparation used in the ancient pharmaceutical practice, and still popular in the preparation of certain herbal remedies. Tinctures are prepared by soaking chopped plant material with a mixture of water and alcohol. This is the basic dosage form for herbal remedies, and the alcohol in the extracting solvent helps in product stability by its antiseptic properties. The ratio of water and alcohol in the solvent mixture depends on the type and nature of the plant material. The drug:menstruum (solvent) ratios for most pharmaceutical tincture are within the range of 1:5 and 1:10 (solvent:plant). The final concentration desired in a tincture can be obtained by

Table 1 (*continued*)

evaporation of the solvent under vacuum. Fluid extracts are produced when the solvent:plant material ratio is 2:1 to 1:1.

Dry extracts: Extracts prepared from tinctures, organic extracts or hydro-alcoholic intermediates by removing the extracting solvent under vacuum. The resultant residue can be formulated as capsules, tablets or other pharmaceutical dosage forms. Solid extracts are dry extracts to which propylene glycol or liquid glycerin is added to produce a semi-solid or semi-solid viscous substance. Powdered extracts are produced in which all the solvent has been removed and/or the residue is diluted with a suitable solid carrier substrate, such as cellulose, malt dextrin, and magnesium carbonate. Dry extracts can be standardized using a chemical marker and, in some cases, also with a biological marker. Dry extracts are usually very hydroscopic, and some of them may be photo- or thermolabile. Care should be taken, therefore, to ensure that the milling and processing of the extract are done in a moisture-free environment and that the final products are packed in a moisture-tight containers (and, when necessary, protected from the light).

Natural product: Entire organisms or parts, extracts and exudates that have not been subjected to any treatment, except perhaps to a simple process of preservation, such as drying.

Pharmacognosy: A term believed to have been introduced by the Austrian physician, Johann Adam Schmidt (1759–1809) to describe the study of natural products used as drugs or for the preparation of drugs. The word 'Pharmacognosy', itself is derived from the two Greek words *pharmakon* = drug (portion) and *gnosis* = knowledge.

References

1. Barsh RL. (1999) Indigenous knowledge and biodiversity. In: Indigenous peoples, their environments and territories, presented by Gray, Andres. In: Posey D, editor. Cultural and spiritual values of biodiversity, a complimentary contribution to the global biodiversity assessment. Nairobi: UNDP, London: Intermediate Technologies, pp. 73–76.
2. Foster GM. (1983) An introduction to ethnomedicine. In: Bannerman RK, Burton J, Wen-Chieh C, editors. Traditional medicine and health care coverage. Geneva: WHO, pp. 17–24.
3. Iwu MM. (1999) Symbols of power and health – forward. In: De Smet PAGM, editor. 'Herbs, health and healers: African art as pharmacological and medical treasury'. The Netherlands: Alphen a.d Rijn.
4. Wiessendanger H, Westmuller L, Reuter K, Walach H. (2001) Chronically ill patients treated by spritual healing improve in quality of life: results of a randomized waiting-list controlled study. J Altern Complement Med 7(1):45–51.
5. Jobst KA. From plants to patients: the alchemy of herbs and healing. J Altern Complement Med 6(2):111–113.
6. Brown CK. (2000) Methodological problems of clinical research into spiritual healing: the healer's perspective. J Altern Complement Med 6(2):171–176.
7. Harkness E, Abbot NC, Ernst E. (2000) A randomized trial of distant healing for skin warts. Am J Med 108(6):448–452.
8. Jonas W. (2001) The middle way: realistic randomized controlled trials for the evaluation of spiritual healing. J Altern Complement Med 7(1):5–7.
9. Astin JE, Harkness E, Ernest E. (2000) The efficacy of 'distant healing': a systematic review of randomized trials. Ann Intern Med 132:903–910.
10. Iwu MM, Gbodossou E. (2000) The role of traditional medicine in drug development. Lancet, Supplement 3.
11. Brekhman II. (1980) Man and biologically active substances. Oxford: Pergamon Press.
12. UNIDO. (1984) Traditional pharmacopoeias revisited. In: Proceedings of the Workshop on the Pharmaceutical industry, Beijing, China. UNIDO/IO/R.121 of 14.
13. Iwu MM. (1993) Handbook of African Medicinal Plants. Boca Raton, FL: CRC Press, p. 435.
14. Alok SK, Akerele O, Heywood V, Synge H (editors). (1991) The conservation of medicinal plants. New York: Cambridge University Press, p. 362.
15. Anon. (1991) Essential drug monitor 11:15–17.

16. Office of Technological Assessment. (1993) Pharmaceutical R & D: costs, risk and rewards. Washington DC: OTA.
17. Bohlin L, Bruhn JG. (1999) Bioassay methods in natural product research and drug development, proceeding of the Phytochemical Society of Europe. Dordrecht: Kluwer Academic, p. 201.
18. Hostettmann K (editor). (1991) Assays for bioactivity, Vol. 6. In: Dey PM, Hraborne JR, editors. Methods in plant biochemistry. London: Academic Press, p. 360.
19. Evans WC (editor). Trease and Evans' pharmacognosy, 14th edition. London: WB Saunders, p. 612.
20. Tyler VE, Brady LR, Robbers JE. Pharmacognosy, 9th edition. Philadelphia, PA: Lea and Febiger.
21. Halliday RG, Drasdo AL, Lumley CE, Walker SR. (1992) Pharmaceut Med 6:281–296.
22. Kravitz R. (1999) Evaluation of Herbal Medicines: Alternatives to Randomized Controlled Trial. In: Eskinazi D, Blumenthal M, Farnsworth N, Riggins CW, editors. Botanical medicine: efficacy, quality assurance, and regulation. New York: Mary Ann Liebert, pp. 59–64.
23. Stewart A, Ware JE (Eds). (1992) Measuring functioning and well-being: the medical outcomes study approach. Durham: Duke University Press.
24. List PH, Schmidt PC. (1989) Phytopharmaceutical technology. Boca Raton, FL: CRC Press, p. 374.
25. Samuelsson G. (xxxx) Drugs of natural origin, a textbook of pharmacognosy. Stockholm: Swedish Pharmaceutical Press, pp. 320 (based on a lecture by Prof. Harnischfeger, D.G. of Schaper and Brummer GmbH & Co. Ringelheim – personal communication).

Iwu and Wootton (eds.), Ethnomedicine and Drug Discovery

CHAPTER 1

Drugs from nature: past achievements, future prospects

GORDON M CRAGG, DAVID J NEWMAN

Abstract

Nature has been a source of medicinal agents for thousands of years, and an impressive number of modern drugs have been isolated from natural sources, many based on their use in traditional medicine. The past century, however, has seen an increasing role played by micro-organisms in the production of the antibiotics and other drugs for the treatment of diseases, ranging from bacterial infections to cardiovascular problems and cancer. Much of the world's biodiversity remains unexplored as a source of novel drug leads, and the search for new bioactive agents from natural sources, including extreme environments, will continue. With less than 1% of the microbial world currently known, advances in procedures for microbial cultivation and the extraction of nucleic acids from environmental samples from soil and marine habitats, and from symbiotic and endophytic microbes associated with terrestrial and marine macro-organisms, will provide access to a vast untapped reservoir of genetic and metabolic diversity. These resources will provide a host of novel chemical scaffolds, which can be further developed by combinatorial chemical and biosynthetic approaches to yield chemotherapeutic and other bioactive agents, which have been optimized on the basis of their biological activities. The investigation of these resources requires multi-disciplinary, international collaboration in the discovery and development process.

Keywords: *traditional medicine, antibiotics, biodiversity, drug leads, bioactive agents, biological activity*

I. Medicinals for the millennia

I.A. Recorded history

Throughout the ages, humans have relied on nature for their basic needs for the production of foodstuffs, shelters, clothing, means of transportation, fertilizers, flavors and fragrances, and, not least, medicines. Plants have formed the basis of sophisticated traditional medicine systems that have been in existence for thousands of years.[1] The first records, written on clay tablets in cuneiform, are from Mesopotamia and date from about 2600 BC; among the substances that they used were oils of *Cedrus* species (cedar) and *Cupressus sempevirens* (cypress), *Glycyrrhiza glabra* (licorice), *Commiphora* species (myrrh), and *Papaver somniferum* (poppy juice), all of which are still in use today for the treatment of ailments ranging from coughs and colds to parasitic infections and inflammation. Egyptian medicine dates from about 2900 BC, but the best known Egyptian pharmaceutical record is the 'Ebers Papyrus' dating from 1500 BC; this documents some 700 drugs (mostly plants), and includes formulas, such as gargles, snuffs, poultices, infusions, pills and ointments, with beer, milk, wine and honey being commonly used as vehicles. The Chinese Materia Medica has been extensively documented over the centuries, with the first record dating from about 1100 BC (Wu Shi Er Bing Fang, containing 52 prescriptions), followed by works such as the Shennong Herbal ($\sim$100 BC; 365 drugs) and the Tang Herbal (AD 659; 850 drugs). Likewise, documentation of the Indian Ayurvedic system dates from about 1000 BC (Susruta and Charaka), and this system formed the basis for the primary text of Tibetan Medicine, Gyu-zhi (Four Tantras) translated from Sanskrit during the eighth century AD.[2]

In the ancient Western world, the Greeks contributed substantially to the rational development of the use of herbal drugs. The philosopher and natural scientist, Theophrastus ($\sim$300 BC), in his 'History of Plants', dealt with the medicinal qualities of herbs, and noted the ability to change their characteristics through cultivation. Dioscorides, a Greek physician (AD 100), during his travels with Roman armies throughout the then 'known world', accurately recorded the collection, storage, and use of medicinal herbs, and is considered by many to be the most important representative of the science of herbal drugs in 'ancient times'. Galen (AD 130–200), who practiced and taught pharmacy and medicine in Rome, and published no less than 30 books on these subjects, is well known for his complex prescriptions and formulas used in compounding drugs, sometimes containing dozens of ingredients ('galenicals'). During the Dark and Middle Ages (fifth to 12th centuries), the monasteries in countries such as England, Ireland, France and Germany preserved the remnants of this Western knowledge, but it was the Arabs who were responsible for the preservation of much of the Greco-Roman expertise. They were also responsible for expanding it to include the use of their own resources, together with Chinese and Indian herbs unknown to the Greco-Roman world. The Arabs were the first to establish privately owned drug stores in the eighth century, and the Persian pharmacist, physician, philosopher and poet, Avicenna, contributed much to the sciences of pharmacy and medicine through works, such as Canon Medicinae, regarded as 'the final codification of all Greco-Roman medicine'.

I.B. Traditional medicine and drug discovery

As mentioned above, plants have formed the basis for traditional medicine systems, which have been used for thousands of years in countries such as China[3] and India.[4] The use of plants in the traditional medicine systems of many other cultures has been extensively documented.[5–10] These plant-based systems continue to play an essential role in healthcare, and it has been estimated by the World Health Organization that approximately 80% of the world's inhabitants rely mainly on traditional medicines for their primary healthcare.[11] Plant products also play an important role in the healthcare systems of the remaining 20% of the population, mainly residing in developed countries. Analysis of data on prescriptions dispensed from community

pharmacies in the United States from 1959 to 1980 indicates that about 25% contained plant extracts or active principles derived from higher plants, and at least 119 chemical substances, derived from 90 plant species, can be considered as important drugs currently in use in one or more countries.[11] Of these 119 drugs, 74% were discovered as a result of chemical studies directed at the isolation of the active substances from plants used in traditional medicine.

The isolation of the antimalarial drug, quinine, from the bark of *Cinchona* species (e.g. *C. officinalis*) was reported in 1820 by the French pharmacists, Caventou and Pelletier. The bark had long been used by indigenous groups in the Amazon region for the treatment of fevers and was first introduced into Europe in the early 1600s for the treatment of malaria. Quinine formed the basis for the synthesis of the commonly used antimalarial drugs, chloroquine and mefloquine. With the emergence of resistance to these drugs in many tropical regions, another plant long used in the treatment of fevers in traditional Chinese medicine, *Artemisia annua* (Quinhaosu), has yielded the agents, artemisinin and its derivatives, artemether and artether, effective against resistant strains.[12] The analgesic, morphine, isolated in 1816 by the German pharmacist, Serturner, from the opium poppy, *Papaver somniferum*, used in ancient Mesopotamia (*vide infra*), laid the basis for alkaloid chemistry, and the development of a range of highly effective analgesic agents.[12] In 1785, the English physician, Withering, published his observations on the use of the foxglove, *Digitalis purpurea*, for the treatment of heart disorders, and this eventually led to the isolation of the cardiotonic agent, digoxin.

Other significant drugs developed from traditional medicinal plants include: the antihypertensive agent, reserpine, isolated from *Rauwolfia serpentina* used in Ayurvedic medicine for the treatment of snakebite and other ailments;[4] ephedrine, first isolated in 1887 from *Ephedra sinica* (Ma Huang), a plant long used in traditional Chinese medicine, and basis for the synthesis of the anti-asthma agents (beta agonists), salbutamol and salmetrol;[12] and the muscle relaxant, tubocurarine, isolated from *Chondrodendron* and *Curarea* species used by indigenous groups in the Amazon as the basis for the arrow poison, curare.[12]

I.C. The golden age of antibiotics

The serendipitous discovery of penicillin from the filamentous fungus, *Penicillium notatum*, by Fleming in 1929, and the observation of the broad therapeutic use of this agent in the 1940s, ushered in a new era in medicine and the 'golden age' of antibiotics, which promoted the intensive investigation of nature as a source of novel bioactive agents. Micro-organisms are a prolific source of structurally diverse bioactive metabolites and have yielded some of the most important products of the pharmaceutical industry. These include: antibacterial agents, such as the penicillins (from *Penicillium* species), cephalosporins (from *Cephalosporium acremonium*), aminoglycosides, tetracyclines and polyketides (all from *Streptomyces* species);[12] immunosuppressive agents, such as the cyclosporins and rapamycin (from *Streptomyces* species);[13] cholesterol lowering agents, such as mevastatin (compactin) and lovastatin (from *Penicillium* species);[12] and anthelmintics and antiparasitic drugs, such as the ivermectins (from *Streptomyces* species).[12] A recent publication reports the isolation of a potential antidiabetic agent from a *Pseudomassaria* fungal species found in the rainforests of the Congo.[14]

I.D. Marine sources

While marine organisms do not have a history of use in traditional medicine, the ancient Phoenicians employed a chemical secretion from marine mollusks to produce purple dyes for woolen cloth, and seaweeds have long been used to fertilize the soil. The world's oceans, covering more than 70% of the earth's surface, represent an enormous resource for the discovery of potential chemotherapeutic agents. All but two of the 28 major animal phyla are represented in aquatic environments, with eight being exclusively aquatic, mainly marine.[15] Prior to the development of reliable scuba diving techniques some 40 years ago, the collection of marine organisms was limited to those obtainable by skin diving. Subsequently, depths from approximately 10 feet to 120 feet became routinely attainable, and the marine environment has been increasingly explored as a source of novel bioactive agents. Deep-water collections can be made by dredging or trawling, but these methods suffer from disadvantages, such as environmental damage and non-selective sampling. These disadvantages can be partially overcome by the use of manned submersibles or remotely operated vehicles (ROVs). However, the high cost of these collecting forms precludes their extensive use in routine collection operations.

The pseudopterosins, isolated from the Carribean gorgonian, *Pseudopterogorgia elisabethae*, possess significant analgesic and anti-inflammatory activity, and defined fractions obtained from extracts of the gorgonian are used topically in skin lotions. Another marine product showing potent anti-inflammatory activity, is manoalide, isolated from the sponge, *Luffarriella variabilis*,[15] which has led to a family of similar compounds via synthesis, some of which have reached clinical trial status. The extremely potent venoms (conatoxins) of predatory cone snails (*Conus* species) have yielded complex mixtures of small peptides (6–40 amino acids), which have provided models for the synthesis of novel painkillers (e.g. Ziconotide).[16]

I.E. Other sources

Teprotide, isolated from the venom of the pit viper, *Bothrops jaracaca*, led to the design and synthesis of the ACE inhibitors, captopril and enalapril,[12] which are used in the treatment of cardiovascular disease, while epibatidine, is isolated from the skin of the poisonous frog, *Epipedobates tricolor*, and has led to the development of a novel class of painkillers.[17]

This interest in nature as a source of potential chemotherapeutic agents continues, and an analysis of the number and sources of anticancer and anti-infective agents, reported mainly in the Annual Reports of Medicinal Chemistry from 1984 to 1995 covering the years 1983 to 1994, indicates that over 60% of the approved drugs developed in these disease areas are of natural origin.[18]

II. Anticancer agents from natural sources

Of the 92 anticancer drugs commercially available prior to 1983 in the United States and approved world-wide between 1983 and 1994, approximately 62% can be related to natural origin.[18]

II.A. Plant sources

Plants have a long history of use in the treatment of cancer,[19] though many of the claims for the efficacy of such treatment should be viewed with some skepticism because cancer, as a specific disease entity, is likely to be poorly defined in terms of folklore and traditional medicine.[20] Of the plant-derived anticancer drugs in clinical use, the best known are the so-called vinca alkaloids, vinblastine and vincristine, which are isolated from the Madagascar periwinkle, *Catharanthus roseus*. *C. roseus* was used by various cultures for the treatment of diabetes, and vinblastine and vincristine were first discovered during an investigation of the plant as a source of potential oral hypoglycemic agents. Therefore, their discovery may be indirectly attributed to the observation of an unrelated medicinal use of the source plant.[20] The two clinically active agents, etoposide and teniposide, which are semi-synthetic derivatives of the natural product epipodophyllotoxin, may be considered being more closely linked to a plant originally used for the treatment of cancer. Epipodophyllotoxin is an isomer of podophyllotoxin, which was isolated as the active antitumor agent from the roots of various species of the genus *Podophyllum*. These plants possess a long history of medicinal use by early American and Asian cultures, including the treatment of skin cancers and warts.[20]

More recent additions to the armamentarium of naturally derived chemotherapeutic agents are the taxanes and camptothecins. Paclitaxel initially was isolated from the bark of *Taxus brevifolia*, collected in Washington State as part of a random collection program by the US Department of Agriculture for the National Cancer Institute [NCI].[21] The use of various parts of *T. brevifolia* and other *Taxus* species (e.g. *canadensis, baccata*) by several Native American tribes for the treatment of some non-cancerous conditions has been reported,[20] while the leaves of *T. baccata* are used in the traditional Asiatic Indian (Ayurvedic) medicine system,[4] with one reported use in the treatment of cancer.[19] Paclitaxel, along with several key precursors (the baccatins), occurs in the leaves of various *Taxus* species, and the ready semi-synthetic conversion of the relatively abundant baccatins to paclitaxel, as well as active paclitaxel analogs, such as docetaxel,[22] has provided a major, renewable natural source of this important class of drug. Likewise, the clinically active agents, topotecan (hycamptamine), irinotecan (CPT-11), 9-amino- and 9-nitro-camptothecin, are semi-synthetically derived from camptothecin, isolated from the Chinese ornamental tree, *Camptotheca acuminata*.[23] Camptothecin (as its sodium salt) was advanced to clinical trials by NCI in the 1970s, but was dropped because of severe bladder toxicity.

II.B. Microbial sources

Antitumor antibiotics are amongst the most important of the cancer chemother-apeutic agents, which include members of the anthracycline, bleomycin, actinomy-cin, mitomycin and aureolic acid families.[24] Clinically useful agents from these families are the daunomycin-related agents, daunomycin itself, doxorubicin, idarubicin and epirubicin; the glycopeptidic bleomycins A_2 and B_2 (blenoxane); the peptolides exemplified by dactinomycin; the mitosanes such as mitomycin C; and the glycosylated anthracenone, mithramycin. All were isolated from various *Streptomyces* species. Other clinically active agents isolated from *Streptomyces* include streptozocin and deoxycoformycin.

II.C. Marine sources

The first notable discovery of biologically active compounds from marine sources was the serendipitous isolation of the C-nucleosides, spongouridine and spongothy-midine, from the Caribbean sponge, *Cryptotheca crypta*, in the early 1950s. These compounds were found to possess antiviral activity, and synthetic analog studies eventually led to the development of cytosine arabinoside (Ara-C) as a clinically useful anticancer agent approximately 15 years later,[15] together with Ara-A as an anti-viral agent. The systematic investigation of marine environments as sources of novel biologically active agents only began in earnest in the mid-1970s. During the decade from 1977 to 1987, about 2500 new metabolites were reported from a variety of marine organisms. These studies have clearly demonstrated that the marine environment is a rich source of bioactive compounds, many of which belong to totally novel chemical classes not found in terrestrial sources.[25]

As yet, no compound isolated from a marine source has advanced to commercial use as a chemotherapeutic agent, though several are in various phases of clinical development as potential anticancer agents. The most prominent of these is bryostatin 1, isolated from the bryozoan, *Bugula neritina*.[15] This agent exerts a range of biological effects, thought to occur through modulation of protein kinase C, and has shown some promising activity against melanoma in Phase I studies.[26] Phase II trials are either in progress or are planned against a variety of tumors, including ovarian carcinoma and NHL.

The first marine-derived compound to enter clinical trials was didemnin B, isolated from the tunicate, *Trididemnum solidum*.[15] Unfortunately, it has failed to show reproducible activity against a range of tumors in Phase II clinical trials, while always demonstrating significant toxicity. Ecteinascidin 743, a metabolite produced by another tunicate *Ecteinascidia turbinata*, has significant in-vivo activity against the murine B16 melanoma and human MX-1 breast carcinoma models, and has advanced to Phase II clinical trials in Europe and the United States (Personal Communication G. Faircloth, PharmaMar). The sea hare, *Dolabella auricularia*, an herbivorous mollusc from the Indian Ocean, is the source of more than 15 cytotoxic cyclic and linear peptides, the dolastatins. The most active of these is the linear tetrapeptide, dolastatin 10, which has been chemically synthesized and is currently in

Phase I clinical trials.[25] Sponges are traditionally a rich source of bioactive compounds in a variety of pharmacological screens,[25] and in the cancer area, halichondrin B, a macrocyclic polyether initially isolated from the sponge, *Halichondria okadai* in 1985, is currently in pre-clinical development by the National Cancer Institute (NCI). Halichondrin B and related compounds have been isolated from several sponge genera, and the present source, a *Lissodendoryx* species, is being successfully grown by in-sea aquaculture in New Zealand territorial waters.[27] The mechanisms of action of discodermolide,[28] isolated from the Caribbean sponge, *Discodermia* sp., and eleutherobin,[29] isolated from a Western Australian soft coral, *Eleutherobia* sp., are similar to that of paclitaxel, with the former now in pre-clinical development with Novartis.

III. Development of molecular targets and high-throughput screens

With the rapid progress in the sequencing of the human genome comprising an estimated 30,000 genes, a vast amount of information is becoming available, which is leading to a better understanding of human diseases in terms of biology, diagnosis, prevention and treatment. Knowledge of the genes associated with the onset of diseases enables the identification of the proteins expressed by these genes. These proteins may serve as molecular targets for the development of high-throughput assays for the testing of thousands of materials, including natural products, some of which may act as inhibitors in the progression of the relevant diseases. It is estimated that about 1–10,000 protein targets may be identified through the human genome project. In addition, the sequencing of the genomes of pathogens and parasites will permit the identification of the genes essential for the survival of the pathogens, and their encoded proteins may serve as molecular targets for drug discovery.

Over the past 20 years, there has been an explosion in the understanding of how cancer cells work. Through the Cancer Genome Anatomy Project (CGAP: www.ncbi.nlm.nih.gov/cgap/), it is the goal of the NCI to identify as many of the human genes associated with cancer as possible. Through gene sequence analysis, numerous mutational sites in cancer cells have been, and are being, identified, some of which are unique to specific types of cancer. This knowledge permits the prediction of the structure of the encoded proteins associated with the malignant process, and the discovery of possible molecular targets affecting important aspects of cancer cell function. Anti-cancer drugs, which have emerged from molecular target approaches, are being evaluated in the clinic, and include inhibitors of angiogenesis, farnesyl tranferase, signal transduction and metalloprotease, protein kinase (PK) antagonists, and modulators of gene expression (antisense oligonucleotides).

The ultimate goals envisaged are the creation of an integrated, cohesive drug-discovery program and early clinical trials system that is founded on mechanistic-based approaches, and to make emerging knowledge of cancer biology the basis for drug discovery, development and testing.

IV. Generation of molecular diversity

IV.A. Exploration of new environments

The potential of the marine environment as a source of novel drugs has already been discussed. The NCI contract collection program has been expanded to the waters off East and Southern Africa, and expansion to under-explored regions, such as the Red Sea, is being considered. These collections are performed in close collaboration with organizations based in the countries controlling the relevant waters.

Exciting untapped resources are the deep-sea vents occurring along ocean ridges, such as the East Pacific Rise and the Galapagos Rift. Exploration of these regions is being performed by several organizations, including the Center for Deep-Sea Ecology and Biotechnology of the Institute of Marine and Coastal Sciences, Rutgers University, using deep-sea submersibles such as Alvin, and their rich biological resources of macro- and micro-organisms are being catalogued.[30,31] Samples are being evaluated by the NCI in collaboration with chemists at Research Triangle Institute.

Despite the more intensive investigation of terrestrial flora, it is estimated that only 5–15% of the approximately 250,000 species of higher plants have been systematically investigated, chemically and pharmacologically.[32] The potential of large areas of tropical rainforests remains virtually untapped and may be studied through collaborative programs with source country organizations, such as those established by the NCI.

Another vast untapped resource is that of the insect world, and organizations, such as the Instituto Nacional de Biodiversidad (INBio) in Costa Rica, are investigating the potential of this resource, in collaboration with some pharmaceutical companies and the NCI.

The continuing threat to biodiversity through the destruction of terrestrial and marine ecosystems lends urgency to the need to expand the exploration of these resources as a source of novel bioactive agents.

IV.B. The unexplored potential of microbial diversity

Until recently, microbiologists were greatly limited in their study of natural microbial ecosystems due to an inability to cultivate most naturally occurring micro-organisms. In a report recently released by the American Academy of Microbiology entitled 'The Microbial World: Foundation of the Biosphere', it is estimated that 'less than 1% of bacterial species and less than 5% on fungal species are currently known', and recent evidence indicates that millions of microbial species remain undiscovered.[33]

The recent development of procedures for cultivating and identifying micro-organisms will aid microbiologists in their assessment of the earth's full range of microbial diversity. In addition, procedures based on the extraction of nucleic acids from environmental samples will permit the identification of micro-organisms through the isolation and sequencing of ribosomal RNA or rDNA (genes encoding for rRNA); samples from soils are currently being investigated, and the methods may be applied to other habitats, such as the microflora of insects and marine animals.[34] Valuable products and information are certain to result from the cloning and understanding of the novel genes, which will be discovered through these processes.

Extreme habitats harbor a host of extremophilic microbes (extremophiles), such as acidophiles (acidic sulfurous hot springs), alkalophiles (alkaline lakes), halophiles (salt lakes), baro- and thermophiles (deep-sea vents),[35] and psychrophiles (arctic and antarctic waters, alpine lakes).[36] While investigations thus far have focused on the isolation of thermophilic and hyperthermophilic enzymes,[37] there are reports of useful enzymes being isolated from other extreme habitats by companies such as Diversa Corporation (www.diversa.com). These extreme environments will undoubtedly yield novel bioactive chemotypes.

As Dr Rita Colwell, Director of the United States National Science Foundation, commenting on the importance of exploration and conservation of microbial diversity, has stated: 'Hiding within the as-yet undiscovered microorganisms are cures for diseases, means to clean polluted environments, new food sources, and better ways to manufacture products used daily in modern society'.[38]

IV.C. Combinatorial biosynthesis

Advances in the understanding of bacterial aromatic polyketide biosynthesis have led to the identification of multifunctional polyketide synthase enzymes. (PKSs) responsible for the construction of polyketide backbones of defined chain lengths, the degree and regio-specificity of ketoreduction, and the regiospecificity of cyclizations and aromatizations, together with the genes encoding for the enzymes.[39] Since polyketides constitute a large number of structurally diverse natural products exhibiting a broad range of biological activities (e.g. tetracyclines, doxorubicin, and avermectin), the potential for generating novel molecules with enhanced known bioactivities, or even novel bioactivities, appears to be high.[40]

The NCI is promoting this area of research through the award of grants to consortia composed of multidisciplinary groups devoted to the application of combinatorial biosynthetic and/or combinatorial chemical techniques to the generation of molecular diversity for testing in high-throughput screens related to cancer.

IV.D. Total synthesis of natural products

The total synthesis of complex natural products has long posed challenges to the top synthetic chemistry groups world-wide and has led to the discovery of many novel reactions, and to developments in chiral catalytic reactions.[41] More recently, the efforts of some groups have been focused on the synthesis and modification of drugs that are difficult to isolate in sufficient quantities for development. In the process of total synthesis, it is often possible to determine the essential features of the molecule necessary for activity (the pharmacophore), and in some instances, this has led to the synthesis of simpler analogs having a similar or better activity. Notable examples in the anticancer drug area are the synthesis of synthetic analogs of the marine organism metabolites, bryostatin 1[42] and ecteinascidin 743.[43]

The synthesis of the epothilones by several groups has permitted the preparation of a large number of designed analogs and detailed structure–activity studies, which are reviewed in Ref. 44. These studies have identified desirable modifications, which might eventually lead to more suitable candidates for drug development, but thus far, none of the analogs have surpassed epothilone B in its potency against tumor cells.

The similarity in the mechanisms of action of paclitaxel, the epothilones, discodermolide and eleutherobin has led to proposals that these structurally dissimilar substances possess common pharmacophores, which could lead to the design and synthesis of analogs having substantially different structures and superior activities.[45]

IV.E. Combinatorial chemistry and natural products

In the study of the structure–activity relationships of the epothilones, solid-phase synthesis of combinatorial libraries has been used to probe regions of the molecule important to retention or improvement of activity.[44] The combinatorial approach, using an active natural product as the central scaffold, can also be applied to the generation of large numbers of analogs for structure–activity studies, the so-called parallel synthetic approach.[46]

The split-and-pool solid-phase synthetic approach has also been used to assemble a library of over two million natural product-like compounds from 18 chiral tetracyclic scaffolds, 30 terminal alkynes, 62 primary amines, and 62 carboxylic acids, using a six-step reaction sequence.[47] This library will be used to probe complex biological processes, including protein–protein interactions, for which no ligands have, as yet, been identified. This approach of probing complex biological processes by altering the function of proteins through binding with small molecules has been called chemical genetics.[48]

V. Collaboration in drug discovery and development

V.A. The NCI role: The Developmental Therapeutics Program (DTP)

Much of the NCI drug discovery and development effort has been, and continues to be, carried out through collaborations between DTP and academic institutions, research organizations and the pharmaceutical industry world-wide. Many of the naturally derived anticancer agents were developed through such efforts. The DTP/NCI thus complements the efforts of the pharmaceutical industry and other research organizations through taking positive leads, which industry might consider too uncertain to sponsor, and conducting the high-risk research necessary to determine their potential utility as anticancer drugs. In promoting drug discovery and development, the DTP/NCI has formulated various mechanisms for establishing collaborations with research groups world-wide.

V.B. Source-country collaboration

Drug Discovery. Memorandum of Understanding. Collections of plants and marine organisms have been carried out in over 25 countries through contracts with qualified botanical and marine biological organizations working in close collaboration with qualified source-country organizations. The recognition of the value of the natural resources (plant, marine and microbial) being investigated by the NCI and the significant contributions being made by source country scientists in aiding the performance of the NCI collection programs have led the NCI to formulate its Letter of Collection (LOC), specifying policies aimed at facilitating collaboration with, and compensation of, countries participating in the drug-discovery program.[49]

With the increased awareness of genetically rich source countries to the value of their natural resources and the confirmation of source country sovereign rights over these resources by the UN Convention of Biological Diversity, organizations involved in drug discovery and development are increasingly adopting policies of equitable collaboration and compensation in interacting with these countries.[50] Particularly in the area of plant-related studies, source-country scientists and governments are committed to performing more of the operations in-country, as opposed to the export of raw materials. The NCI has recognized this fact for several years and has negotiated Memoranda of Understanding (MOU) with a number of source country organizations suitably qualified to perform in-country processing. In considering the continuation of its plant-derived drug-discovery program, the NCI has de-emphasized its contract collection projects in favor of expanding closer collaboration with qualified source country scientists and organizations. In establishing these collaborations, the NCI undertakes to abide by the same policies of collaboration and compensation as specified in the LOC. A number of other organizations and companies have implemented similar policies.[50] Through this mechanism, collaborations have been established with organizations in Bangladesh, Brazil, China, Costa Rica, Fiji, Iceland, Korea, Mexico, New Zealand, Nicaragua, Pakistan, Panama, South Africa, and Zimbabwe.

V.C. Drug development

The Calanolides. In 1988, an organic extract of the leaves and twigs of the tree, *Calophyllum lanigerum*, collected in Sarawak, Malaysia in 1987, through the NCI contract with the University of Illinois at Chicago (UIC) in collaboration with the Sarawak Forestry Department, showed significant anti-HIV activity. Bioassay-guided fractionation of the extract yielded (+)-calanolide A as the main in-vitro active agent.[51] Attempted recollections in 1991 failed to locate the original tree, and collections of other specimens of the same species gave only trace amounts of calanolide A. In 1992, UIC and botanists of the Sarawak Forestry Department undertook a detailed survey of C. lanigerum and related species. As part of the survey, latex samples of *Calophyllum teysmanii* were collected and yielded extracts showing significant anti-HIV activity. The active constituent was found to be (−)-calanolide B, which was isolated in yields of 20–30%. While (−)-calanolide B is slightly less active than (+)-calanolide A, it has the advantage of being readily available from the latex, which is tapped in a sustainable manner by making small slash wounds in the bark of mature trees without causing any harm to the trees. A decision was made by the NCI to proceed with the pre-clinical development of both the calanolides, and in June of 1994, an agreement based on the NCI Letter of Collection was signed between the Sarawak State Government and the NCI. Under the agreement, a scientist from the University of Malaysia Sarawak was invited to visit the NCI laboratories in Frederick to participate in the further study of the compounds.

The NCI obtained patents on both calanolides, and in 1995, an exclusive license for their development was awarded to Medichem Research, Inc., a small pharmaceutical company based near Chicago. Medichem Research had developed a synthesis of (+)-calanolide A[52] under a Small Business Innovative Research. (SBIR) grant from the NCI. The licensing agreement specified that Medichem Research negotiate an agreement with the Sarawak State Government. Meanwhile, by late 1995, the Sarawak State Forestry Department, UIC, and the NCI had collaborated in the collection of over 50 kg of latex of *C. teysmanii*, and kilogram quantities of (−)-calanolide B have been isolated for further development towards clinical trials. Medichem Research, in collaboration with the NCI through the signing of a Cooperative Research and Development Agreement (CRADA) by which NCI is contributing research knowledge and expertise, has advanced (+)-calanolide A through pre-clinical development, and was granted an INDA for clinical studies by the US Food and Drug Administration (FDA). The Sarawak State Government and Medichem Research formed a joint-venture company, Sarawak Medichem Pharmaceuticals Incorporated (SMP), in late 1996, and SMP has sponsored Phase I clinical studies with healthy volunteers. It has been shown that doses exceeding the expected levels required for efficacy against the virus are well tolerated. Trials using patients infected with HIV-1 were initiated in early 1999.

The development of the calanolides is an excellent example of collaboration between a source country (Sarawak, Malaysia), a company (Medichem Research, Inc.) and the NCI in the development of promising drug candidates, and illustrates the effectiveness and strong commitment of the NCI to policies promoting the rights of source countries to fair and equitable collaboration and compensation in the drug discovery and development process. The development of the calanolides has been reviewed as a 'Benefit-Sharing Case Study' for the Executive Secretary of the Convention on Biological Diversity by staff of the Royal Botanic Gardens, Kew.[53]

V.D. Screening agreement

In the case of organizations wishing to have pure compounds tested in the NCI drug-screening program, such as pharmaceutical and chemical companies or academic research groups, the DTP/NCI has formulated a screening agreement, which includes terms stipulating confidentiality, patent rights, routine and non-proprietary screening and testing versus non-routine, and levels of collaboration in the drug-development process. Individual scientists and research organizations wishing to submit pure compounds for testing generally consider entering into this agreement with the NCI. Should a compound show promising anticancer activity in the routine screening operations, the NCI will propose the establishment of a more formal collaboration, such as a Cooperative Research and Development Agreement (CRADA) or a Clinical Trial Agreement (CTA).

V.E. DTP WWW homepage

The NCI DTP offers access to a considerable body of data and background information through its www homepage: http://dtp.nci.nih.gov. Publicly available data include results from the human tumor cell line screen and AIDS antiviral drug screen, the expression of molecular targets in cell lines, and 2D and 3D structural information. Background information is available on the drug screen and the behavior of standard agents, NCI investigational drugs, analysis of screening data by COMPARE, the AIDS antiviral drug screen, and the 3D database. Data and information are only available on so-called open compounds, which are not subject to the terms of confidential submission.

In providing screening data on extracts, they are identified by code numbers only; details of the origin of the extracts, such as source organism taxonomy and location of collection, may only be obtained by individuals or organizations prepared to sign agreements binding them to terms of confidentiality and requirements regarding collaboration with, and compensation of, source countries. Such requirements are in line with the NCI commitments to the source countries through its LOC.

VI. Conclusion

As illustrated in the foregoing discussion, nature is an abundant source of novel chemotypes and pharmacophores. However, it has been estimated that only 5–15% of the approximately 250,000 species of higher plants have been systematically investigated for the presence of bioactive compounds,[32] while the potential of the marine environment has barely been tapped.[15] The *Actinomycetales* have been extensively investigated and have been, and remain, a major source of novel microbial metabolites;[54] however, less than 1% of bacterial and less than 5% of fungal species are currently known, and the potential of novel microbial sources, particularly those found in extreme environments, seems unbounded. To these natural sources can be added the potential to investigate the rational design of novel structure types within certain classes of microbial metabolites through genetic engineering, as has been elegantly demonstrated with bacterial polyketides. The proven natural product drug-discovery track record, coupled with the continuing threat to biodiversity through the destruction of terrestrial and marine ecosystems, provides a compelling argument in favor of expanded exploration of nature as a source of novel leads for the development of drugs and other valuable bioactive agents. It is apparent that nature can provide the novel chemical scaffolds for elaboration by combinatorial approaches (chemical and biochemical), thus leading to agents that have been optimized on the basis of their biological activities.

References

1. Herbalgram. (1998) 42:33–47.
2. Fallarino M. (1994) Herbalgram 31:38–44.
3. Chang HM But, PP-H. (1986) Pharmacology and applications of Chinese Materia Medica, Vols 1 and 2. Singapore: World Scientific Publishing.
4. Kapoor LD. (1990) CRC handbook of Ayurvedic medicinal plants. Boca Raton, FL: CRC Press.
5. Schultes RE, Raffauf RF. (1990) The healing forest. Portland, OR: Dioscorides Press.
6. Arvigo R, Black M. (1993) Rainforest remedies. Twin Lakes, WI: Lotus Press.
7. Gupta MP. (1995) 270 Plantas Medicinales Iberoamericanas. Bogota: Talleres de Editoral Presencia.
8. Ayensu ES. (1981) Medicinal plants of the West Indies. Algonac, MI: Reference Publications.
9. Iwu MM. (1993) Handbook of African medicinal plants. Boca Raton, FL: CRC Press.
10. Jain SK. (1991) Medicinal plants of India, Vols 1 and 2. Algonac, MI: Reference Publications.
11. Farnsworth NR, Akerele O, Bingel AS, Soejarto DD, Guo Z. (1985) Medicinal plants in therapy. Bull WHO 63:965–981.
12. Buss AD, Waigh RD. (1995) In: Wolff ME, editor. Burgers medicinal chemistry and drug discovery 5th edition, Vol. 1, Chapter 24, pp. 983–1033. New York: Wiley.
13. Buss AD, Waigh RD. (1995) In: Wolff ME, editor. Burgers medicinal chemistry and drug discovery 5th edition, Vol. 1, Chapter 20. New York: Wiley.
14. Zhang B, Salituro G, Szalkowski D, Zhang Z, Li Y, Royo I, Vilella D, Diez MT, Pelaez F, Ruby C, Kendall RL, Mao X, Griffin P, Calaycay J, Zierath J, Heck JV, Smith RG, Moller DE. (1999) Discovery of a small molecule insulin mimetic with anti-diabetic activity in mice. Science 284: 974–976.
15. McConnell O, Longley RE, Koehn FE. (1994) In: Gull VP, editor. The discovery of natural products with therapeutic potential. Boston, MA: Butterworth-Heinemann, pp. 109–174.
16. Olivera BM. (1997) Conus venom peptides, receptor and ion channel targets, and drug design: 50 million years of neuropharmacology. Mol Biol Cell 8:1–9.
17. Daly JW. (1998) Thirty years of discovering arthropod alkaloids in amphibian skin. J Nat Prod 61:162–172.
18. Cragg GM, Newman DJ, Snader KM. (1997) Natural products in drug discovery and development. J Nat Prod 60:52–60.
19. Hartwell JL. (1982) Plants used against cancer. Lawrence, MA: Quarterman.
20. Cragg GM, Boyd MR, Cardellina II, JH, Newman DJ, Snader KM, Cloud TG. (1994) In: Chadwick DJ, Marsh J, editors. Ethnobotany and the search for new drugs. Ciba Foundation symposium. Wiley, Chichester, UK, Vol. 185, pp. 178–196.
21. Cragg GM, Schepartz SA, Suffnes M, Grever MR. (1993) The taxol supply crisis. New NCI policies for handling the large-scale production of novel natural product anticancer and anti-HIV agents. J Nat Prod 56:1657–1668.
22. Cortes JE, Pazdur R. (1995) Docetaxel. J Clin Oncol 13:2643–2655.
23. Potmeisel M, Pinedo H. (1995) Camptothecins. New anticancer agents. Boca Raton, FL: CRC Press.
24. Foye WO. (1995) Cancer chemotherapeutic agents, ACS professional reference book. Washington, DC: American Chemical Society.
25. Carte BK. (1996) Biomedical potential of marine natural products. Bio-Science 46:271–286.
26. Philip PA, Rea D, Thavasu P, Carmichel J, Stuart NSA, Rockett H, Talbot DC, Ganesan T, Pettit GR, Balkwill F, Harris AL. (1993) Phase I Study of bryostatin 1: assessment of interleukin 6 and tumor necrosis factor alpha induction in vivo. The Cancer Research Campaign Phase I Committee. J Natl Cancer Inst 85:1812–1818.
27. Cragg GM, Newman DJ, Weiss RB. (1997) Coral reefs, forests, and thermal vents: the worldwide exploration of nature for novel antitumor agents. Semin Oncol 24:156–163.
28. Haar E ter, Kowalski RJ, Lin CM, Longley RE, Gunasekera SP, Rosenkranz HS, Day BW. (1996) Discodermolide, a crytotoxic marine agent that stabilizes microtubules more potently than taxol. Biochemistry 35:243–250.

29. Long BH, Carboni JM, Wasserman AJ, Cornell LA, Casazza AM, Jensen PR, Lindel T, Fenical W, Fairchild CR. (1998) Eleutherobin, a novel cytotoxic agent that induces tubulin polymerization, is similar to paclitxel. Cancer Research 58:1111–1115.
30. Lutz RA, Shank TM, Fornari DJ. (1994) Rapid growth at deep sea vents. Nature 371:663–664.
31. Lutz RA, Kennish MJ. Ecology of deep-sea hydrothermal vent communities: A review. Rev Geophys 31:211–242.
32. Baladrin MF, Kinghorn AD, Farnsworth NR. (1993) Plant-derived natural products in drug discovery and development. An overview. In: Kinghorn AD, Balandrin MF, editors. Human medicinal agents from plants. American Chemical Society symposium series 534, Washington, DC: American Chemical Society, pp. 2–12.
33. Young P. (1997) Major microbial diversity initiative recommended. ASM News 63:417–421.
34. Handelsman J, Rondon MR, Brady SF, Clardy J, Goodman RM. (1998) Molecular biological access to the chemistry of unknown soil microbes: a new frontier for natural products. Chem Biol 5:R245–R249.
35. Persidis A. (1998) Extremophiles. Nat Biotechnol 16:593–594.
36. Psenner R, Sattler B. (1998) Life at the freezing point. Science 280:2073–2074.
37. Adams MW, Kelly RM. (1998) Finding and using hyperthermophilic enzymes. Trends Biotechnol 16:329–332.
38. Colwell RR. (1997) Microbial diversity: the importance of exploration and conservation 302. J Ind Microbiol Biotechnol 18:302–307.
39. Hutchinson CR. (1999) Microbial polyketide synthases: more and more prolific. Proc Natl Acad Sci USA 96:3336–3338.
40. Gokhale RS, Tsuji SY, Cane DE, Khosla C. (1999) Dissecting and exploiting intermodular communication in polyketide synthases. Science 284:482–485.
41. Service RF. (1999) Race for molecular summits. Science 285:184–187.
42. Wender PA, DeBrabander J, Harran PG, Jiminez JM, Koehler MFT, Lippa G, Park CM, Siedenbiedel C, Pettit GR. (1998) The design, computer modeling solution structure, and biological evaluation of synthetic analogs of bryostatin 1. Proc Natl Acad Sci USA 95:6624–6629.
43. Martinez EJ, Owa T, Schreiber SL, Corey EJ. (1999) Phthalascidin, a synthetic antitumor agent with potency and mode of action comparable to ecteinascidin 743. Proc Natl Acad Sci USA 96:3496–3501.
44. Nicolaou KC, Roschangar F, Vourloumis D. (1998) Angew Chem Int Ed 37:2014–2045.
45. Borman S. (1999) Four types of natural products that stabilize cell microtubules share structural features. C&EN April 26:35–36.
46. Nicolaou KC, Kim S, Pfefferkorn J, Xu J, Oshima T, Hosokawa S, Vourloumis D, Li T. (1998) Synthesis and biological activity of sarcodictyins. Angew Chem Int Ed 37:1418–1421.
47. Schreiber SL, Tan DS, Foley MA, Shair MD. (1998) Stereoselective synthesis of over two million compounds having structural features both eminiscent of natural products and compatible with miniaturized cell-based assays. J Am Chem Soc 120:8565.
48. Schreiber SL. (1998) Chemical genetics resulting from a passion for synthetic organic chemistry. Bioorg Med Chem 6:1127–1152.
49. Mays TD, Mazan KD, Cragg GM, Boyd MR. (1997) In: Hoagland KE, Rossman AY, editors. Global genetic resources: access, ownership, and intellectual property rights. Washington, DC: Association of Systematics Collections, pp. 279–298.
50. Baker JT, Borris RP, Carte B, Cragg GM, Gupta MP, Iwu MM, Madulid DR, Tyler VE. (1995) Natural product drug discovery and development: new perspectives on international collaboration. J Nat Prod 58:1325–1357.
51. Kashman Y, Gustafson KR, Fuller RW, Cardellina II JH, McMahon JB, Currens MJ, Buckheit RW, Hughes SH, Cragg GM, Boyd MR. (1992) The calanolides, a novel HIV-inhibitory class of coumarin derivatives from the tropical rainforest tree, Calophyllum lanigerum. J Med Chem 35:2735–2743.
52. Flavin MT, Rizzo JD, Khilevich A, Kucherenko A, Sheinkman AK, Vilaychack V, Lin L, Chen W, Greenwood EM, Pengsuparp T, Pezzuto J, Hughes SH, Flavin TM, Cibulski M, Boulanger WA, Shone RL, Xu Z-Q. (1996) Synthesis, chromatographic resolution, and anti-human immunodeficiency virus activity of (+ / −)-calanolide A and its anantiomers. J Med Chem 39:1303–1313.
53. Ten Kate K, Wells A. (1998) Benefit-sharing case study. The access and benefit-sharing policies of the United States National Cancer Institute: a comparative account of the discovery and development of the drugs Calanolide and Topotecan. Submission to the Executive Secretary of the Convention on Biological Diversity by the Royal Botanic Gardens, Kew.
54. Horan AC. (1994) Actinomycetes. In: Gullo VP, editor. The discovery of natural products with therapeutic potential with therapeutic potential. Boston, MA: Butterworth-Heinemann, pp. 3–30.

Iwu and Wootton (eds.), Ethnomedicine and Drug Discovery

CHAPTER 2

Natural products for high-throughput screening

ALAN L HARVEY

Abstract

Not all traditional remedies are likely to be proven to be efficacious in scientific studies. This means that natural product researchers are going to accumulate negative results along with considerable information about the chemistry and biology of components found in the ingredients of the traditional leads. In any sizeable or long-term research programe involving natural product leads, there will be a large collection of background information and, potentially, a large number of apparently inactive extracts and compounds. Can something of value be done with such material? Since the vast majority of pharmaceutical drug discovery does not involve ethnopharmacological leads but is centered on high-throughput screening of the widest possible chemical diversity, the answer to this question has to be 'yes'. The difficulty in realizing the value of natural products for industrial drug discovery is in the area of communications between industrial and natural product researchers. There needs to be education in both directions: about issues relating to the Convention on Biological Diversity to industry, and about the technological demands of high-throughput screening to natural product researchers.

Keywords: *natural products, drug discovery, high-throughput screening, UN Convention on Biological Diversity, combinatorial chemistry, pharmaceutical industry*

I. Where do drugs come from?

Historically, ethnopharmacology was the origin of all medicines. Therefore, natural products were the most important source of drugs. In the 20th century, synthetic chemistry and then biotechnology offered alternatives to natural products. Due to technological convenience, the efforts of drug-discovery scientists have tended to be directed at synthetic rather than natural products.

Ethnopharmacology has provided some very notable past successes, including morphine (isolated in 1804), quinine (isolated in 1820), digitoxin (isolated in 1841), ephedrine (isolated in 1897), and tubocurarine (isolated in 1935). These compounds, or their analogs and derivatives, are still in widespread use. More recent developments undergoing trials and with an ethnopharmacological association include artemisinin for malaria, components from marigolds for psoriasis, flavones as anti-anxiety compounds, prostratin as an anti-viral, and the South African appetite suppressant being developed by Phytopharm.[1]

It is often claimed that using ethnopharmacological information will greatly increase the chances of discovering new drugs. However, it is not clear from published information whether this assertion is valid. Two large-scale studies provide some pointers to success rates with natural product screening. The National Cancer Institute (NCI) anti-cancer screening program revealed that screening of random samples gave a hit rate of 10.4%, while plants collected on the basis of some ethnopharmacological information gave a hit rate of 19.9%. However, plants that were known to be poisonous had a much higher initial hit rate, 50.0%. The second example comes from the Central Drug Research Institute in Lucknow, India. On a wide range of assays, randomly collected plants had a hit rate of 18.9%, whereas plants with associated ethnopharmacological uses had a hit rate of 18.3%.

One problem with assessing the relative merits of different plant selection strategies is that most of the published studies involve relatively few species against a small number of specific assays. In addition, negative results tend not to get published, and hence, overall failure rates are underestimated or unknown.

Looking at commonly accepted statistics of failure rates in drug-discovery and development programes of the pharmaceutical industry:

Discovery to development: 4 in 10 proceed
Toxicology and ADME: 1 in 2 proceed
Clinical Phase I: 2 in 3 proceed
Clinical Phase II: 1 in 2 proceed
Clinical Phase III: 3 in 4 proceed

That is, there is a 5% chance of a discovery project going from the preclinical stage to a marketed product. This would suggest the conclusion that most traditional leads will not make the transition into a pharmaceutical medicine.

II. What to do when traditional leads fail?

What can natural product researchers do when their traditional leads fail? The alternatives generally chosen are either to move on to another traditional lead, or to publish the phytochemical information without reference to biological activity. However, there is an alternative strategy. This involves embracing the concept that natural products are a superbly diverse chemical collection that can be productively applied to random screening approaches to drug discovery. This involves switching from traditional to modern approaches to drug discovery.[2]

Modern routes to drug discovery are dominated by molecular approaches, e.g. with cloned receptors and high throughput screening. Although molecular strategies also include so-called rational drug design and various genetic approaches, random screening is probably the most prevalent activity, and also the most relevant for natural products researchers. Random screening is likely to be more productive than other approaches because of the continuing technical challenges faced by the other technologies. For example, with rational drug design, there is a need for accurate three-dimensional information about the drug's binding target. Genomics gives an abundance of predicted protein sequence data, but potential therapeutic targets are not linear proteins. Therefore, structural information is required, but currently, X-ray crystallography is too slow to keep pace with the production of sequence information or not feasible because the protein targets are integral membrane proteins, and de-novo prediction of structures from linear protein sequences is still an inexact art. In the case of genetic approaches to treatment of disease, gene therapy will have to overcome major hurdles to go from dealing with single to multiple gene defects. Likewise, antisense RNA therapeutics have to demonstrate convincingly that they can provide the required specificity in a form that allows appropriate delivery to the body, to cells, and to the ultimate target at a cost that is affordable. By contrast, most of the technical problems associated with random screening appear to be solved, or at least solvable.

Random screening is the rapid testing of large numbers of compounds on molecular targets, and is referred to as 'high-throughput screening' (or HTS). It is built on molecular biology to provide the assays and robotics and information technology (IT) to provide the throughput (e.g. up to 500,000 points/week/assay). However, the success of any HTS program depends on the chemical diversity that is applied to the assay. This is because success depends on detecting a specific binding interaction between the target and one of the test compounds: the greater the variations in the three-dimensional shapes of the compounds and in their chemical characteristics, the greater the chance of a detectable interaction with the target. Compound supply, therefore, becomes critical for successful HTS programs, and three-dimensional diversity of compounds is even more important than the number of compounds.

Compounds for HTS can come from two main sources: natural products and synthetic chemistry. Some large collections of synthetic chemicals are available, e.g. from years of in-house synthesis in major pharmaceutical companies, or from collections of the output of many academic laboratories. However, these numbers are still small when the HTS system can assay half a million compounds each week. Because of this, many companies have turned to synthetic chemistry based on automated parallel syntheses, using so-called combinatorial chemistry. These methods can provide very high numbers of compounds, but they are generally lacking in three-dimensional diversity. Combinatorial chemistry also suffers from a number of practical issues relating, for example, to the purity of the screening samples, to their stability, and to the reproducibility of the syntheses on scale-up.

In contrast, natural products offer a much greater structural diversity. In a recent comparison of published databases, 40% of chemical scaffolds from natural products were found to be absent from synthetic libraries.[3] Additionally, very few of the world's natural products have been tested in HTS programs,[4] yet even today, nine out of the top 20 best-selling medicines are based on leads from natural products.[5]

III. What holds back the use of natural products in high-throughput screening?

What, then, is preventing the greater use of natural products in HTS? There are several real and perceived limitations of natural products: their chemical complexity; the difficulty of screening mixtures of compounds in natural product extracts; the time-consuming nature of natural products chemistry; the belief that screening of natural products gives rise to large numbers of artefacts; the supposedly common occurrence of synergistic actions between different components in an extract; the fear of poor reproducibility between different batches of extracts, possibly from seasonal effects on plant secondary metabolism; the uncertainty of being able to obtain supplies of an interesting extract in large quantities; and the general political problems of access to biodiversity and the implications of the United Nations Convention on Biological Diversity.

Given the continued successes deriving from natural product lead compounds, it seems worthwhile for the screening departments of pharmaceutical companies to seriously consider natural products as a source of compounds for screening. The structural diversity is acknowledged to be high, so can the other technical barriers be reduced?

Some of the comparisons of the analysis of databases of structures of natural products and synthetic compounds indicate that natural products are not necessarily much more complex chemicals than the synthetic products.[3] Also, there have been many advances in separation chemistry and in techniques for analysis and structural elucidation of natural products. Therefore, following up initial hits made with extracts is not necessarily any longer or more difficult than scaling up and reconfirming hits made from a combinatorial library. Additionally, the higher resolution of current techniques means that structures can be obtained from much smaller quantities of natural products than before, opening the way to early production of synthetic material and of analogs for optimization studies.

The major hindrance to the wider use of natural products in HTS systems appears to relate to the difficulties of access to biodiversity. Most companies with interests in screening natural products are aware of the United Nations Convention on Biological Diversity (CBD), but few companies are very clear on its impact.

The CBD has been ratified by most countries in the world in recognition of a need to encourage the preservation of the world's biodiversity. In addition to its conservation purposes, the CBD provides a framework for the sustainable exploitation of the genetic resources contained in biodiversity. Under the CBD, countries are recognized as having sovereign rights over the biodiversity within their national boundaries. Countries have to develop appropriate regulations to facilitate access to biodiversity for research and development purposes. Access has to be under principles of prior informed consent, and it must involve fair and equitable benefit-sharing. The involvement of source countries in research on their biodiversity and technology transfer to the bioresource-rich countries are expected under the CBD.

This raises many practical problems for companies wishing to access samples of natural products for screening. Although many countries have ratified the CBD several years ago, very few have introduced the necessary laws and regulations governing access. Also, it is daunting to companies seeking the broadest range of biodiversity because they would need to establish links with several different countries throughout the world.

Conversely, it is also daunting for research groups working on traditional medicinal leads to know how to gain additional value from natural products that are no longer actively being pursued in the traditional medicines validation program. Natural-product researchers may not find it easy to make appropriate contacts with counterparts in HTS companies. Also, the numbers of samples that they can contribute may be too small to interest a company.

For these reasons, it would be useful for groups of natural products researchers in different countries to operate in networks that pool their resources. If such networks also have a single contact for commercialization of natural products available for random screening, they will facilitate the development of collaborations with industry.

IV. SIDR's natural product network

An example of such a collaborative research network is that organized by Strathclyde Institute for Drug Research. (SIDR). It is long established, being operative for more than 10 years and self-sustaining through commercial interactions. The network was originally based on research contacts between the Phytochemistry Research Group of the University of Strathclyde and several departments and institutes in developing countries. The network is based on a collaborative approach to natural product-based drug research, wherever possible. Each institution in the network signs a legally binding agreement with the University of Strathclyde covering conditions of collecting (sustainable, no endangered species, compatible with local laws and regulations) and a commitment to training, technology transfer and involvement in research projects. Benefit-sharing from commercial exploitation coordinated by SIDR is on a 60:40 basis, with 60% of income going to the overseas institution. Currently, groups in 20 different countries are involved. The natural product library that has been created is largely based on higher plants.

Because of the geographical spread of the participants, samples from 90% of the families of higher plants are included in the collection. Such biodiversity guarantees exceptionally high chemical diversity, making the network's library particularly attractive for HTS purposes.

V. Conclusion

From the average numbers of compounds that drop out of industrial drug development during toxicological, pharmacokinetic and clinical testing, not all traditional leads will succeed as pharmaceutical medicines. However, natural products can be applied to HTS where their exceptional chemical diversity can provide new lead structures against molecular targets. Despite this, the pharmaceutical industry is not generally enthusiastic about using natural products in HTS systems. The advances in the processing of natural products need to be more widely appreciated, and there is a need for networks of natural products researchers to provide convenient and politically appropriate access to the world's biodiversity.

References

1. Harvey AL, Waterman PG. (1998) The continuing contribution of biodiversity to drug discovery. Curr Opin Drug Disc Dev 1:71–76.
2. Harvey AL. (1998) Advances in drug discovery techniques. Chichester, UK: Wiley.
3. Henkel T, Brunne RM, Muller H, Reichel F. (1999) Statistical investigation into the structural complementarity of natural products and synthetic compounds. Angewandte Chemie Int Ed 38:643–647.
4. Harvey AL. (2000) Strategies for discovering drugs from previously unexplored natural products. Drug Disc Today 5:294–300.
5. Harvey AL. (2001) Natural product pharmaceuticals: a diverse approach to drug discovery. Richmond, UK: PJB Publications.

CHAPTER 3

Medical ethnobotanical research as a method to identify bioactive plants to treat infectious diseases

THOMAS JS CARLSON

Abstract

In tropical rural communities, most people do not have access to pharmaceuticals. However, tropical countries contain the most biologically and culturally diverse traditional medicine systems in the world. Ethnobotanical diversity from tropical ecosystems supplies affordable medicines and foods to rural communities. As ecosystems are degraded and languages are lost, these traditional botanical systems and the healthcare they provide become significantly diminished. It is valuable to have interdisciplinary research teams of ethnobiologists with training in botany, anthropology, medicine, pharmacology, and field linguistics to understand how these traditional medical systems contribute to public health. Experimental biology research on tropical medicinal plants has demonstrated bioactivity for the treatment of malaria, infections of the skin, lungs, and gastrointestinal tract and other common diseases. The red stembark latex (sangre de drago) of *Croton lechleri* Müell.-Arg., a common medicinal plant in the western Amazon basin of South America, is used by local communities for a variety of ailments, including respiratory infections, gastric ulcers, diarrhea, herpes simplex lesions and skin infections, and as a wound healer. Based on medical ethnobotanical research, hypotheses were generated on the therapeutic benefits of this phytomedicine. Preclinical experimental biology studies were designed and conducted on different formulations derived from the stembark latex of this plant, and bioactivity was confirmed for the treatment of wound healing, herpes simplex virus, respiratory viruses, gastric ulcers, and secretory diarrhea. In human clinical studies, formulations of the stembark latex demonstrated activity for the topical treatment of herpes simplex virus and the oral treatment of both acute and chronic watery diarrhea.

Keywords: *ethnobotany*, Croton lechleri, *SP-303, SB-300, viral infections, standardized botanical extract, type 2 diabetes mellitus*

I. Introduction

When plants have been evaluated for medicinal qualities by large pharmaceutical companies in recent decades, the typical approach has been to collect large numbers of plant species randomly or taxonomically to feed the standardized high-throughput in vitro screening systems. In this high-throughput, random approach, plant species are screened for activity against a variety of diseases without consideration of whether or not the plant is used ethnomedically to treat any of the maladies. While this approach can lead to identification of safe and effective bioactive molecules,[1] a more efficient method is the ethnomedically directed screening of plants for medicinal bioactivity. Plant species are selected by conducting careful ethnobotanical field research with traditional botanical healers. Utilization of sound medical ethnobotany field research methods generates a high percentage of biologically active plants. An example of this is 40 out of 70 plants (57%) plants used ethnomedically to treat Type 2 diabetes demonstrated statistically significant antihyperglycemic activity in the db/db in vivo diabetic mouse model compared to a vehicle control.[2] This paper describes a spectrum of antiviral activities of medicinal plants. There is also a detailed description of the pre-clinical and human clinical therapeutic activity of formulations of *Croton lechleri* for the treatment of a variety of infectious disease induced forms of diarrhea.

II. Bioactivity of ethnobotanical plant collections compared to randomly collected plants

Several studies have demonstrated that there is a significantly higher rate of pharmacological activity in plant extracts used ethnomedically compared to extracts from randomly collected plants. An early comparison survey publication[3] evaluated multiple published studies on plant extracts assessed in anti-cancer in vitro assays at the NCI (National Cancer Institute). This survey reported anti-cancer in vitro activity in 52.2% of plants used as dart or arrow poisons, 38.6% of plants used as fish poisons, and 29.3% plants used to treat nematodes, while 10.4% of the randomly collected plants demonstrated activity. Numerous studies have been conducted comparing anti-infective activity of ethnomedically collected plants versus randomly collected plants. In Brazil, a study demonstrated anti-malarial activity in 0.07% (two of 273) of randomly collected plants compared to 18% (four of 22) of the plants used ethnomedically for treatments.[4] Anti-HIV activity was assessed in NCI (National Cancer Institute) assays comparing extracts of randomly collected plants from Belize to extracts of 'powerful plants' used by a traditional Mayan healer in Belize. In many traditional botanical medicines, tannins are present. When the tannins in the plant extracts were not selectively removed from the Belize plants, 25% (five of 20) of the ethnobotanically indicated plants were active compared to 6% (one of 18) of the randomly collected plants.[5] A larger group of plants from Belize were assessed in a similar way in the same NCI anti-HIV assay, and this study showed that 1.6% (one of 61) of the randomly collected plants were active compared to 15% (11 of 73) of the plants presented by the traditional Mayan healer.[6]

Another study assessing antiviral activity found that 25% of the ethnobotanical plant collections demonstrated anti-viral activity compared to 5% of the randomly collected plants.[7] The frequency of isolating antiviral compounds from natural products that were randomly collected was compared to medicinal plants identified through ethnobotanical research to treat viral infections.[2] In these studies, the randomly collected natural products were screened against an antiviral assay for herpes simplex, whereas the plants collected ethnobotanically were assessed against three different antiviral assays (cytomegalovirus, influenza, and respiratory syncytial virus). The screening programs for each of the four different viruses were supervised by the same virologist. The results are as follows: 15,000 randomly collected natural products screened against herpes simplex virus yielded two active compounds with 0.013% isolation frequency; 231 plants used ethnomedically to treat viral infections screened against cytomegalovirus yielded five active compounds with 2.2% isolation frequency; 123 plants used ethnomedically to treat viral infections screened against influenza virus yielded two active compounds with 1.6% isolation frequency; 97 plants used ethnomedically to treat viral infections screened against cytomegalovirus yielded eight active compounds with 8.2% isolation frequency. The frequency of isolating pure compounds with anti-viral activity from plants used ethnomedically to treat viral infections is, respectively, 123 times, 169 times, or 630 times higher, depending, respectively, on the virus (influenza, cytomegalovirus, or respiratory syncytial virus) being tested when compared to the frequency of isolating anti-herpes simplex virus compounds from randomly collected natural products. While different viruses were screened in these studies, these data suggest that being guided by ethnobotanical knowledge of local people can enable laboratory scientists to isolate pure pharmacologically active anti-viral compounds from plants more successfully.

III. Ethnobotanical uses of *Croton lechleri*

A medium-sized tree, *Croton lechleri* L., in the family Euphorbiaceae, is a medicinal plant species used widely by multiple ethnolinguistic groups in the western Amazon basin to treat a variety of ailments. Common names for this plant include sangre de drago, sangre de grado, conéwé (*waorani* in Ecuador), lan huiqui (*quichua* in Ecuador), uruchnum (*untsuri shuar* in Ecuador), and masujuain (*cofan* in Ecuador/ Colombia). The red bark latex of this tree is used in South America to treat a variety of diseases,[8] including: oral administration for diarrhea, stomach ulcers, hepatitis, lung problems, flu, cough, and diabetes; topical administration to treat herpes simplex lesions, cuts, and for wound healing. Additional ethnomedical reports on the use of this plant latex include: orally to treat rheumatism in Peru;[9] orally to treat leucorrhea, fractures, and piles in Peru;[10] topically as a hemostatic and wound healer, as a vaginal bath before childbirth, and orally for intestinal and stomach ulcers in Peru;[11] to treat cancer and applied topically for wound healing in Brazil and Ecuador;[12,13] topically as a wound healer in South America;[14] topically to treat inflammations in Ecuador.[13]

Other species (*Croton draco* and *C. urucurana*) with red latex have also been reported to be used as medicines. *Croton draco* is used in the following ways: latex is applied topically to treat bleeding in Mexico;[15] infusion of dried aerial parts is taken orally to treat fevers in Mexico;[16] dried aerial parts are used to treat cancer in Panama.[17] *C. urucurana* has been reported to treat the following diseases: stembark sap is applied topically as a wound healer, and to treat cancer, and taken orally to treat rheumatism in Brazil.[18]

Our primary ethnobotanical field research in South America has demonstrated the use of this latex to treat a variety of adult and child diseases of the gastrointestinal tract, respiratory system, skin, and mouth.[19]

IV. *Croton lechleri* uses to treat diarrhea

Diarrhea is a major cause of both morbidity and mortality world-wide, especially in tropical countries.[20] It is estimated that six million people die of diarrhea each year.[20] Diarrhea is emerging in developing countries as the major cause of death in children under five years of age.[20,21] In developed and developing countries, diarrhea is a growing problem in people living with HIV. Botanical medicines are used to treat diarrhea throughout the tropical world. Primary medical ethnobotany field research conducted in Peru in 1993 and 1994 found that the bark latex from *Croton lechleri* L. is taken orally as a medicine by indigenous people and mestizos in numerous South American countries to treat a variety of types of diarrhea and dysentery in children and adults including cholera,[19] since cholera is a type of secretory diarrhea. Based on our medical ethnobotanical research observations that the latex from *Croton lechleri* is used to treat cholera, we formulated the hypothesis that this botanical medicine works by an anti-secretory mechanism. To test this hypothesis, pre-clinical and clinical studies were designed and conducted.

V. Pre-clinical safety and efficacy studies of *Croton lechleri* formulations in treatment of diarrhea

An oligomeric proanthocyanidin compound, SP-303 (crofelemer) was extracted from the latex of *Croton lechleri*, an Amazonian rainforest tree in the plant family Euphorbiaceae.[8] The molecular weight of this compound is 2100. A standardized botanical extract, SP-300, was formulated from the latex of *Croton lechleri*. This standardized botanical extract contains the principal ingredient, SP-303. The SP-300 formulation is a 350 mg tablet containing 250 mg of SP-303.

There has been a strong medical need for a safe and effective anti-diarrhea agent with a new mechanism of action. Opiate-derived anti-diarrhea medicines, e.g. imodium, lomotil, and tincture of opium all act by reducing intestinal motility. Unfortunately, a common side-effect of these medicines is constipation. Studies in rodents demonstrated that SP-303 does not affect the intestinal motility. Safety was demonstrated in SP-303 in chronic consumption and LD_{50} studies in rodents and dogs. Safety was also demonstrated with SP-300 in LD_{50} studies in rats. To test our hypothesis that SP-303 and SP-300 act through an antisecretory mechanism, studies were conducted to assess activity of SP-303 and SP-300 in models of secretory diarrhea. Mechanism-of-action studies were conducted in a Ussing chamber, and in vivo efficacy studies were conducted in a cholera mouse model. The research in the cholera mouse model demonstrated that SP-303 has a significant inhibitory effect on fluid secretion into the intestinal lumen when given simultaneously with cholera toxin as well as when given 3 h after the cholera toxin was introduced.[22] In the in vitro Ussing chamber model, cAMP-mediated chloride ion secretion is measured as current across human colonic epithelial cells. Studies in the Ussing chamber demonstrate that SP-303 inhibits chloride-ion secretion, which mediates some forms of secretory diarrhea in humans.[22] A subsequent study compared the SP-303 compound to a SP-300 extract and found that both had a comparable significant inhibition of cAMP-medicated chloride ion secretion in the Ussing chamber.[23] These data support the finding that both the SP-303 compound and the SP-300 standardized botanical extract are potent inhibitors of chloride ion secretion.

VI. Human clinical studies of SP-303 from *Croton lechleri* in traveler's diarrhea

Phase I human clinical studies demonstrated safety in both infants and adults. Pharmacokinetic studies in infants and adult humans demonstrated that there is no significant absorption of the SP-303 compound into the bloodstream. A double-blind, randomized, placebo-controled study was conducted on tourists from the USA with traveler's diarrhea in Jamaica or Mexico. There was a statistically significant reduction in diarrhea in the 48 people in the 250 mg SP-303 treatment group compared to the 44 people in the placebo group.[24] The results of this study are as follows: compared to the placebo-control group, the treatment group that received 250 mg of SP-303 for 2 days significantly reduced the time to the last unformed stool during the 48-h ($P = 0.0004$), and the 72-h ($P = 0.0002$) treatment/surveillance period in patients with traveler's diarrhea; partial or complete improvement of diarrhea in the first 24 h was demonstrated in more than 90% of people in the treatment group which was statistically significant compared to the placebo group ($P = 0.003$); the treatment regimens were tolerated well by the patients. There was also a significant improvement in the treated group in subjective symptom scoring, such as relief from cramping and urgency, compared to the placebo group.

VII. Human clinical studies of *Croton lechleri* formulations in people with HIV and chronic diarrhea

A large percentage of people with HIV with or without AIDS have frequent bouts of watery diarrhea. This is caused by the disease state and can also be a side-effect of anti-HIV medications. For example, nelfinavir, a popular anti-HIV protease inhibitor medication, has been shown to cause secretory diarrhea in both in vitro and human clinical studies.[25] So, chronic watery diarrhea is a significant problem in HIV positive people whether or not they are treated with HAART (highly active anti-retroviral therapy).

VIII. Studies of SP-303 compound in people with HIV and chronic diarrhea

A randomized, double-blind, placebo-controlled, phase II safety and efficacy study was conducted to assess the safety and efficacy of orally administered SP-303 for the symptomatic treatment of diarrhea in people with HIV. This study was conducted at the Palo Alto Veterans' Administration Hospital at Stanford University Medical Center and the San Francisco General Hospital at the University of California at San Francisco Medical Center. A daily measures analysis of treatment with 500 mg of SP-303 taken orally four times a day for 4 days in 51 people (25 placebo/26 SP-303) revealed the following results:[26] a significant reduction in stool weight in treatment group compared to the placebo group ($P = 0.008$) (see Figure 1); a significant reduction in stool frequency in treatment group compared to the placebo group ($P = 0.04$); treatment regimens were tolerated well by patients. A Phase III

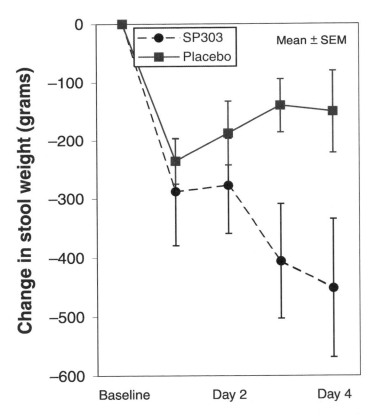

Fig. 1. Double-blind, placebo-controled study in people with HIV and chronic diarrhea demonstrating a significant reduction in stool weight in the SP-303 treatment group compared to the placebo group ($P = 0.008$). From Ref. 26.

randomized, double-blind, placebo-controled 7-day efficacy and safety study was conducted in which orally administered 500 mg SP-303 was given four times a day for 7 days in people with HIV and chronic diarrhea. The SP-303 treatment group (98 people), compared to the placebo group (100 people), revealed a statistically significant reduction in stool weight $P = 0.033$ in all people and $P = 0.008$ in people with severe diarrhea, i.e. > 1000 g stool/day. SP-303 was well tolerated by the people in the Phase II and Phase III studies.

IX. Studies of SP-300 standardized botanical extract in people with HIV and chronic diarrhea

The anti-diarrhea effect of a standardized botanical extract, SP-300, was assessed in a 20-patient open label safety and efficacy study on HIV+ people with chronic diarrhea. The open label study was conducted at San Francisco General Hospital at the University of California, San Francisco Medical Center. People in the study received one or two 350-mg tablets of SP-300 four times a day by mouth for 2 weeks. The pre-treatment 2-day baseline stool weights compared to the stool weights on day 13 and day 14 of treatment demonstrated a statistically significant reduction in stool weight $(P = 0.0026)$.[27] In the study, the SP-300 was safe and well tolerated.

A study was also conducted on quality-of-life issues in a combined group of 42 people in the SP-300 study and the Phase III SP-303 study. This study[28] demonstrated a significant improvement in the following measures of quality of life: ability to leave home $(P = 0.03)$; time spent resting $(P = 0.03)$; sexual activity. (0.01); and sum score of activities of daily living (0.024).

Conclusion

Affordable, safe, and effective locally available botanical medicine treatments based on ethnobotanical knowledge are used in tropical communities around the world. Multiple experimental biology studies in a variety of different disease categories described earlier in this paper demonstrate a higher rate of bioactivity in plants used medicinally by local people compared to plants collected randomly for the treatment of malaria and a spectrum of different viral infections. These studies confirm the wisdom of the local ethnomedical and ethnolinguistic systems that have identified and developed safe and therapeutic medicinal plants that grow within the different ecosystems.[29]

The ethnomedical uses, experimental biology studies, and human clinical studies on sangre de drago, the stembark latex of *Croton lechleri*, are described in detail in this paper. *Croton lechleri* latex is taken orally by South American peoples to treat different types of watery diarrhea, including cholera. The ethnomedical use of *Croton lecheri* provided information that generated a hypothesis on a potential antisecretory mechanism of action for treatment of diarrhea. This hypothesis was tested, and different formulations of *Croton lechleri* (SP-303 compound and SB-300 extract) have demonstrated potent antisecretory activity in a cellular anti-diarrhea model, while the SP-303 compound also showed strong antisecretory activity in a cholera-infected mouse diarrhea model. The mechanism of action of most types of watery diarrhea is secretory. Human clinical studies showed that *Croton lechleri* formulations are effective in treating traveler's diarrhea (SP-303 compound) and HIV-associated diarrhea (SP-303 compound and SB-300 extract), formulations derived from *Croton lechleri* latex that are now available to people in urban areas for the treatment of watery diarrhea. The *Croton lechleri* tree grows abundantly in the western Amazon basin and provides local people with an inexpensive, safe, and effective treatment for a variety of ailmants, including watery diarrhea and herpes simplex skin infections.

References

1. Cragg GM, Newman DJ, Snader KM. (1997) Natural products in drug discovery and development. J Nat Prod 60:52–60.
2. Carlson TJ, Cooper R, King SR, Rozhon EJ. (1997) Modern science and traditional healing. Special publication. R Soc Chem, 200 (Phytochemical Diversity), 84–95.
3. Spjut RW, Perdue RE Jr. (1976) Plant Folklore: A tool of predicting sources of antitumor activity? Cancer Treat Rep 60:979–985.
4. Carvalho LH, Krettli AU. (1991) Antimalarial chemotherapy with natural products and chemically defined molecules. Memorias do Instituto Oswaldo Cruz, Rio de Janeiro 86 (Suppl 11):181–189.
5. Balick MJ. (1990) Ethnobotany and the identification of therapeutic agents from the rainforest. In Chadwick DJ, Marsh J, editors. Bioactive compounds from plants, CIBA Foundation Symposium 154, pp. 22–39.
6. Balick MJ. (1994) Ethnobotany, drug development and biodiversity conservation: exploring the linkages. In: Chadwick DJ, Marsh J, editors. Ethnobotany and the search for new drugs, CIBA Foundation Symposium 185, pp. 4–24.
7. Vlietnick AJ, Vanden Berghe DA. (1991) Can ethnopharmacology contribute to the development of antiviral agents? J Ethnopharmacol 32(1–3):141–154.
8. Ubillis R, Jolad SD, Bruening RC, Kernan MR, King SR, Sesin DF, Barrett M, Stoddart CA, Flaster T, Kuo J, Ayala F, Meza E, Castanel M, McMeekin D, Rozhon E, Tempesta MS, Barnard D, Huffman J, Smee D, Sidwell R, Soike K, Brazier A, Safrin S, Orlando R, Kenny PTM, Berova N, Nakanishi K. (1994) Phytomedicine 1:77–106.
9. Persinos GJ, Blomster RN, Blake DA, Farnsworth NR. (1979) South American Plants. II. Taspine, Isolation and Antiinflammatory Activity. J Pharm Sci 68(1):124–126.
10. Rutter RA. (1990) Catalogo de plantas utiles de la Amazonia Peruana. Instituto Linguistico de Verano. Yarinacocha, Peru, p. 349.
11. Duke JA, Vasquez R. (1994) Amazonian ethnobotanical dictionary. Boca Raton, FL: CRC Press, p. 58.
12. Pieters L, De Bruyne T, Mei G, Lemiere G, Vanden Berghe D, Vlietinck AJ. (1992) In vitro and in vivo biological activity of South American dragon's blood and its constituents. Planta Med Suppl 58(1):A582–A583.
13. Cai Y, Chen ZP, Phllipson J. (1993) Diterpenes from Croton lechleri. Phytochemistry 32(3):755–760.
14. Pieters L, Bruyne TD, Claeys M, Vlietnick A. (1993) Isolation of a dihydrobenzofuran lignan from South American dragon's blood (*Croton* spp.) as an inhibitor of cell proliferation. J Nat Prod 56(6):899–906.
15. Zamora-Martinez MC, Pola CNP. (1992) Medicinal plants used in some rural populations of Oaxaca, Puebla and Veracruz, Mexico. J Ethnopharmacol 35(3):229–257.
16. Hernandez J, Delgado G. (1992) Terpenoids from aerial parts of *Croton draco*. Fitoterapia 63(4):377–378.
17. Gupta MP, Monge A, Karikas GA, Lopez De Cerain A, Solis PN, De Leon E, Trujillo M, Suarez O, Wilson F, Montenegro G, Noriega Y, Santana AI, Correa M, Sanchez C. (1996) Screening of

panamanian medicinal plants for brine shrimp toxicity, crown gall tumor inhibition, cytotoxicity and DNA intercalation. Int J Pharmacog 34(1):19–27.
18. Peres MTLP, Monache FD, Cruz AB, Pizzolatti MG, Yunes RA. (1997) Chemical composition and antimicrobial activity of *Croton urucurana* Baillon (Euphorbiaceae). J Ethnopharmacol 56(3):223–226.
19. Carlson TJS, King SR. (2000) Sangre de Drago (*Croton lechleri* Müell.-Arg.) – a phytomedicine for the treatment of diarrhea. Health Notes: Rev Complement Integr Med 7(4):315–320.
20. Guerrant RL. (1995) Principles and syndromes of enteric infection. In: Mandell GL, Bennett JE, Dolin R, editors. Principles and practice of infectious diseases, 4th edition. New York: Churchill Livingstone, pp. 945–962.
21. Guerrant RL, Hughes JM, Lima NL, Crane J. (1990) Diarrhea in developed and developing countries: magnitude, special settings, and etiologies. Rev Infect Dis 12 (Suppl 1):S41–S50.
22. Gabriel SE, Davenport SE, Steagall RJ, Vimal V, Carlson T, Rozhon EJ. (1999) A novel plant-derived inhibitor of cAMP-mediated fluid and chloride secretion. Am J Physiol 276 (Gastrointest Liver Physiol 39):G58–G63.
23. Illek B, Machen TE, Widdicombe JH, Carlson T, Chow J. (2000) A Novel Botanical Extract, DS4/44F2 from the Bark Latex of *Croton lechleri* Inhibits cAMP-mediated chloride secretion in human colonic epithelial cells, the extraction process of DS4/44F2 was later optimized to produce the botanical extract for SB-Normal Stool Formula℠. Poster presented to 17th Annual Aids Investigators' Meeting and 3rd Annual Conference on AIDS Research in California, 2/25/00 South San Francisco, CA.
24. Dicesare D, DuPont HL, Mathewson JJ, Ericsson CD, Ashley D, Martinez-Snadoval FG, Pennington JE, Porter SB. (1998) A double blind, randomized, placebo controled study of SP-303 in the symptomatic treatment of acute diarrhea among travelers to Mexico and Jamaica. Abstract presented in 1998 in Denver at Infectious Diseases Society of America, 36th Annual Meeting, Denver, 1998.
25. Andrade A, Sears C, Rufo P, Lencer W, Flexner C. (2000) Nelfinavir (NFV)-associated diarrhea: secretory versus osmotic. Poster presented at the 7th Conference on Retroviruses and Opportunistic Infections, Jan 30–Feb 2, 2000, San Francisco, CA.
26. Holodniy M, Koch J, Mistal M, Schmidt JM, Khandwala A, Pennington JE, Porter SB. (1999) A double blind, randomized, placebo-controlled, phase II study to assess the safety and efficacy of orally administered SP-303 for the symptomatic treatment of diarrhea in patients with AIDS. Am J Gastroenterol 94(11):3267–3273.
27. Koch J, Tuveson J, Carlson T, Schmidt JM. (1999a) Human studies on SB-300 in HIV-associated diarrhea SB-300: a new and effective therapy for HIV-associated diarrhea poster presented at the Seventh European Conference on Clinical Aspects and Treatment of HIV-Infection, October 23–27, 1999, Lisbon.
28. Koch J, Tuveson J, Carlson T, Schmidt JM. (1999b) SB-300: A new and effective therapy for HIV-associated diarrhea improves quality of life. Poster presented at the Seventh European Conference on Clinical Aspects and Treatment of HIV-Infection, October 23–27, 1999, Lisbon.
29. Carlson TJS. (2001) Language, ethnobotanical knowledge, and tropical public health. In: Maffi L, editor. On biocultural diversity: linking language, knowledge, and the environment. Washington, DC: Smithsonian Institute Press.

Iwu and Wootton (eds.), Ethnomedicine and Drug Discovery

CHAPTER 4

Development of HerbMed®: an interactive, evidence-based herbal database

JACQUELINE C WOOTTON

Abstract

HerbMed® is an interactive, electronic herbal database providing hyperlinked access to the scientific data underlying the use of herbs for health. Hyperlinks and dynamic links are made to PubMed and other electronic resources, providing evidence-based information for healthcare professionals, pharmacists, researchers and healthcare consumers. The features of this resource are discussed from the perspective of ethnobotanists and field biologists, and of chemists and biochemists interested in drug development for infectious diseases. Searching the database yields new information that cannot be obtained from PubMed alone. The resource has a breadth and comprehensiveness that enable creative cross-referencing and unexpected links to provide fresh insights. HerbMed® is freely available on the World Wide Web at: www.herbmed.org

Keywords: *HerbMed®, herbal database, evidence-based resources, infectious diseases, hepatitis, malaria, leishmaniasis*

I. Introduction

HerbMed® is an interactive, electronic herbal database providing hyperlinked access to the scientific data underlying the use of herbs for health. It is an evidence-based information resource, designed and built by the Alternative Medicine Foundation to serve the needs of healthcare professionals, pharmacists, researchers, and consumers. The database is freely available on the World Wide Web at http://www.herbmed.org.

HerbMed® was specifically researched, designed and developed to provide:

- healthcare professionals with convenient access to the supporting data underlying the use of herbs for health;
- pharmacists and research scientists with convenient access to information on herb actions and mechanisms of action;
- healthcare consumers with responsible and reliable information about the herbal products they purchase for medicinal or preventive use.

In designing the resource, the emphasis was on speed and useful content, so minimal graphics are used. The database will be re-engineered to improve the interface for compilers, editors and users, but the same design principles will be retained.

II. Description

Each herb is identified by common name and Latin binomial. Information on individual herbs is sorted into clear categories. The information is as impartial and comprehensive as possible. Some selection is necessary where there is a very large amount of information. This judgment is made by the Research Pharmacologist, on the basis of relevance and completeness. Information on individual herbs is hyperlinked directly to the scientific literature using PubMed, the freely available interface for MEDLINE, and to other research resources. Dynamic links are used to create automatic updating of information, and a search engine allows for simple searches by own choice keywords, such as disease category.

There are six main categories of information for each herb entry:

- **Evidence for activity**: This includes data from controled clinical trials, observational data and case studies, and empirical knowledge from long-standing folk or traditional use.
- **Warnings**: These include links to published information on contraindications, toxic and adverse effects, and drug interactions.
- **Preparations**: These give data on commercial methods of preparation and suppliers, along with information on folk methods of preparation suitable for a home or office dispensary.
- **Mixtures**: These may be novel commercial mixtures or blends from traditional or folk systems where the multiple components are reputed to have a synergistic effect.
- **Mechanisms of action**: The primary constituents, the biochemistry of action, and miscellaneous information on the scientific basis of action.
- **Other**: For links to herb pictures, related keyword search strategies that are automatically updated from the PubMed database, conservation and cultivation issues, and related Web sites.

III. Ethnobotanical and field biology uses of HerbMed®

Figure 1 provides a sample standard format using the example of Garcinia kola, also known by the common name, bitter kola. By clicking under 'Evidence' on 'clinical trials', few entries are found, reflecting the fact that there is a dearth of clinical data at present. However, many entries are listed under 'Biochemical Mechanisms of Activity'. Clicking on 'Related Keyword PubMed Links', a dynamic link function, automatically updates the resource listing the latest peer-reviewed research, a clear indication that there is a great deal of interest and new research in this area.

Under the category title, 'Warnings', the Food and Drug Administration (FDA) link is used dynamically to provide automatic updating in the event of new reported

Alternative Medicine Foundation
HerbMed ®

common name: Bitter Kola

scientific name: **Garcinia kola**

Evidence for Activity:

Human Clinical Data | Case Reports | Traditional & Folk Use

Warnings:

Contraindications | Toxic & Adverse effects | Interactions

Preparations:

Commercial | Suppliers | Folk & Traditional

Mixtures:

Modern | Traditional herbals

Mechanism of action:

Constituents | Biochemistry | Miscellaneous

Other:

Pictures | Related Keyword Pubmed Links | Other Web Site Links

Copyright c 1998 Alternative Medicine Foundation

Fig. 1. Standard HerbMed® format for a herb record using the example of *Garcinia kola*.

information on adverse events, contraindications, or toxicity information for any of the herbs listed.

Figure 2 demonstrates the search function and shows the results of searching on 'Tubercul* TB'. The 'wild card', *, is used to maximize the search. Each of the herb entries retrieved can be individually searched using the 'Find' function, under 'Edit' on the top bar, for the pertinent reference or references to tuberculosis.

Ethnobotanists, field biologists, chemists and biochemists interested in drug development for infectious diseases will find that searching on the database yields new information that cannot be obtained from PubMed or other bibliographic databases alone. The resource has a breadth and comprehensiveness that enable creative cross-referencing and unexpected links to provide fresh insights for new approaches to diseases of the developing countries.

A current (January 2000) search on 'Hepatitis' yields a listing of 26 herbs with full information on the current state of the research. A search on malaria produces seven hits. A search on Leishmania*, using the asterisk as a wild card, comes up with five

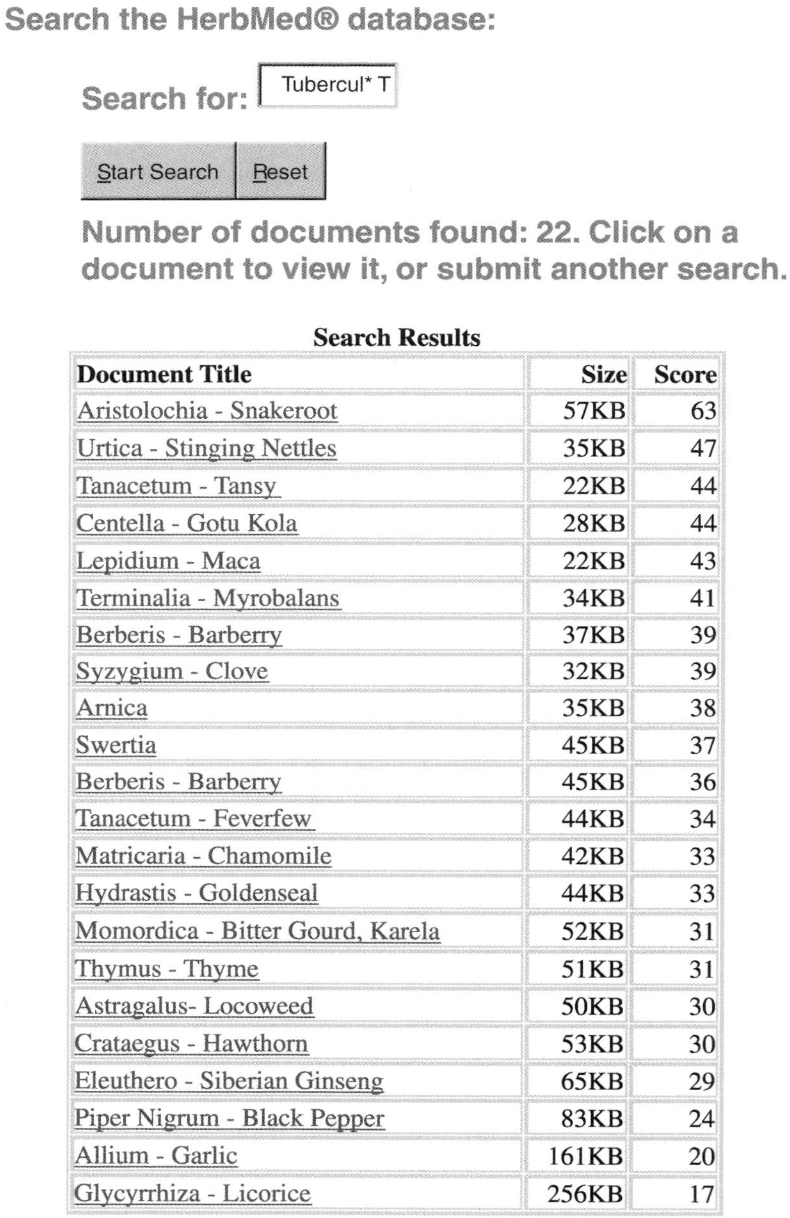

Search the HerbMed® database:

Search for: | Tubercul* T |

| Start Search | Reset |

Number of documents found: 22. Click on a document to view it, or submit another search.

Search Results

Document Title	Size	Score
Aristolochia - Snakeroot	57KB	63
Urtica - Stinging Nettles	35KB	47
Tanacetum - Tansy	22KB	44
Centella - Gotu Kola	28KB	44
Lepidium - Maca	22KB	43
Terminalia - Myrobalans	34KB	41
Berberis - Barberry	37KB	39
Syzygium - Clove	32KB	39
Arnica	35KB	38
Swertia	45KB	37
Berberis - Barberry	45KB	36
Tanacetum - Feverfew	44KB	34
Matricaria - Chamomile	42KB	33
Hydrastis - Goldenseal	44KB	33
Momordica - Bitter Gourd, Karela	52KB	31
Thymus - Thyme	51KB	31
Astragalus- Locoweed	50KB	30
Crataegus - Hawthorn	53KB	30
Eleuthero - Siberian Ginseng	65KB	29
Piper Nigrum - Black Pepper	83KB	24
Allium - Garlic	161KB	20
Glycyrrhiza - Licorice	256KB	17

Fig. 2. Demonstration of the HerbMed® search function using the search terms: 'Tubercul* TB'.

hits: *Berberis* sp. (Barberry), *Echinacea*, *Swertia* sp., *Hydrastis canadensis* L. (Goldenseal) and *Guatteria gaumeri* Greenman. A typical result from the Guatteria gaumeri entry is shown in Figure 3. Once again, there is no clinical evidence, but

Alternative Medicine Foundation
HerbMed®

common name: .

scientific name: **Guatteria gaumeri Greenman**

Evidence for Activity:

Human Clinical Data   Case Reports   Traditional & Folk Use

Warnings:

Contraindications   Toxic & Adverse effects   Interactions

Preparations:

Commercial   Suppliers   Folk & Traditional

Mixtures:

Modern   Traditional herbals

Mechanism of action:

Constituents   Biochemistry   Miscellaneous

Other:

Pictures   Related Keyword Pubmed Links   Other Web Site Links

---

Copyright c 1998 Alternative Medicine Foundation

Plain English summaries of major research
articles from Medline abstracts with hyperlink
to original. (PubMed is the public version of Medline
from U.S. National Library of Medicine.)

Traditional & Folk Use Reports

Leishmaniasis traditional remedies includes Guatteria foliosa,
which contains isoguattouregidine (an isoquinoline alkaloid)
Akendengue 1999 .

EthnobotDB .

top of page

Fig. 3. HerbMed® data record for *Guatteria gaumeri* – showing only the category 'Evidence for Activity'.

under 'Traditional and Folk Use', there is a review article by Akendengue (1999) on the use of natural products against Leishmaniasis. By clicking on the article, and then on related articles, a host of more relevant articles can be located.

New entries are added to the database monthly, so the potential for fresh findings and creative use of cross-referencing continues to grow.

IV. Academic model

The Alternative Medicine Foundation has used what may be termed a standard academic model to develop this database. The model is characterized by the following features:

● open, global access to basic research material;
● neutral presentation of relevant information and hyperlinks;
● current and comprehensive content, frequent updates;
● no frills, minimum cost, sustainable, technically feasible and robust operation;
● priority given to logic of presentation for researcher needs;
● editorial rather than promotional role of designers, developers and compilers.

HerbMed® epitomizes high leverage for minimal investment – appropriate for an underfunded research area. The intention is to continue to develop the resource through world-wide collaborations funded by charitable donations, research grants, and possibly also money from industry, but only on the proviso that there could be no direct sponsorship of specific parts of the database, or of particular herb entries, so ensuring that the database retains its status as an impartial resource.

Acknowledgment

The author gratefully acknowledges Soaring Bear, a post-doctoral research pharmacologist and biochemist, with a background in herbalism, for his work in compiling the herb entries.

Reference

1. Akendengue B et al. (1999) Recent advances in the fight against leishmaniasis with natural products. Parasite 6(1):3–8.

Iwu and Wootton (eds.), Ethnomedicine and Drug Discovery

CHAPTER 5

Natural products: a continuing source of inspiration for the medicinal chemist

S Mbua Ngale Efange

Abstract

The study of naturally occurring compounds has led to innumerable advances in pharmacology and medicine. For the medicinal chemist, a biologically active natural compound provides a useful template for designing novel molecules, which can be used to probe the mechanism of action of the parent compound, or to develop drugs with improved pharmacological profiles. The discipline of medicinal chemistry is therefore crucial to modern drug development. Developing countries are richly endowed with flora containing many useful natural products. Unfortunately, little effort has been made in these same countries to develop the expertise (in areas such as medicinal chemistry) required to mount a credible drug-development effort. The lack of expertise in modern drug-development methods has relegated most developing countries to the role of raw material supplier. To address this inadequacy, training programs in modern drug design and development should become an integral part of national policy in the South. In the subsequent presentation, two case studies are provided to illustrate the use of natural products in drug design.

Keywords: *pharmacology, medicinal chemistry, drug-development, natural products*

I. Introduction

Modern medicine traces its origins to Man's use of plant and animal parts in alleviating human suffering caused by disease. Some of the earliest compilations of the medicinal uses of plants include the De materia medica of Dioscorides (AD 78) and the Egyptian Papyrus Ebers (1550 BC) (reviewed in Ref. 1). It is estimated that as recently as 1985, over 120 pharmaceutical products in use were derived from plants.[2] Moreover, up to 75% of these agents appear to have been discovered by examining the use of these plants in traditional medicine. According to some estimates, there are at least 250,000 plant species in existence on this planet (most of which are found in the tropical rain forest). Therefore, it is fair to conclude that plants remain an important source of new drugs.

The benefits of plant products extend far beyond their immediate use as medications. A review of the scientific literature shows that natural products have contributed immensely to the advancement of pharmacology and physiology. Historically, validation of the medicinal uses of a given plant has been invariably followed by attempts to isolate and determine the structure of the active ingredient(s). In the ensuing investigation of the isolated compound, new insights into the mechanism of drug action have been revealed, new areas of pharmacology have been born, and new drugs have been developed.[3] In this connection, the annals of modern pharmacology have recorded impressive gains from natural products such as heroin, nicotine, acetylcholine, penicillin.

Since molecular structure is intricately linked to biological activity, for the medicinal chemist, the natural product represents a treasure trove of possibilities. When the mechanism of action of a compound is unknown, the synthesis and study of carefully designed analogs of the lead compound can provide deeper insights into the interactions of the compound with the biological target. If the mode of action of the compound is known, modification of the lead structure can be used to fine-tune the drug-molecular target interaction so as to produce the desired biological response. Finally, lead modification is also employed simply to alter the physico-chemical characteristics of a given molecule so that the latter can be amenable to formulation (e.g. as an oral dosage form).

In the following examples, we demonstrate (a) how lead modification was used to simplify molecular structure while maintaining the desired pharmacological characteristics, and (b) how knowledge of the mechanism of action of a natural product was used in the design of a subsequent generation of pharmacologically active agents.

II. Example 1: Design of ibogaine analogs

Ibogaine (1, Figure 1) is the major constituent of the root of Tabernanthe iboga, a shrub commonly found in West and Central Africa.[4,5] Ibogaine has been shown to reduce cocaine and morphine self-administration in laboratory animals (reviewed in Ref. 6), and to attenuate alcohol intake by alcohol-preferring rats.[7] The compound also reduces the rewarding effects of nicotine in animals.[8,9] In addition, the compound reduces craving in humans.[10–13] Ibogaine is thus a potentially useful anti-addictive agent.

The goal of this study was to develop synthetic ibogaine-like compounds that might be more effective than the parent compound. Due to the fact that ibogaine is a relatively complex structure, our immediate objective was to identify the simplest ibogaine fragment that retains anti-addictive activity. The fragmentation approach has been used extensively in medicinal chemistry to probe ligand–receptor interactions and to generate novel molecular entities.[14] A close inspection of ibogaine reveals that the latter is composed of two major fragments, 2 and 3 (Figure 1), derived from 5-methoxytryptamine and isoquinuclidine. An intermediate structure, which incorporates elements of both fragments, is the tricyclic heterocycle 4 (Figure 1). In previous studies of conformationally restricted tryptamines, compounds derived from this basic skeleton had been found to antagonize aggressive behavior in animals[15] and to display potential antidepressant activity.[16] The current study investigated the biological activities of the corresponding azepino[4,5-b]bezothiophenes.

The target compounds (12a–d) were synthesized as shown in Figure 2. When tested in vitro (Table 1), all four compounds displayed moderately high affinity for

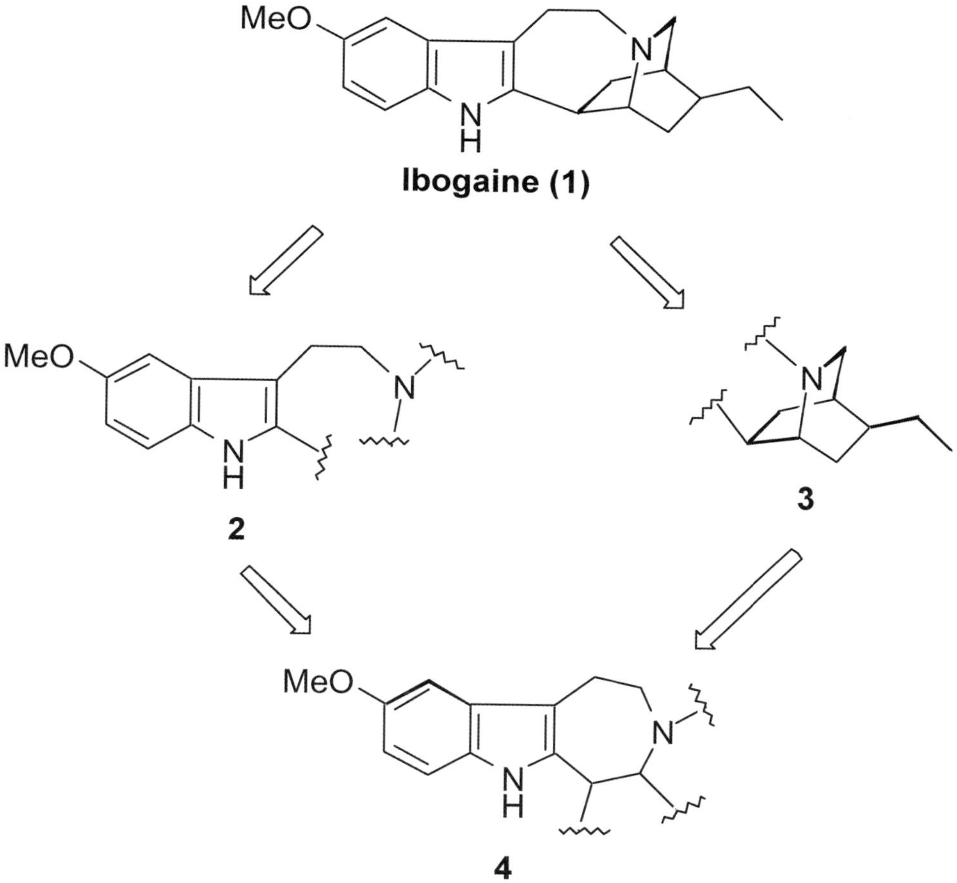

Ibogaine (1)

2

3

4

Fig. 1. Ibogaine and selected ibogaine fragments.

dopamine and serotonin transporters,[17] two molecular targets implicated in the mode of action of cocaine[18,19] and ibogaine.[20,21] The compounds also interact with other molecular targets with varying degrees of effectiveness and thus appear to mimic the parent compound. In subsequent studies, compounds 12a, 12c and 12d were found to dose-dependently reduce cocaine self-administration in rats without affecting general locomotor activity (Efange et al., unpublished). Taken together, the foregoing investigations confirmed that the tricyclic ibogaine fragment 4 retains anti-addictive activity and is thus a useful platform for developing new anti-addictive agents.

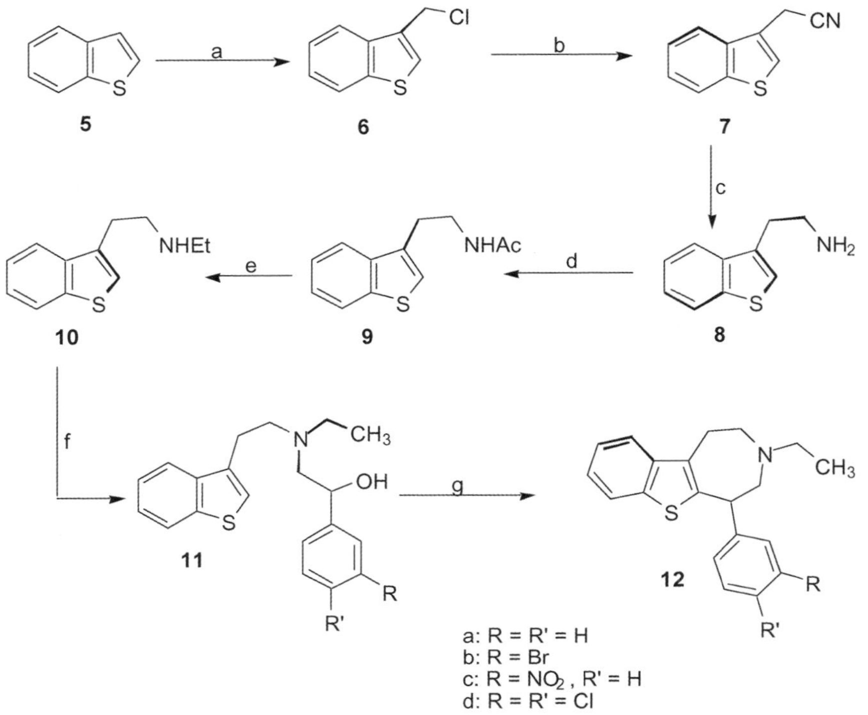

a: CH_2O (aq.), HCl(g), conc. HCl; b: KCN, PTC,H_2O, 90-95 °C; c: $LiAlH_4$, ether, $AlCl_3$, reflux;
d: Ac_2O, Et_3N; e: BH_3.THF, reflux; f: styrene oxide, EtOH, reflux; g: CF_3CO_2H, conc. H_2SO_4 (cat.).

Fig. 2. Synthesis of substituted hexahydroazepino[4,5-b]benzothiophenes.

Table 1
Relative Affinities (IC$_{50}$ ± SD, μM) of hexahydroazepino[4,5-b]benzothiophenes and Reference
Compounds at Selected Molecular Targets*

Target	DAT (WIN35,428)	SERT (RTI-55)	D1 (SCH23390)	D2 (YM-091512)	D3 (7-OHDPAT)
12a	0.14 ± 0.003	0.30 ± 0.001	3 ± 0.05	5.0 ± 0.07	0.6 ± 0.01
12b	0.21 ± 0.02	4.30 ± 0.01	5 ± 0.05	3.5 ± 0.28	2.0 ± 0.28
12c	0.25 ± 0.01	1.10 ± 0.002	5 ± 0.06	3.1 ± 0.14	0.6 ± 0.04
12d	0.25 ± 0.001	1.20 ± 0.002	70 ± 0.09	5.0 ± 0.21	2.0 ± 0.14
Ibogaine	4.11 ± 0.45[a]	0.59 ± 0.09[a]	>10[a]	>10[a]	>10[a]
Noribogaine	3.35 ± 0.50[a]	0.04 ± 0.01[a]	>10[a]	>10[a]	>10[a]

*Data taken from ref. 17. Values shown for reference compounds are Ki ± SD, μM.

III. Example 2: Design of dual modulators of serotonergic and opioidergic function

At pharmacologically relevant doses, ibogaine interacts with several molecular targets implicated in drug dependence, including serotonin transporters,[20,21] mu and kappa opioid receptors,[22-24] NMDA,[25-29] sigma[30,31] and nicotinic acetylcholine receptors.[32,33] These observations have led to the suggestion that the anti-addictive properties of this molecule derive from its simultaneous interaction with multiple molecular targets. To investigate this possibility, we chose to develop compounds that can modulate central nervous system function by interacting with both serotonin reuptake sites and opioid receptors, two molecular targets implicated in psychostimulant dependence. Previously, Hitzig[34] had observed synergistic effects in the treatment of alcohol dependence with a combination of dopamine and serotonin agonists. Therefore, it was expected that dual modulators of serotonergic and opioidergic function would display more useful anti-addiction profiles than agents directed solely at the individual molecular sites.

Inspiration for the design of this class of molecules came from an inspection of the structures of ibogaine and morphine, two important natural products. As shown in Figure 3, a tryptaminyl (2, Figure 3) or tryptamine-like (13, Figure 3) fragment can be found embedded in both natural compounds. While ibogaine recognizes both serotonin reuptake sites and opioid receptors, morphine only recognizes opioid sites. For this exercise, it was postulated that a suitably designed tryptamine-like compound could be made to recognize both serotonin reuptake sites and one or more classes of opioid receptors. After further analysis and modeling, the spiro compound 14 (Figure 3) emerged as the consensus structure for this drug discovery effort. Two noteworthy design features are: a) incorporation of the flexible aminoethyl side chain of the tryptaminyl fragment within a five-membered spirofused heterocycle; and b) the introduction of a C3-aryl group. Conformational restriction in part [a] was employed to hold the aminoethyl position in a suitable orientation for opioid receptor binding and to reduce the number of molecular targets that can recognize this structure. Indeed, spiro[indan-1,3'-pyrrolidine] derivatives, synthesized as conformationally restricted profadol analogs, had been shown to alleviate pain in animals, presumably via opioid receptors.[35,36] The C3-aryl group was introduced because 3-phenyl-1-indamines are known to inhibit monoamine reuptake.[37] The presence of the C3 substituent introduces a second chiral center. Therefore, two sets of diastereomers are possible. For the initial effort, the syn diastereomers 21a–e were synthesized as shown in Figure 4 and tested in vitro (Table 2).[38]

The compounds were tested for binding at selected binding sites with the aid of radioligand binding techniques. The spiroindanamines were tested as the hydrochlorides in racemic form. SERT refers to the cocaine binding site on the serotonin transporter. DAT is the dopamine transporter, and mu and kappa refer to the mu and kappa opioid receptors. Affinities are described as high (< 0.1 μM); moderate (0.1–1 μM), weak (1–10), and poor (> 10 μM).

Of the four compounds tested, the *N*-cyclopropylmethyl analog 21e consistently displayed the lowest affinity for monoamine transporters. The other three

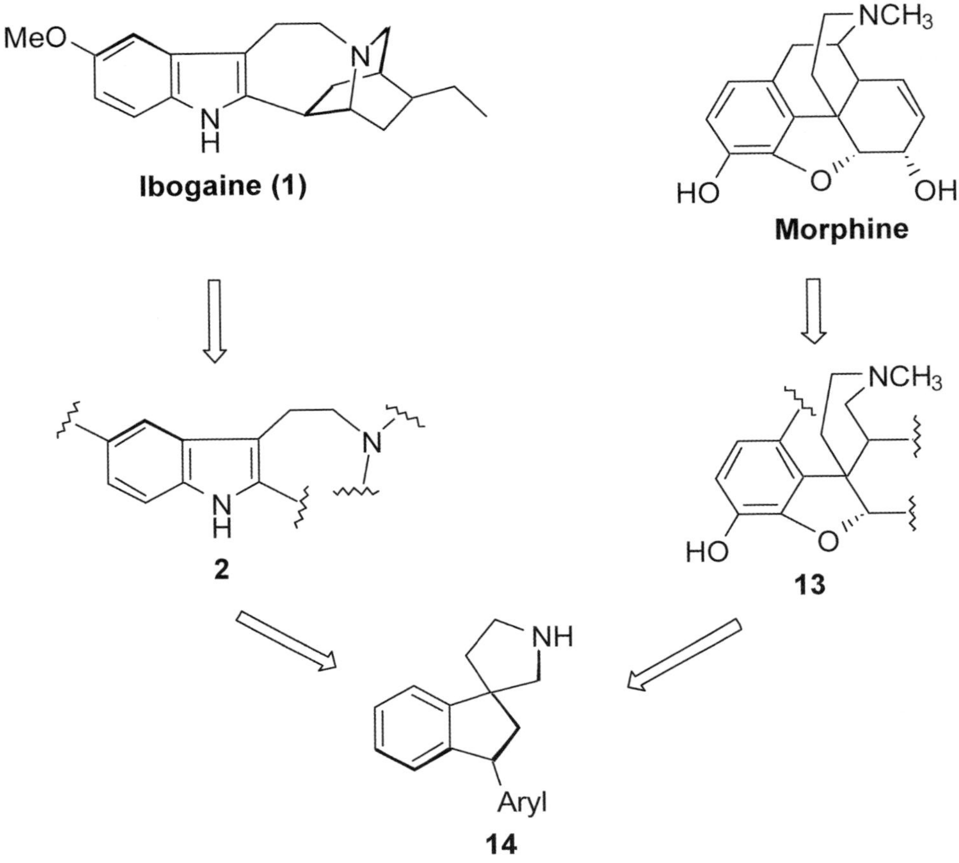

Fig. 3. Design of 3-arylspiro[indan-1,3′-pyrrolidine] analogs.

Table 2
Relative Affinities (IC$_{50}$, μM) of spiro[indan-1,4′-pyrrolidine] Analogues for Monoamine Transporters and Opioid Receptors*

Target	SERT(RTI-55)	DAT([³H]WIN)	μ-OPIOID([¹²⁵I]DAMGO)	κ-OPIOID([³H]U69593)
21a	0.01 μM	0.2 μM	0.5 μM	10 μM
21c	0.002	0.15	5.0	80
21d	0.75	0.73	2.9	1.52
21e	2.41	2.9	0.98	1.0
Fluoxetine	0.034	—	—	—
GBR12935	—	0.002	—	—
Naloxone	—	—	0.002	—
U50488	—	—	—	0.003

*Data taken from ref. 38. Ki values of reference compounds are provided for comparison. Relative affinities were determined by previously published methods using the radioligands shown in the table above.

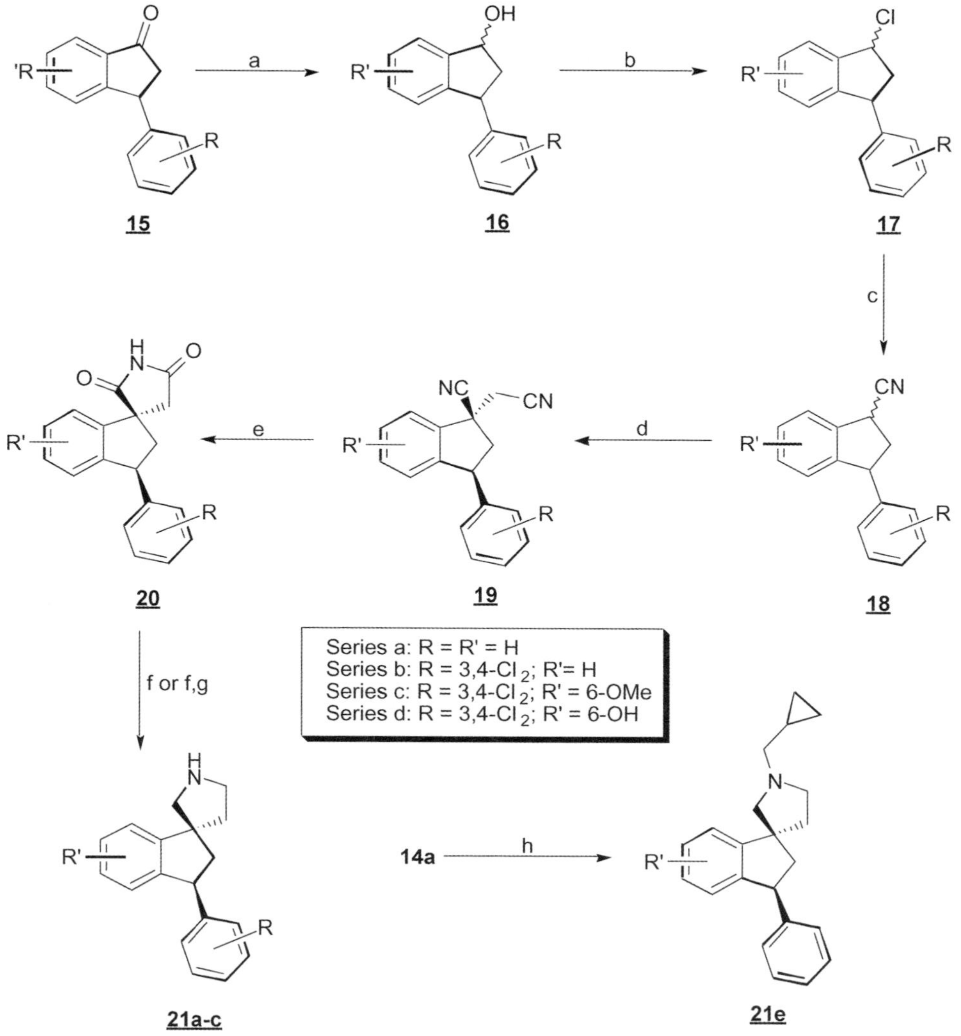

Fig. 4. Synthesis of 3-arylspiro[indan-1,3′-pyrrolidine] analogs.

a: NaBH$_4$, MeOH, r.t.; b: SOCl$_2$, pyridine, 0°C to r.t.; c: NaCN, DMF, KI (cat.), 70 °C;
d: BrCH$_2$CN, LDA,THF, -78° to -60 °C to r.t.; e: 78% H$_2$SO$_4$-HOAc (w/v), 125 °C;
f: BH$_3$.THF, reflux; g: 50% HBr-HOAc, reflux; h: cyclopropylmethyl bromide, K$_2$CO$_3$, DMF, r.t.

compounds (21a, c, d) showed moderate to high affinity for these transporter sites. The compounds also displayed some affinity, albeit moderate, for opioid receptors. In this respect, compound 21e was the most potent at mu opioid receptors, while 21e was the most potent ligand at kappa opioid receptors. We conclude, then, that compounds in this class recognize both serotonin transporters and opioid receptors. Additional investigations will reveal the intrinsic activity of these compounds at opioid receptors and their potential utility as pharmacologic agents.

IV. Conclusion

Tropical forests harbor a vast abundance of plant life and continue to serve as a useful source of new natural products. Unfortunately, the source countries have done little to capitalize on the enormous potential of natural products as lead compounds in a broader drug development effort. The lack of participation in this value-added area of drug development has relegated most developing countries to the role of raw material supplier. For the reason that lead compounds can inspire the development of novel molecular entities, it is safe to say that many developing countries now import pharmaceutical products that were inspired by natural products obtained from the very same countries. The development of expertise and infrastructure that can capitalize on valuable natural resources should, therefore, become a priority at the dawn of the Third Millennium.

References

1. Robbers JE, Speedie MK, Tyler VE. (1996) Pharmacognosy and pharmacobiotechnology. Philadelphia, PA: Williams & Wilkins.
2. Farnsworth NR, Akerele O, Bingel AS, Soejarto DD, Guo Z. (1985) Medicinal plants in therapy. Bull. WHO 63:965–981.
3. Buss AD, Waigh RD. (1995) Natural products as leads for new pharmaceuticals. In: Wolff ME. Burger's medicinal chemistry. NewYork: Wiley, pp. 983–1033.
4. Dybowski J, Landrin E. (1901) Sur l'iboga, sur ses propriétés excitantes, sa composition et sur l'alcaloïde nouveau qui renferme. Compt Rend 133:748–750.
5. Haller A, Heckel E. (1901) Sur l'iboga, principe actif d'une plante du genre Tabernanaemontana, originaire du Congo. Compt Rend 133:850–853.
6. Popik P, Layer RT, Skolnick P. (1995) 100 years of ibogaine: neurochemical and pharmacological actions of a putative anti-addictive drug. Pharmacol Rev 47:235–253.
7. Rezvani AH, Overstreet DH, Lee Y-W. (1995) Attenuation of alcohol intake by ibogaine in three strains of alcohol-preferring rats. Pharmacol Biochem Behav 52:615–620.
8. Benwell ME, Holtom PE, Moran RJ, Balfour DJ. (1996) Neurochemical and behavioral interactions between ibogaine and nicotine in the rat. Br J Pharmacol 117:743–749.
9. Maisonneuve IM, Mann GL, Deibel CR, Glick SD. (1997) Ibogaine and the dopamine response to nicotine. Psychopharmacology 129:249–256.
10. Lotsof HS. (1985) Rapid method for interrupting the narcotic addiction syndrome. US Patent No. 4,499,096.
11. Lotsof HS. (1987) Rapid method for interrupting the cocaine and amphetamine abuse syndrome. US Patent No. 4,587,243.
12. Kaplan CD, Ketzer E, De Jong J, De Vries M. (1993) Reaching a stage of wellness: multistage explorations in social neuroscience. Soc Neurosci Bull 6:6–7.
13. Sisko B. (1993) Interrupting drug dependency with ibogaine: a summary of four case histories. Multidisc Assoc Psychedelic Stud 4: 15–24.
14. Cannon JG. (1995) Analog Design. In: Wolff ME, editor. Burger's medicinal chemistry. NewYork: Wiley, pp. 783–802.
15. Hester JB, Tang AH, Keasling HH, Veldkamp W. (1968) Azepinoindoles. I. Hexahydroazepinol[4,5-b]indoles. J Med Chem 11:101–106.

16. Elliot AJ, Gold EH, Guzik H. (1980) Synthesis of some 5-phenylhexahydroazepino[4,5-b]indoles as potential neuroleptic agents. J Med Chem 23:1268–1269.
17. Efange SMN, Mash DC, Khare AB, Ouyang Q. (1998) Modified ibogaine fragments: synthesis and preliminary pharmacological characterization of 3-ethyl-5-phenyl-1,2,3,4,5,6-hexahydroazepino[4,5-b]benzothiophenes. J Med Chem 41:4486–4491.
18. Woolverton WL, Johnson KM. (1992) Neurobiology of cocaine abuse. Trends Pharmacol Sci 13:193–200.
19. Nestler EJ. (1993) Cellular responses to chronic treatment with drugs of abuse. Crit Rev Neurobiol 7:23–39.
20. Mash DC, Staley JK, Baumann MH, Rothman RB, Hearn WL. (1995) Identification of a primary metabolite of ibogaine that targets serotonin transporters and elevates serotonin. Life Sci 57:PL45–50.
21. Staley JK, Ouyang Q, Pablo J, Hearn WL, Flynn DD, Rothman RB, Rice KC, Mash DC. (1996) Pharmacological screen for activities of 12-hydroxyibogamine: a primary metabolite of the indole alkaloid ibogaine. Psychopharmacology 127:10–18.
22. Glick SD, Maisonneuve IM, Pearl SM. (1997) Evidence for roles of kappa-opioid and NMDA receptors in the mechanism of action of ibogaine. Brain Res 749:340–343.
23. Codd EE. (1995) High affinity ibogaine binding to a mu opioid agonist site. Life Sci 57:PL315–320.
24. Pablo JP, Mash DC. (1998) Noribogaine stimulates naloxone-sensitive [35S]GTPgS binding. NeuroReport 9:109–114.
25. Mash DC, Staley JK, Pablo JP, Holohean AM, Hackman JC, Davidoff RA. (1995) Properties of ibogaine and its principal metabolite (12-hydroxyibogamine) at the MK-801 binding site of the NMDA receptor complex. Neurosci Lett 192:53–56.
26. Popik P, Layer RT, Fossom LH, Benveniste M, Geter-Douglass B, Witkin JM, Skolnick P. (1995) NMDA antagonist properties of the putative anti-addictive drug, ibogaine. J Pharmacol Exptl Ther 275:753–760.
27. Sershen H, Hashim A, Lajtha A. (1996) The effect of ibogaine on sigma- and NMDA-receptor-mediated release of [3H]dopamine. Brain Res Bull 40:63–67.
28. Chen K, Kokate TG, Donevan SD, Carroll FI, Rogawski MA. (1996) Ibogaine blockade of the NMDA receptor: in vitro and in vivo studies. Neuropharmacology 35:423–431.
29. Layer RT, Skolnick P, Bertha CM, Bandarage UK, Kuehne ME, Popik P. (1996) Structurally modified ibogaine analogs exhibit differing affinities for NMDA receptors. Eur J Pharmacol 309:159–165.
30. Bowen WD, Vilner BJ, Williams W, Bertha CM, Kuehne ME, Jacobson AE. (1995) Ibogaine and its congeners are sigma-2 receptors-selective ligands with moderate affinity. Eur J Pharmacol 279:R1–3.
31. Mach RH, Smith CR, Childers SR. (1995) Ibogaine possesses a selective affinity for _2 receptors. Life Sci 57:PL57–62.
32. Schneider AS, Nagel JE, Mah SJ. (1996) Ibogaine selectively inhibits nicotinic receptor-mediated catecholamine release. Eur J Pharmacol 317:R1–2.
33. Badio B, Padgett WL, Daly JW. (1997) Ibogaine: a potent noncompetitive blocker of ganglionic/neuronal nicotonic receptors. Mol Pharmacol 51:1–5.
34. Hitzig P. (1993) Combined dopamine and serotonin agonists: a synergistic approach to alcoholism and other addictive behaviors. Med Med J 42:153–156.
35. Crooks PA, Szyndler R. (1980) Synthesis and analgesic properties of some conformationally restricted analogs of profadol. J Med Cem 23:679–682.
36. Crooks PA, Sommervillle R. (1982) The synthesis and analgesic activities of some spiro[indan-1,3'-pyrrolidine] derivatives designed as rigid analogs of profadol. J Pharm Sci 71:291–294.
37. Bogeso KP, Christensen AV, Hyttel J, Liljefors T. (1985) 3-Phenyl-1-indanamines. Potential antidepressant activity and potent inhibition of dopamine, norepinephrine, and serotonin uptake. J Med Chem 28:1817–1828.
38. Efange SMN, Mash DC. (1999) Spiroindanamines and spiroindanamides. US Patent No. 5,948,807.

CHAPTER 6

Integrating African ethnomedicine into primary healthcare: a framework for South-eastern Nigeria

CHIOMA OBIJIOFOR

Abstract

The health sectors in most developing countries, including South-eastern Nigeria, face formidable hurdles with improving the quality of life of their people with increasingly dwindling resources. The ability to prevent and treat diseases at minimal cost is fundamentally dependent on the integration of Western medicine and traditional ethnomedicine. The purpose of this study was to design a model of an 'integrated clinic' as well as determine the necessity and acceptability of these clinics in South-eastern Nigeria. The study was conducted by surveying 800 patients in randomly selected hospitals and herbal clinics, who voluntarily agreed to participate. The results of this study are discussed as well as the recommended conceptual model of an 'Integrated clinic', which will complement and become part of the formal national health program.

Keywords: *traditional medicine, integrated clinics, Southeastern Nigeria, herbal clinics, hospitals*

I. Introduction

Culture is defined as that complex of activities, which include the practice of the arts and of certain intellectual disciplines. 'Culture comprises a people's technology, its manners and customs, its religious beliefs and organization, its systems of valuation, whether expressed or implicit'.[1] Indigenous knowledge in Africa is the information base unique to a given culture and forms the basis for local-level decision making in healthcare, agriculture, food preparation, natural resource management, way of life and other social and rural activities. This knowledge, in most cases, is handed down from one generation to another. Traditional medicine or ethnomedicine is one such indigenous system of knowledge, which has affected the lives of people around the globe. This includes the healing methods and health practices, which are dynamic and are continually influenced by internal creativity and experimentation as well as by the interplay of a complex of social, economic and political factors. The international community formally adopted this view of culture and ethnomedicine in the UNESCO-sponsored definition in the Declaration of MondiaCult:

Culture and ethnomedicine are the whole complex of spirit, material, intellectual and emotional features that characterize a society or social group. It includes not only arts and letters, but also modes of life, the fundamental rights of the human being, value systems, traditions and beliefs.[2]

Iwu (1986, 1993) noted that any attempt to understand the fundamental conceptions of health and disease among any people must deal with the broader attitudes that they have regarding life itself. In Africa, the preservation, restoration, and enhancement of health necessarily involve the whole human community, the apparent dead, spirit and gods, the natural environment and God. This, in part, forms the foundation of traditional systems of medicine all over the world.

In Western Europe, the professional use of herbs and phytomedicines enjoys a relatively strong integration with conventional medicine. In some European countries, herbal medicines are generally sold in pharmacies as licensed non-prescription or prescription drugs. Botanical medicines are part of larger markets, referred to in the United States as 'dietary supplements', and the industry is experiencing rapid growth world-wide. Annual growth rates are between 10 and 20% in most countries.[3] But in Africa, Western biomedicine and traditional medicine are often viewed as two parallel entities, without any formal linkages or complementarity.

II. Federal Republic of Nigeria

The Federal Republic of Nigeria is Africa's most populous nation. The estimated total of over 100 million inhabitants in 1995 represents roughly 14% of the entire population of sub-Saharan Africa.[4] Nigeria is a diverse West African nation located on the shores of the Atlantic Ocean's Gulf of Guinea, bordering Benin to the west, Niger to the north, Lake Chad to the northeast and Cameroon to the east and southeast. The country's land area of 923,768 km^2 (356,669 mi^2), places it 11th in size among sub-Saharan African countries. Nigeria is made up of 30 federated states and one federal territory. States in turn are composed of local government areas. (LGAs). The states are primarily divided into four regions – Southeast, Southwest, Northeast and Northwest. Southeast, which is the study area for this project, is made up of 10 states (Enugu, Anambra, Imo, Ebonyi, Abia, Cross River, Akwa Ibom, Benue, Bayelsa and Rivers) and is the ethnic home of the 'Igbos'. The Nigerian population is concentrated in the southern part of the country (Southeast and Southwest). Immigration into Nigeria from other West African countries has increased since 1990, especially with the influx of refugees from war-torn Sierra Leone and Liberia. This has in turn impacted the practice of traditional medicine in the country.

III. Traditional or 'alternative' medicine?

Traditional medicine, as described in the introductory chapter in this volume, is a system of health practice based on indigenous knowledge. In most industrialized countries, because traditional medicine had become largely a very minor component of health practice, it is often described as 'folk medicine'. There has been, however, a resurgence of interest in non-Western medicine during the last three decades that such medical practices have now come to be recognized as credible forms of healthcare. The term most often used to describe this form of healthcare in most industrialized countries is 'alternative medicine'. The synonyms used for alternative medicine include 'complementary medicine', a British term, that emphasizes the joint use of conventional and alternative therapies, and 'Holistic', which is derived from holos, a Greek word meaning 'whole' was coined in 1926 and revived in the 1970s. It denotes an approach that addresses the uniqueness of each individual that sought to understand whole people in their total environments and that employed a wide range of conventional and alternative therapies. Most recently, the term 'new medicine' has been used because it suggests a synthesis of the wisdom of ancient healing traditions, such as classical Chinese medicine and Indian Ayurveda medicine, and the critical perspective and technology of modern science. This new medicine includes an appreciation of the power of modern biomedicine and an understanding that it is one step in, and not the end point of, medical evolution.

Murray and Rubel[5] defined alternative medicine as a set of practices offered as an alternative to conventional medicine for the preservation of health and the treatment of health-related problems. Eisenberg et al.[6] classified alternative medicine as a broad range of modalities, including relaxation technique, chiropractic, massage, imagery, spiritual healing and commercial weight-loss programs as well as the use of herbs, vitamins, diets, hypnosis, energy, biofeedback, acupuncture, homeopathy, folk remedies, and self-help.[7] Since the word 'alternative' is simply a descriptive term used to denote non-Western medicine relative to the dominant modern healthcare system, it would be inappropriate to describe 'traditional medicine' as alternative medicine in countries and communities where traditional medicine is the primary healthcare system that is readily available to the population. According to the World Health Organization, 70–80% of Africans today depend either totally or partially on traditional or alternative medicine. This form of treatment, which is referred to as ethnomedicine, is sometimes the only kind of healthcare available to the rural population. The use of the terms 'alternative' or 'complementary' in this chapter when dealing with traditional medicine is purely normative and meant to situate these healthcare systems within what is known as 'complementary and alternative medicine' (CAM).

IV. Alternatives to modern biomedicine

In recent years, many non-Western healthcare practices have become more visible as an alternative to modern biomedicine. Herbal remedies and acupuncture, which have been used for treatment of various ailments in traditional societies for hundreds of years, have now gained increasing acceptance in the US and other Western countries as a treatment option for certain illness conditions. At the same time, professionals in healthcare and allied fields are seeing greater numbers of immigrants and ethnic minorities whose belief about health and illness differ from those of Western people.[8]

Many publications on African traditional medicine[9-11] note that in spite of the influence of Western civilization, many Africans seek recourse to spiritual diagnosis for any serious or chronic illness, irrespective of the social status held or the religion professed by the individual. The priests, medicine men and herbalists are thus an integral part of the cultural history of the African. Some have speculated that these beliefs and practices may pose health risks[12,13] or barriers to utilization of mainstream healthcare systems. Others consider that quality healthcare requires a cultural sensitivity to the people being cared for.[14] This has aroused a renewed interest in knowing more about health beliefs and practices rooted within non-Western cultures as a way to address emerging concerns of indigenous people.[8] Since modern biomedicine has been shown to be inadequate in addressing the healthcare needs of traditional societies, most people are eclectic in their choice of healthcare system, using modern medicine or traditional medicine as they consider these more suitable for a given illness or situation.

V. Integration

Imagine a clinical demonstration unit in which clinical care integrates practitioners such as rheumatologists, orthopedists, physical therapists, chiropractors, acupuncturists, massage therapists, *'dibias'*, *'babalawos'*, seers and practices such as exercise, meditation, rituals, magic and incantations; and the team demonstrates its utility in providing better care in terms of clinical outcomes, health status and patient satisfaction. Can we imagine doctors who are bilingual, who are capable of thinking in more than one system, who are capable of breaking down barriers?

In recent decades, the Nigerian healthcare sector has witnessed a rapid decline of once flourishing research institutes, universities and medical institutions dedicated to health improvement and provision of care as a result of the economic crisis currently affecting most of the continent. The effects of this economic crisis have led to erosion of health infrastructures, decline in the quantity and quality of access to health services, an increase in cases of malnutrition, inadequate supply of drugs, unsatisfactory rate of infant mortality and neglect of health information collection and dissemination by national programs. With a ratio of people to physicians of 4496 to one, Nigeria has a relatively high number of trained healthcare personnel compared to other Sub-Saharan nations (median of over 14,000:1).[15] The African traditional healer differs from the contemporary doctor trained in the ways of Western science in their conception of health, the cause of ill health, and the methods of treatment.

The Western-trained doctor is generally analytic in their methodology, concentrating efforts on units, while the native healer focuses on the whole human being, and recognizes the units as parts that can never equal the sum – all treatment, therefore, is holistic. In the traditional African society, the native doctor is a person of immense social standing and significance; they are considered as the greatest gift from God, the most useful source of help and succor in an otherwise harsh environment.[10] Their judgment and counsel are usually accepted without question or proof. Both, however, are concerned for the improvement of the community health, and they make genuine efforts to cure and prevent diseases, and maintain good health. With such common objectives, it would be desirable to have a pooling of resources by the two groups, to achieve their goals, but unfortunately, what exists presently is mutual antagonism, distrust and contempt. No dialogue exists between the two professional groups, and the desirable complementary role of the herbalist in the health system is non-existent.

Despite widespread utilization, most traditional health systems are outside the formal health sector or have marginal status within it. Accordingly, there is relatively little research done on these traditions. Many international organizations have published reports on traditional systems of healthcare in recent years, all making strong recommendations for research into integrated or complementary medicine. To date, no coordinated effort has existed to bring together research in this field, which reflects a meaningful and well-implemented agenda.

Western medical systems need to heed the wisdom of traditional healers and incorporate the psychosocial aspects of the patient's illness into the patient's care.[16] The standard medical system treats diseases, rather than the people who happen to be ill. The integration of mind and spirit with body can be accomplished by improving the poor communication that exists between patients and health providers. Good communication requires non-judgmental mannerisms, empathy for the patient, and good listening skills. The doctor must respect the patient's autonomy, concerns, and priorities. Treatment decisions should be made collaboratively. The World Health Organization examines the research issues and methodological problems that need to be resolved in order to exploit the therapeutic potential of medicinal plants and herbs.[17] The objective of this exercise was to establish possible means of integrating traditional knowledge into the mainstream of medical practice, whether through the development of new drugs or through quality control and standardization of traditional remedies. The potential uses of medicinal herbs in primary healthcare, criteria for the selection of plants for further research or immediate use, and the use of specific remedies in different countries should be studied along with different research methodological approaches for the development of herbal products and models for clinical evaluation. This must be done if we hope to strengthen the role of herbal remedies in primary healthcare and address questions of standardization and regulation.

The widespread use of complementary and alternative medicine techniques, often explored by patients without discussion with their primary care physician, illustrates a request from patients for care as well as cure.[18] Despite this era of rapidly advancing medical technology, there are obvious reasons for the growth of, and interest in, complementary and alternative medicine. For instance, there is evidence of the efficacy of supportive techniques such as group psychotherapy in improving adjustment and increasing survival time of cancer patients.

Recently, the Integrative Medicine Clinic has been perceived as the most effective model of healthcare delivery in the developed world. Its philosophic and consistent means of delivering complementary medicine ensures open communication and integration of Western and complementary techniques. Physicians, in this system, incorporate complementary practices and practitioners into a single setting or related settings. Examples of such programs include The King County natural Medicine Clinic in Kent, Washington; American Holistic Centers in Chicago, Denver and Boston; UCLA's center for East–West Medicine and Spense Centers for Women's Health, Maryland. There are also many hospital-based programs, which vary in size and scope from single-philosophy programs in mind–body therapies, spirituality, healing touch, and acupuncture to multidisciplinary clinical, educational, or referral-based programs. Examples are California Pacific Medical Center's Institute for Health and Healing; Sisters of Charity of Nazareth Health System, Kentucky; Tzu Chi Institute for Complementary and Alternative Medicine, Vancouver Canada and Griffith Hospital, Connecticut. Other independent and affiliated programs are The Mind/Body Medical Institute of Deaconess Hospital and Harvard Medical School; Georgetown University's Department of Psychiatry and Community Medicine in conjunction with Dr Andrew Weils' Program in Integrative Medicine; and the Complementary Medicine Clinic at Stanford.

A recent article by Marion Meines agrees that healthcare providers and patients are becoming increasingly aware of the options available with alternative medicine or unconventional therapies.[7]

Patients seek alternative medicine for a variety of reasons. With all the media hype, skepticism, and desire for cure, the medical community must take the lead in assisting patients to travel along a safe path to get the treatment they want and the information they need. Validation, self-determination and autonomy are key factors in the process. Patients want the best care they can receive, and sometimes their choices may bring conflict with their care providers and healthcare personnel. Therefore, developing a rapport that establishes trust with both sides will benefit all involved.

The purpose of this research study was to establish lines of action for bringing the wealth of traditional knowledge into the mainstream of medical practice through 'integrated clinics' as well as provide the much needed demography information on the use of traditional medicine in South-eastern Nigeria. A model of an integrated clinic was recommended to the government, and an evaluation technique was developed to assess the success and failure of this approach to healthcare delivery.

VI. Research problems/issues

What percentage of the population surveyed utilizes traditional medicine? Can the integration of African and Western medicine be achieved through integrated clinics? What type of 'integrated clinics' models will be suitable for South-eastern Nigeria? What 'types' of services should a clinic have? What framework of action should be adopted? How do you deal with the related issues of intellectual property rights, access and ownership to traditional knowledge?

This project is unique because the 'integrated clinics' will combine existing resources to achieve a wholesome healthcare system, which can be available at reduced cost. Traditional medicine makes significant contributions to the healthcare and health status of the population, particularly in rural communities in South-eastern Nigeria. The services they offer are demand-driven. Despite the extensive use of traditional medicine, very few public resources are made available to support them. The integrated clinics will emphasize the promotion of the best practice as the national standard. In the 'integrated clinics', there will be physicians with various specialties, traditional practitioners with varied practices, levels of skills, competence and values as well as nurses and health educators. The size of these clinics will be based on available resources and the consumer population.

VII. Survey

In this study, 668 residents of South-eastern Nigeria were surveyed by the administration of questionnaires. A total of 73.05% (488) participants reported using a form of alternative therapy or procedure. The three most commonly used therapies were herbal therapy (53.89%), prayer (49.4%) and vitamin therapy (35.03%). Participants were allowed to choose more than one option. The least commonly used are acupuncture, yoga, homeopathy, hypnosis and chiropractic.

Among the respondents, 65.12% had never been treated outside a hospital, which indicates that a considerable percentage of alternative therapy users obtained their treatment in their homes. Out of the alternative therapy users, 65.12% were uninformed about integrated clinics.

However, 83.98% said they would attend an integrated clinic if it were available. The services most preferred were prayer (79.79%), herbal therapy (36.08%), family practice (25.90%), vitamin therapy (17.07%), massage (14.52%), surgery (14.37%), internal medicine/cardiology (13.32%), pediatric services (11.38%) and chiropractic (10.03%). It was also interesting to note that 49.40% of the respondents explained that they would inform their primary care providers if they were using an alternative therapy.

A model of an integrated clinic that is closely incorporated with existing hospitals and home health services is recommended. The majority of the residents (98.8 %) use the hospital system. Home healthcare services are not routinely offered in hospitals in the area, so this may account for the better service that users reported (31.14%) from alternative therapy providers. The integrated clinic should have at least one doctor, two healers, three nurses and one health educator. Services provided should include prayer, herbal therapy, vitamin therapy, massage, surgery, family practice, and internal medicine, pediatrics and chiropractic services. To achieve a dialogue between Western and traditional systems of health, it would be desirable to have a pooling of resources. Dissemination and sharing of information is of utmost importance in achieving this objective. The activities reported by survey participants that will enhance the success of these integrated clinics include introducing courses on integrative medicine in medical schools (38.17%), the education of healers, the establishment of prayer ministries on Western medicine (20.1%), the education of physicians, nurses and other healthcare workers (21.86%) and the circulation of fliers to inform the general public (18.11%).

VIII. Framework for action

- *Recognition and identification* of traditional practices and healers should be accorded respect and recognition. The potential disappearance of many indigenous practices could have a negative effect primarily on those who have developed them and who make a living through them.[19] A greater awareness of the important role that traditional medicine can play in the development process in Africa is likely to help preserve valuable medical skills, technologies and problem solving strategies.
- *Recording and documentation* is a major challenge because of the tacit nature of traditional medicine. The recording may require audio-visuals, taped narrations and information on the source. This information may need to be validated in terms of its significance and authenticity. Preferably, this should be done through associations of traditional medicine practitioners (e.g. National Union of Herbal Medical Practitioners). Collected information should be stored in retrievable repositories. This would involve categorization, indexing and integrating it into other health information with a user-friendly technology. This could be further developed into a database of traditional practices, lessons learned, sources, etc.
- *Dissemination* activities should include public awareness campaigns, advertisements, seminars, workshops, distribution of information materials, publications and the incorporation of integrated clinics into existing primary and public health programs. Electronic networking on this issue should also be encouraged among researchers and civil society. The government could encourage the process by creating a favorable economical, political and legal framework.
- *National policies in support of TM development*. Knowledge as an instrument of development has not received the needed attention in Africa and other developing countries.[19] In the past, country policies would typically focus on the adoption of 'Western' practices with a view to modernizing the society and transforming the health sector (op. cit., 1998). As a result, there has been very little effort put forth to merge both health systems. The government should begin elaborating specific policies in support of this acquiring knowledge (e.g. accessing and adapting successful global models of integrated clinics), absorbing knowledge in terms of training and learning processes (e.g. incorporating traditional medicine into college curriculum) and harnessing the potential use of new information technology to bring access and knowledge to the healers.
- *Building partnerships*. Meaningful collaboration should be established between government, hospital communities, healers associations, non-governmental organizations and local communities. These partners can play active technical, financial and supportive roles in developing state and national health policies. By

leveraging limited resources of various partners, a greater development impact can be made.

This framework could be a powerful tool to facilitate the integration of traditional medicine into national health sectors. External support to build on the local capacities can also be sought.

References

1. Trilling L. (1978) Beyond culture. New York: Harcourt Press, p. 16.
2. Culture and Development in Africa (1994) Proceedings of an International Conference held at the World Bank, Washington DC, p. 7.
3. ten Kate K, Laird S. (1999) The commercial use of biodiversity – Access to genetic resources and benefit sharing. London: Earthscan Publications.
4. CIHI (1998) Country Health Profile Series (1996) Center for International Health Profile on Nigeria, Washington, DC, CIHI Press.
5. Murray and Rubel. (1992) Healing without Doctors. Am Demogr 15(7):46–49.
6. Eisenberg DM, Kessler RC, Foster C, Norlock FE, Calkins DR, Delbanco TL. (1993) Unconventional medicine in the United States: prevalence, costs and patterns of use. N Engl J Med 328(4):246–252.
7. Meines M. (1998) Should Alternative treatment be integrated into mainstream medicine? Nursing Forum 33:11.
8. Chau K. (1997) Bridging socio-cultural practices into healthcare and related services among ethnic populations. In: Wozniak DA, Yuen S, Garett M, Shuman T, editors. Proceedings of the International Symposium on Herbal Medicine – a holistic approach. Honolulu: University of San Diego Press, pp.145–148.
9. Iwu MM. (1986) African ethnomedicine. Nsukka, Nigeria: University Press.
10. Iwu MM. (1993) Handbook of African Medicinal Plants, Boca Raton, FL: CRC Press.
11. W.H.O. (1978) Traditional medicine. Geneva: World Health Organization.
12. Chan C., Tomlinson L., Critchley P. (1993) Chinese herbal medicine revisited: a Hong Kong perspective. Lancet (England), 342(8886–8887):1534–1534.
13. De Smet P. (1999) Herbs health healers – Africa as ethnopharmacological treasury. Berg en Dal, The Netherlands: Afrika Museum.
14. Hatch L. (1985) Reducing barriers to utilization of health services by racial and ethnic minorities. In: Watkins E, Johnson A, editors. Removing cultural and ethnic barriers to healthcare. Washington, DC: National Center of Education in Maternal Child Health.
15. World Almanac. (1998) URL: www.worldbank.org
16. Merwe JV van der. (1995) Physician–patient communication using ancestral spirits to achieve holistic healing (Transactions of the Thirteenth Annual Meeting of the American Gynecological and Obstetrical Society). Am J Obstet Gynecol 172:1080.
17. Chaudhury RR, Herbal medicine for human health. WHO Regional Publications, Southeast Asia Series, No. 20.
18. Spiegel D., Stroud P., Fyfe A. (1998) Complementary medicine. Western J Med 168:241.
19. World Bank. (1998) Indigenous knowledge for development: a framework for action knowledge and learning center. Africa Region publication.

Iwu and Wootton (eds.), Ethnomedicine and Drug Discovery

Current initiatives in the protection of indigenous and local community knowledge: problems, concepts and lessons for the future

ROBERT LETTINGTON

Abstract

This paper examines a range of current initiatives being undertaken for the protection and promotion of indigenous and local community knowledge. Particular attention is paid to the proposed Peruvian draft law, Philippines legislation and the Organization of African Unity Model Law. However, reference is also made to the activities of the World Intellectual Property Organization, the World Trade Organization and the United Nations Convention on Biological Diversity. This international perspective is considered vital on the basis that the failure to establish some minimal international regulation of indigenous and local community knowledge will be likely to undermine the effectiveness of any national systems that are developed. This systemic risk exists because the main demand for access to, and capacity to commercially develop, indigenous and local community knowledge tends to be present in states and regions other than the points of origin of that knowledge. Many of the fundamental problems in addressing the protection of indigenous and local community knowledge involve accommodating the world views according to which the holders and potential users of such knowledge conduct their daily lives. It is argued here that the current state of international debate on these questions indicates that, at least for the time being, understanding of the most basic of points is not significantly increasing. This paper asserts that the protection and promotion of knowledge cannot be approached as an isolated goal in the indigenous and local community context.

Keywords: *intellectual property, indigenous community knowledge, world views, belief systems*

I. Introduction

The exact meaning of the phrase 'indigenous and local community knowledge' is still a subject of continuing debate. In this chapter, no attempt has been made to offer a precise definition or to join in the debate on the exact meaning of that phrase. This is not to suggest that this is not a worthwhile discussion but rather that it is one for a different context than the objective of this chapter. For the purposes of this discussion, the term will be left somewhat ambiguous and regarded as relating approximately to that knowledge which is developed in an informal context. This generally, but not exclusively, means that which is developed and applied away from the modern context of laboratories, universities, research-oriented hospitals and the like. A useful, although still largely inadequate, description might be 'village level knowledge'. Given that the current types of formal legal protection for knowledge consist of the bundle of rights known as 'intellectual property rights', a first step in seeking protection for indigenous and local community knowledge is to see to what extent it might be considered 'intellectual property'. This paper argues that there is no doubt that indigenous and local knowledge falls within accepted definitions of 'intellectual property' and that the problems that exist arise exclusively from the question of 'rights'. The hypothesis here is that difficulties in the protection of indigenous and local community knowledge have arisen largely because of the fact that the dominant intellectual property rights paradigm was designed for the protection of types of knowledge developed through a radically different process, and in an equally different context, from that of indigenous and local community knowledge. Consequently, the error that has been made to date has been to assume that the dominant paradigm would be capable of protecting such knowledge rather than a failure to make it do so adequately. However, recognizing that this somewhat academic debate does not provide much immediate practical assistance to those dealing with the pressing problems of epidemics such as HIV/AIDS and malaria, amongst others, in Africa today, the paper proceeds to look at ways of applying the existing intellectual property rights system for the benefit of traditional healers and those who work with them.

II. Is indigenous and local community knowledge intellectual property?

- TRIPs Article 1(2):[1]

  For the purposes of this Agreement, the term 'intellectual property' refers to all categories of intellectual property that are the subject of Sections 1 through 7 of Part II.

- Article 2(viii) of the Convention Establishing the World Intellectual Property Organization (WIPO), 1967:

  ... resulting from intellectual activity in the industrial, scientific, literary or artistic fields.

At the international level, there are two bodies that have recognized competence in the field of intellectual property, the World Intellectual Property Organization (WIPO) and the Council for the Uruguay Round Agreement on Trade Related

---

[1] General Agreement on Tariffs and Trade (GATT) Uruguay Round Agreement on Trade Related Aspects of Intellectual Property (TRIPs), 1994.

Aspects of Intellectual Property Rights (TRIPs Council). As the recognized international bodies, it is logical to turn to the two of these for clarification on the issue of whether they consider the field of traditional knowledge as falling within that of intellectual property. Neither of these bodies has a particularly high profile to those not working in fields directly related to intellectual property. The reasons for this anonymity are twofold. First is the simple fact that intellectual property law is a fairly arcane and inaccessible field that historically has not lent itself to popular interest. The second is the technical fact that intellectual property law as such does not exist at the international level. All states implement, or in many cases decline to implement, intellectual property law in national legislation. WIPO has traditionally provided a forum for some procedural matters, such as facilitating simultaneous applications to multiple states, and has acted as the secretariat for a series of loose, principle establishing treaties, such as the Berne and Paris Conventions (addressing copyright and industrial property respectively, the two of which make up the established body of intellectual property rights). The TRIPs Council oversees the TRIPs Agreement, which also does not create any law directly but rather establishes a set of minimum standards that national laws must uphold. The big distinction between WIPO and the TRIPs Council, and, many argue, the reason for the existence of TRIPs, is that through the processes of the World Trade Organization (WTO) Dispute Settlement Panels (DSPs), TRIPs has real enforceability due to the threat of trade sanctions as a consequence of violation. A combination of interpretations, particularly by the North, and the impact of the ever-present threat of DSPs has led some commentators to suggest that while, in theory, TRIPs does not create law at the national level, in practice, it is doing exactly that.

The most commonly recognized of these two bodies is the TRIPs Council. This is largely true due to the high profile of its parent, the WTO. Article 1(2) of TRIPs, quoted above, deals with the definition of intellectual property. The first thing that should be noted here is the fact that this article is highly unusual within the TRIPs structure because it is a form of definition. TRIPs avoids defining most critical terms, such as 'inventor' or 'invention', and yet it does provide a definition of 'intellectual property'. However, there are two particular points that should be noted with Article 1(2).

First is the fact that, while the Article initially seems to take a pragmatic and straightforward approach by simply referring to its substantive sections and stating that whatever is covered by them constitutes intellectual property, it actually complicates the issue. The complication lies in the fact that the substantive sections of TRIPs are not constructed so as to refer to various fields of intellectual property; rather they focus on various types of intellectual property *right*. The WTO operates on a rule of literal interpretation when its texts are being examined, normally in the context of disputes. Consequently, the most obvious meaning here would seem to be that intellectual property *only* consists of that information that is provided protection by the mechanisms covered by TRIPs. This ultimately leaves a flexible situation where whatever national intellectual property authorities decide to offer protection for may be considered as intellectual property. However, there is a strong argument that this is illogical as intellectual property is clearly a wider field, as many of the valuable products of intellectual activity are not necessarily offered protection. This has clearly been recognized over the years as new fields have developed systems of rights that might be defined as intellectual property rights, whether within, or outside of, the TRIPs framework, such as the protections offered to integrated circuits, or more recently databases.

The second point that should be noted regarding Article 1(2) is the fact that it restricts itself by stating 'for the purposes of this Agreement', a standard phrase in many legal texts. It is clear that TRIPs does not claim to cover all intellectual property, or even potentially all intellectual property *rights* as it only encompasses those that are related to trade. Consequently, it could be argued that Article 1(2) has no bearing at all on definitions of intellectual property or in the alternative that it only acts in defining intellectual property rights that are grounded in trade issues. Whichever argument one chooses to pursue, it is clear that one could present a solid case that Article 1(2) of TRIPs does not affect the status of traditional knowledge as intellectual property.

Overall, any discussion of TRIPs Article 1(2) can be no more than academic debate as it is only a dispute settlement panel of the member states that have the power to provide definitions of the treaty. It is unlikely that Article 1(2) will become the subject of a dispute settlement panel anywhere in the near future. As for the member states, the TRIPs Council, until now, has been unable to reach any reliable conclusions on the status of traditional knowledge. Indeed, it has been the subject of considerable, often heated, debate in the corridors around TRIPs Council meetings and in NGO discourse. The reason for this problem is that as part of the WTO the TRIPs Council is extremely limited in its ability to define any terms let alone the scope of concepts as ambiguous as that of 'intellectual property' itself. For example in its six years of existence the TRIPs Council has declined to provide any definition even for what constitutes 'an invention'!

In contrast to the situation with TRIPs, what constitutes intellectual property is not a highly controversial question in the context of WIPO. In the 1967 WIPO Convention, intellectual property is defined as any product resulting, *from intellectual activity in the industrial, scientific, literary or artistic fields*. There is no doubt that according to this definition, traditional knowledge, and particularly elements such as medicinal or agricultural knowledge that are more practically oriented in the Western sense, fall within the boundaries of intellectual property.

The previous discussion focuses on the *de jure*, or legal, aspect. If one looks at the *de facto*, or existing, then the situation is even clearer as WIPO has clearly placed traditional knowledge within the ambit of its Global Issues Division, and the newly formed Inter-Governmental Committee while traditional knowledge has become a controversial fringe issue in the TRIPs Council since the Seattle Third Ministerial Conference of 1999. Consequently, there is little doubt that indigenous and local community knowledge is intellectual property, and the question then moves on to why the intellectual property rights system is inadequate to serve the needs of this sector.

III. If indigenous and local community knowledge is intellectual property, why are intellectual property rights inadequate?

Patents:

- 20-year time limit
- Expensive to apply and enforce
- Technically not suited
  - Novelty
  - Non-obviousness
  - Industrial applicability

Trade secrets:

- Indefinite duration is useful
- Not really intellectual property rights so do not protect the invention
- Weak for knowledge that is widespread.

While the field of intellectual property rights is far more limited than that of intellectual property, it is still a large and varied field that has been expanding further in recent years. Consequently, it would be impossible to examine all of even the potentially most relevant rights here. Two of the best-known types of intellectual property rights, patents and trade secrets, are examined in terms of particular advantages and disadvantages that they present for indigenous and local community knowledge-holders.

Patents are the strongest form of intellectual property right currently available. They provide a 20-year limited monopoly for the patent holder after payment for initial examinations and processes and, in many cases, upon the continued payment of annual maintenance fees. While these fees can be significant (frequently amounting to a total of $US 30,000 or more) they are strictly of an administrative nature. It should also be remembered that the infringement of patent rights is a civil rather than criminal matter, and thus the enforcement of rights is normally to be achieved through the courts rather than by the government or by means of an administrative process. A final significant point is that one should remember the history of patents, with its roots in the development of mechanical inventions in Fifteenth Century Venice.

Patents are a tremendously complex field, or at least courts and patent offices have managed to make them such, and there are multiple volumes dealing even with the relationship between them and indigenous and local community knowledge, let alone in general. However, here, focus will be placed on the three main questions mentioned above: time, cost and technical relevance.

In terms of time, 20 years with a monopoly on a product would seem to be a generous period. However, it is considered to be of a sufficient length to allow an inventor to recoup any investment he or she may have made in developing the product in question and to also make a reasonable level of profit as a reward for their creativity. The time period is particularly relevant to the medical field as very often, at least several years of the 20 are lost due to the process of bringing the product to market.[2] The idea is that there is a balance between the length of the period and the strength of the right. In the case of patents, this is held to be a relatively short period,[3] which allows for a strong right. There are two major problems that this generally raises in the context of indigenous and local community knowledge. The first is that knowledge-holders often see their knowledge as culturally or spiritually significant in some manner. In such a situation, they maintain that it is impossible to place time limits on rights over that knowledge. This is, however, often balanced by the fact that knowledge-holders who maintain such beliefs are not generally averse to the use of their knowledge on certain terms and in certain conditions. Consequently, they are seeking a longer-term, but weaker, right. For those seeking to commercialize their knowledge and who do not have significant non-material interest in it, the 20-year period would not seem to present major problems. If one wishes to take advantage of the market, one must accept the rules of that market. The second problem that knowledge holders must be aware of is that part of the policy foundation of intellectual property rights is that they make information widely available. As a part of this theory, many jurisdictions require that a patent holder make use of their rights if they are to be enforceable. This can be interpreted in many ways in different jurisdictions, but at a minimum, it tends to mean that you have to make your product available on the market to be able to maintain your rights to it.[4] Making your product available costs money, and these funds have to be at hand in advance.[5] Equally, one has to be prepared for others to be able to make use of your invention for their own research purposes. Not all knowledge-holders feel comfortable with this as one does not have any power to limit the type of research: it could well be something such as human genetic research that may offend their personal beliefs. Overall, the 20-year period, and its associated restrictions, can be considered effective if a knowledge-holder's main purpose in seeking patent protection is commercialization and the associated financial gain. If their purposes are different, or financial concerns are only incidental, it will often not be appropriate.

The costs involved with the process of patenting are often misunderstood as the actual process of patenting is not always clearly understood, and many of the costs

---

[2] Pharmaceutical companies often claim that they lose up to 10 years of their monopoly rights, as they need to apply for a patent early in the product life cycle to protect their rights. This is usually before they have submitted their product to clinical tests, etc. that can frequently take years to complete.

[3] At least compared to the standard period of life of the author plus 50 years for copyright or indefinite time for trade secrets.

[4] However, it should be noted that generally, a patent holder will not automatically forfeit their rights in a situation where they are not making use of them; rather, it means that they risk losing them if a challenge is presented by means of a mechanism such as compulsory licensing or governmental use provisions. Also, the cost and difficulties involved with making a product 'available' can be frequently mitigated by the selling of a voluntary license to a local partner who will make it available and thus allow a patent-holder to maintain their basic right.

[5] Although the Paris Convention and TRIPs do require that states give patent-holders a 3- to 4-year grace period on these provisions.

cannot be accurately predicted in advance. Patents potentially incur costs in two principal ways: fees and enforcement costs. Fees are limited but can still be prohibitive for individuals or institutions with limited means, something that is particularly true when considering the protection of an invention with wide application geographically. This tends to be where the process of patenting is misunderstood. Despite the fact that people usually speak of holding 'a patent', this is not what generally occurs. Patents are still granted and maintained on a national basis.[6] Thus, one needs to submit a separate application in *every* country where one wishes to be protected. Since fees have to be paid for all of these applications, the total figure can mount up very quickly; hence the above quoted figure of $US 30,000 is not unusual. This becomes more extreme when one considers that many intellectual property offices require that patent holders pay a recurring maintenance fee throughout the 20-year period. Cutting back on these costs is often not an option as submitting fewer applications can frequently lead to weak protection. In the medicinal field, it is essentially required to seek patents in the United States, Europe, India, Argentina, Brazil, Thailand and a few other countries as these all have active pharmaceutical industries that would be easily capable of exploiting an unprotected invention. The same is largely true, even within Africa as Kenya, South Africa and Nigeria, amongst others, all have active generic manufacturers that are quite capable of reproducing even some of the more complex products on offer. If one takes an alternative approach and targets countries that constitute potential markets, the situation is often worse. In the case of traditional medicines for HIV/AIDS and malaria, an effective treatment for either would have to be considered to have a potential market in at least 70 or 80 countries, if not considerably more.

Unfortunately, the costs involved with patenting do not end there. As mentioned earlier, patent infringement cases are legally considered as civil matters and thus are not handled by criminal law or even normally by an administrative mechanism. This is important as, in a criminal case, the state assumes the burden of investigating and prosecuting violations on behalf of its citizens or, in this case, its patent holders. All the individual needs to do is to submit their complaint to the appropriate authority. In an administrative process, the individual will have to prove a case of infringement of their rights, but while expensive, this expense is generally limited by the rules of the administrative body concerned. Where the infringement concerned is another patent, administrative processes can be used to attack the potentially infringing patent in most jurisdictions, but this requires that the individual seeking to protect their rights be extremely active in doing so. Generally, administrative mechanisms are most available while a patent application is still under examination where an individual can register an objection to the granting of a patent. However, this requires that the individual track what applications are being submitted around the world, something that can be a full time job for several people if it is to be done effectively. A second administrative process has been used on several occasions

---

[6] The two exceptions to this rule are Western Europe and the bulk of Francophone West Africa where regional agreements provide for single patents covering the whole region.

recently, that of requesting a re-examination of the grounds for the grant of a patent after it has been issued.[7] The most severe problems arise where one is not challenging the validity of a patent but rather an infringing activity, product or practice. In most countries, such a situation goes straight to the courts. Since patent rights are often most valuable in developed countries, this can entail enormous litigation expenses, with no guarantee of recovery. In the US, admittedly a country with some of the highest litigation costs in the world, the average cost of a straightforward patent litigation has been estimated at anywhere from $US 300,000 to $US 500,000 upwards. Undoubtedly, these figures are somewhat speculative, but one can be sure that whatever the exact figure is, it is significant.

In considering patents as an option, an individual must be sure that they have sufficient resources available to apply for and maintain that patent. This may be possible for some indigenous and local community knowledge holders in Africa, particularly where they have a very promising product and are able to make a strategic partnership for commercialization. However, a potential applicant also needs to be sure that they will be able to defend that patent once it has been issued. This can mean having access to at least several hundred thousand dollars *with no guarantee of a return on those funds*. Certainly, a patent does act as a form of deterrent to potential infringers, but without the resources to go to court, this deterrent is based on a bluff, and, considering the investment required for applications, it is an expensive bluff.

The third aspect of patents considered here is that of their technical requirements and the degree to which these are suitable to the case of indigenous and local community knowledge. There are three basic requirements for the grant of a patent: novelty, non-obviousness and industrial applicability. These three standards are sometimes known by different terms, but the meaning remains the same throughout the world.[8] However, the fact that the basic meaning remains the same does not necessarily mean that all intellectual property offices interpret that meaning to produce the same results.

Novelty, or a requirement that an invention is new, is generally accepted to mean that any invention that is the subject of a patent application must be sufficiently different from any previous invention to merit protection. This is often argued to preclude the protection of indigenous and local community knowledge on the basis that such knowledge is the product of the accretion of community knowledge and thus does not actually constitute anything sufficiently new in and of its own right. However, if one considers how often an invention is created that is not based on the accretion of previous knowledge, virtually nothing would be patented. In the pharmaceutical field, almost all new products are developed by means of practices

---

[7] This was successfully done in the case of Ayahuasca by the center for International Environmental Law (CIEL) on behalf of various South American Indian Organizations and is currently being pursued by CIAT against a US patent holder in the case of the Mexican Yellow Bean.

[8] For instance, TRIPs addresses these requirements in Article 27 (1): patents shall be available for any inventions, whether products or processes, in all fields of technology, provided that they are new, involve an inventive step and are capable of industrial application.

and equipment that have been in existence for years; they are developed from the accretion of knowledge in the chemical and biological fields.

Non-obviousness, or the requirement for an inventive step, presents a similar problem to that of novelty. In basic terms, this requires that an invention, even though it may be novel, should not be obvious to an 'ordinary person skilled in the art'. This ordinary person is what is known as a legal fiction as we are required to imagine an individual that is skilled but who possesses no imagination. The context of indigenous and local community knowledge is a good one in this context – is it possible to picture a traditional healer who has no imagination? Admittedly, this holds true for most fields of expertise, and the fiction has been considered worthwhile nonetheless. However, the real problem here is the fact that the ordinary person considered is usually not a contemporary of the knowledge holder but rather an individual skilled in the modern version of the particular field in question. Thus, in the context of traditional medicines, the question of obviousness is decided in the context of modern pharmaceuticals. In many instances, this may not create significant difficulties, but in some instances, it is bound to, simply because of the fundamental fact that one cannot judge one system by the rules and standards of a completely alien system and expect satisfactory results.

'Industrial applicability', or 'usefulness', is the idea that for an invention to be worthy of patent protection, it must serve some useful purpose, something that is often defined in terms of commercial activity. In most instances, this is interpreted very broadly and can include 'potential' usefulness as much as that which has already been realized. It is in this manner that many gene sequences that currently have no known use, indeed whose function has not been identified, have received patent protection. Despite this generally broad approach, many elements of indigenous and local community knowledge have been said to fall short of this standard. The basis for this argument is usually along the lines of the idea that something such as a traditional medicine is a raw material that is not 'useful' in and of itself, but rather requires the refinements of modern science to be so.

Trade secrets law is another form of rights involving knowledge that is often cited as having potential for protecting indigenous and local community knowledge. The basic idea of trade secrets law is that as long as you treat something as a secret, the law will regard it as such and punish others who misappropriate it. There are clear advantages in this strategy. First and foremost is the fact that no explicit actions are required on the part of knowledge-holders beyond that they treat their knowledge as a secret. This in turn means that no costs are incurred in receiving protection. Treating one's knowledge as a secret does not mean that one cannot share it with others, whether for a fee or otherwise, but rather means that when sharing it, one should clearly communicate that it is a secret and that in sharing one is not permitting further dissemination. In some situations, it may be difficult to make indigenous and local communities aware of what they need to do, but it should not be an unreasonable barrier where that community is aware of the need to protect their knowledge or where they may wish to commercialize it. In many ways, a large number of knowledge holders, particularly healers, already operate in a manner that is broadly of the same nature as trade secrets. Furthermore, the fact that trade secrets protection is not subject to any fixed time limits may serve the interests of many knowledge-holders whose principal concern is ensuring respect for their knowledge as a cultural rather than a material asset. In sharing a trade secret, there is no reason why a knowledge holder cannot make stipulations as to how it should, and how it should not, be treated – this is basically forming a contract for the sharing of the knowledge.

However, there are some major drawbacks with trade-secrets protection. The main problem lies in the fact that trade secrets is not really a form of intellectual property right but rather a branch of unfair competition law. As such, it does not actually protect the knowledge or invention but rather prevents others from discovering it unfairly. If the knowledge or invention is discovered unfairly, the remedy is monetary compensation, the status as a trade secret cannot be restored, and thus one loses control. Equally, trade secrets do not prevent the discovery of knowledge or inventions by fair means, which is understood to include separate discovery and reverse engineering. In the case of traditional medicines, this could be particularly difficult as it will often not be a problem for a modern pharmacist or chemist, and in some cases even a botanist, to identify the active elements of a remedy and then reproduce it. In this situation, where the knowledge itself is not protected, trade secrets are also vulnerable where particular knowledge or, in some cases even a system of knowledge, is widespread. In such a situation, one needs to be able to form an agreement among all the holders of that knowledge that it will be treated as a secret. Even one dissenter could defeat the protection provided that they could demonstrate that they legitimately held the knowledge in question. Projects such as that developed by Ecociencia and CARE in Ecuador are based on this idea of developing a 'cartel' to claim trade secrets protection as a group.[9]

The unsuitability of patent protection to indigenous and local community knowledge and the difficulties in using trade-secrets protection to provide a comprehensive solution highlight the overall problem with the technical standards and practices of intellectual property rights when applied to indigenous and local community knowledge, particularly traditional medicines. The intellectual property rights system grew out of a Northern European philosophical paradigm over a period of some 400–500 years, and indeed as this paper hopefully suggests, it is still evolving today. The systems and traditions that provide the context for the generation of indigenous and local community knowledge have developed along a parallel philosophical paradigm that often stretches far further back than a few hundred years. However, in suggesting that indigenous and local community knowledge can be adequately protected by the intellectual property rights system, one is assuming that the philosophies of indigenous and local communities can be somehow 'fitted in' to the dominant Northern European philosophical paradigm. Such an approach clearly poses significant risks to the future dynamism of indigenous and local community knowledge, and this paper would argue that it is thus a nonsensical path to follow whether from a practical or ethical point of view.

---

[9] See www.thebiodiversitycartel.com

This should not necessarily be interpreted as a wholesale attack on the patent or intellectual property rights systems but rather an attack on the idea that they can provide lasting solutions of benefit to indigenous and local communities. Indeed, many of the reasons that this paper is being written at all are rooted in the fact that the intellectual property rights system is being used to encroach on indigenous and local community rights. What needs to be developed is a mechanism, or mechanisms, that will restrict the predatory behavior of those who have learnt to manipulate the intellectual property rights system to their exclusive advantage. This should not be overly difficult to realize if the underlying policy justifications for the intellectual property rights system are stuck to. One of the main justifications is that intellectual property rights should encourage creativity. There should not be any subjective element here that places the value of creativity resulting from one philosophical paradigm above that resulting from another. How one arrives at one's creative result should be equally respected. Indigenous and local community knowledge may well seem difficult to protect appropriately at the current time but that does not mean that it should not be. It is clear that the appropriate protection of indigenous and local community knowledge will, in many instances, lead to the wider dissemination of that knowledge while also respecting and promoting the interests of the knowledge generators and holders. In terms of the policy basis of intellectual property rights, this is an ideal outcome as it only results in social positives, a win–win situation for society as a whole.

IV. Where do existing initiatives fit into this pattern?

African Model Legislation for the Protection of the Rights of Local Communities, Farmers and Breeders, and for the Regulation of Access to Biological Resources (OAU, 1999):

- strong principles
- not model legislation
- weak details

The Indigenous Peoples Rights Act of 1997. Philippines Republic Act No. 8371.

- strong principle of local control
- complex system
- transaction costs

Proposal for a Regime for the Protection of the Collective Knowledge of Indigenous Communities and Access to Genetic Resources (Peru, 1999):

- strong on recognition, respect and preservation
- weak on local control
- remote from community

Convention on biological diversity:

- ambiguous principles
- remote from community
- sectoral.

A variety of initiatives have been developed in the past decade in an attempt to address the lack of effective protection for indigenous and local community knowledge. These have taken place in various parts of the world at various levels and embody a range of approaches. A cross-section of these initiatives is briefly examined here. This is not to devalue the efforts made by other institutions and states but rather to give an idea of the current state of regulatory development in the field and to highlight areas where further work needs to be done.

At the international level, one of the best-known instruments that addresses the question of indigenous and local community knowledge is the United Nations Convention on Biological Diversity (CBD). The CBD addresses the question directly in Article 8(j) and indirectly through other provisions such as articles 10(c) and 15. The CBD is a very useful instrument in that it is widely adopted, with over 170 countries having ratified it to date, and has developed a high-profile riding on the development of environmental consciousness around the world. It is also one of the most accessible international fora where NGOs and indigenous groups regularly participate with little interference.[10] As a framework agreement, the CBD is also useful as it allows for considerable flexibility in implementation at the national level while ensuring that certain basic principles are observed. However, during its 7-year life, a variety of problems have arisen in terms of the CBD's ability to provide an adequate umbrella for the protection of indigenous and local community knowledge rights. The most obvious of these is the fact that some of what seemed like fairly straightforward terms, such as 'prior informed consent' and 'mutually agreed terms', have made very limited progress. The main reason for this is that while most delegations and observers believed that they understood what these meant, their detailed iteration has proved almost impossible. Questions such as 'whose prior informed consent' and 'what does it actually take to be informed' are yet to be answered. In the absence of answers to these and similar questions, the usefulness of the CBD at the practical level is yet to be proven.

A second problem that has arisen is that despite its relatively inclusive nature, the CBD has proved to be fairly distant from the interests and concerns of communities at ground level. This has manifested itself in two ways. The first is that indigenous and local community groups have expressed considerable concern that while the CBD does address their rights, it does so in a manner that places responsibility for realizing these rights in the hands of national governments. The objection is that it is frequently national governments that place the greatest obstacles in the way of

---

[10] Although there are exceptions to this such as at the 1998 4th Conference of the Parties (COP4) when indigenous groups and NGOs were ejected from discussions of Article 8(j) at the instigation of the Brazilian delegation.

indigenous and local community empowerment such that in some cases, the CBD may actually be a step backwards from earlier agreements such as the International Labor Organization's Convention 169 (ILO 169). The second manner in which this distance has been manifested has been in the proceedings of the meetings of the Conferences of the Parties (COPs). An interim working group on Article 8(j) met in early 2000 and produced a report that was intended to form the basis of a decision by COP5 that would move the implementation of the Article forward considerably. This working group had a broad cross-section of participants, including considerable indigenous participation, and its conclusions were almost unanimously adopted with only one country (Colombia) maintaining a reservation. However, when the report came up for discussion at COP5 as the basis of a decision, at least half of its provisions were immediately abandoned and the prioritization of activities within it altered considerably. This happened despite the objections of indigenous groups and several developing countries.

The final problem with the CBD to be addressed here is possibly the most telling. As the indigenous and local community perspective on the question of rights to their knowledge becomes clearer, it is also becoming clearer that the sectoral nature of the CBD is likely to create ever-greater problems. The CBD, quite logically, only addresses knowledge that is related to biological resources. Since the lifestyles of indigenous and local communities tend to be more intertwined with their environments than those of modern communities, this might not seem to be a problem. This is particularly true in the case of traditional medicines that are almost always plant-based remedies. However, indigenous and local community knowledge is just as multi-faceted as any other community's body of knowledge and thus includes many elements that are deserving of protection but that are not biological resource-related. To what degree can it be justified to protect some forms of knowledge, because the outside world has become interested in their usefulness for conservation, while ignoring others that are equally important to the knowledge holders? Since many indigenous and local communities, and particularly indigenous communities, reject the division of fields of knowledge as an alien concept, the problem becomes even more severe.

The CBD may in time prove useful for advancing the rights of indigenous and local communities, but at the current time, this is far from clear. What is clear is that the chances of the CBD providing a comprehensive solution to the problem of community rights over their knowledge is unlikely at best, simply because of its mandate if nothing else. Possibly, the agreement has already served its greatest role by raising the profile of the issue to make it a serious subject of discussion at all levels from the community to international.

At a regional level, there is the draft African Model Legislation for the Protection of the Rights of Local Communities, Farmers and Breeders, and for the Regulation of Access to Biological Resources (OAU Model Law). This document has been endorsed by the Heads of State of the Organization of African Unity but is yet to be implemented in any of the member states. The first thing to be noticed about the OAU Model Law is the length of its title. The length and complexity of the title are fully reflected in the substance of the law itself. By this, it is meant that it addresses at least three or four regulatory questions: indigenous and local community knowledge and resource rights, farmers' rights, plant breeders' rights and access to genetic resources. It is commonly acknowledged that these fields are interrelated, particularly in the developing country context, but it has also thus far been acknowledged that they are sufficiently distinct to deserve individual consideration. The main reason for this individual consideration is the fact that each of the questions is enormously complex, and trying to deal with them *en masse* inevitably leads to a situation where clarity is lost and any subsequent law loses effectiveness. The OAU Model Law consists of 68 articles. Very often, a law on any one of the questions it addresses runs to such a length; as an example, just look at the plant variety protection (plant breeders' rights) legislation that many African states have already implemented. Of course, the length of a document is not a reliable indicator of its quality. However, in the case of the OAU Model Law, further inspection reveals that it leaves many of the details of each of the four questions unanswered, details that would be required in national legislation. Having said this, the OAU Model Law does clearly iterate the basic principles that should be applied to various areas. This combination of strong principles and weak details would seem to suggest that the OAU Model Law should not be considered by states as model legislation but rather as a regional framework treaty consisting of obligations that must be further iterated at the national level.

Where states do proceed to develop national legislation, whether based on, or merely deriving principles from, the OAU Model Law, they should still be careful in how they implement the details. A good example of this are the provisions on prior informed consent (PIC), something that has come to be considered as fundamental to protecting the rights of indigenous and local communities. In Article 5(3) the minimum requirements for PIC are laid out. Although the subject is addressed elsewhere and 5(3) appears to be the ultimately controlling provision. This Article contains no mandatory mechanism for direct contact between an applicant seeking to work with a community or their knowledge and the relevant community. What is required is that a prospective researcher contacts a designated national authority that is in turn required to 'consult' with the relevant community. This implies that the designated national authority does not actually have to follow the decisions of the community. Instead, it is free to decide what constitutes the grant of consent and is also free to ignore this anyway as access to knowledge is only invalid *without consultation with the concerned community or communities*, regardless of whether they give their consent. Consequently, the OAU Model Law does not actually include a requirement that communities in which research is to be undertaken give their prior informed consent at all, only that a central regulatory authority does.

The OAU Model Law is an international document intended to be equally applicable across one of the most diverse continents on Earth, and there is a strong argument that prior informed consent procedures need to be tailored to particular national, and even local, circumstances. Consequently, it is right that the OAU Model Law should leave some details to be addressed at the national level and in supplementary regulations. However, the example of PIC does highlight the dangers of this kind of international initiative. In its inability to provide the requisite detail, it risks building systemic flaws into national systems. If national policy-makers do not recognize the gaps in the OAU Model Law's approach to prior informed consent, they risk leaving implementers and prospective users of the Law with no clearer way forward than they had before but with a multiplicity of new obligations. This would probably not present a problem to those acting in bad faith, but it would make the lives of those acting in good faith, whether government regulator, prospective researcher or indigenous or local community substantially harder.

The OAU Model Law was developed by a group of African experts over several years. The majority of these experts, while acting in their personal capacities, were government representatives, and the document is clearly intended to be an inter-governmental one. The resources available for the development process were limited, and thus the final product is based upon the experience of those involved in its drafting rather than on any specifically targeted primary research. The centralizing and high-level approach of the OAU Model Law probably derives from this history.

While the experience of the drafting group would be hard to match, the process has still inevitably been a top down one that seeks to find solutions derived from what the experts know, and thus to some degree based on a theory of amending existing legal systems, rather than creating tailored solutions. As a result, in its details, the OAU Model Law is a useful advance in thinking but is unlikely to provide a comprehensive solution, whether for the protection of indigenous and local community knowledge or any of the other fields it covers.

Peru has been developing one of the first specifically tailored national regimes for the protection and promotion of indigenous and local community knowledge. The draft law is still under consideration after having gone through a period of being available for public comment for more than 6 months, but it is expected to be finalized in the near future. The brief outline of the proposed system is that it is a centrally maintained register that will preserve knowledge while providing a mechanism that will prevent piracy. This will be achieved simply by means of the fact that anything that is contained in the register obviously constitutes prior art and consequently could be used to defeat any subsequent attempts at privatization by means of intellectual property rights in Peru and in most other countries. The existence of the register is based on the idea that indigenous and local knowledge is a useful asset rather than simply a curiosity and as such is positive in the recognition and status that it provides. This is partly because the intellectual property office (INDECOPI) will host the register. This stands in distinction to the approach often seen elsewhere that leaves indigenous and local community knowledge as the domain of authorities responsible for cultural matters, thereby devaluing its practical value. However, the drawbacks of the system as proposed are already becoming apparent, as indigenous and local communities in Peru have been reluctant to engage in the process of developing the law. Some of this seems to be to do with the question of perceived benefit for communities from the new law and some resulting from suspicion of central authorities. INDECOPI has made efforts to create a participatory process, but limited resources and the fact that the process has been seen as being driven from the top have hampered these. This situation has come to be reflected in the current draft where there is little provision for local control of the system; it is more along the lines of the state protecting the rights of its communities rather than empowering the communities to protect themselves. The registry system is strong on the basis that it is practical in terms of management and recognizes solid principles of community rights, but it has weaknesses in that it is dominated by central authorities and does not clearly establish that its results will match the needs and aspirations of the communities it is intended to serve.

The Philippines has been an innovator in all areas related to indigenous and local community knowledge, but more particularly related environmental areas such as access to genetic resources, since the entry into force of the Convention on Biological Diversity in 1993. They have developed two particular legal mechanisms for protecting and promoting indigenous and local community knowledge with a particular focus on traditional medicine. The approach taken has been to provide a means whereby communities can have their customary systems and practices recognized by law. The basic requirement is that the community concerned record these systems and practices and communicate this to the central authorities, in many ways a classic decentralization approach. The second mechanism is the establishment of an institute for traditional medicine to conduct research on traditional remedies and look to their wider application. Such an institute is not an uncommon feature in Africa. What is uncommon in Africa is the accompanying legal recognition of customary systems and practices as regulating the field. There are clear advantages to this approach. The strong principle of local control of knowledge and resources means that regulation will be appropriate to local circumstances and in harmony with community aspirations.[11] Local control is also important to ensure that a system functions effectively. Marginalized sectors of society and, throughout most of the African Continent, traditional healers have been highly marginalized since independence, tend to be suspicious of centralized initiatives and, as a consequence, will defeat them simply by ignoring them. If knowledge-holders are not prepared to participate in a regulatory system, it is not worth the paper it is written on. Given this discussion, it might seem that the Philippines has found the answer to the problem of establishing indigenous and local community rights to their knowledge. However, there are significant drawbacks to the Philippines approach. Principal among these is the fact that the system that is developed becomes highly complex, certainly to the degree that it is beyond the management capacity of most developing country governments. Complexity, particularly where there is not sufficient capacity to handle it, leads to high transaction costs. By this, it is meant that it is difficult to find out exactly what the system is, and for an outside researcher to engage in it is a daunting prospect. If proposed research covers more than one ethnic region, the problems grow exponentially. Another aspect of high transaction costs is that the financial burden of implementing a system, even inadequately, can be prohibitive,

---

[11] Of course, there is always the problem of internal community dynamics and whether those with power truly represent the community, but there can be no doubt that whatever is developed by the community will, in the majority of cases, be an improvement upon whatever a central government might wish to impose, particularly in a culturally diverse state.

sometimes to the degree that it cripples the system. The ultimate result can be a situation where researchers prefer to go elsewhere, communities derive little benefit and governments are left with a big bill, i.e. nobody is happy. The Philippines approach is based on some very solid and realistic principles, but for it to function effectively and work for those it is intended to serve, the implementation will need a more creative solution.

V. What do knowledge holders want?

- recognition/respect
- a future for their tradition
- wider application – to help
- control (but not necessarily in the western sense)
- greater opportunities
- financial returns.

The intellectual property rights that are currently established in law, whether ancient, such as patents, or modern, such as those for integrated circuits, have had one major common characteristic. This is that they have almost exclusively been developed through pressure from the knowledge holders rather than through the initiative of policy-makers. One of the first ever patents was issued for a form of armoured vehicle in 15th century Venice. The reason it was issued was that the inventor refused to hand over the details of his invention until he was given such a monopoly over it. Equally, the 1994 TRIPs Agreement was largely the result of the US chemical and pharmaceutical industries lobbying their government to introduce the treaty into the General Agreement on Tariffs and Trade (GATT) framework. Given the problems that have been experienced in seeking to develop systems for the protection of indigenous and local community knowledge, it is surprising that this history has been forgotten. It is entirely possible that these problems have arisen because we have ignored history and approached the situation the wrong way. If one wishes to create a system of rights, one must be clear on what rights the intended beneficiaries are seeking, rather than trying to guess at what might be useful to them. The above list is one that has been developed from some limited experience of working with key knowledge-holders, indigenous and local communities in general and also with some extrapolations thrown in. It is unlikely to be exhaustive but will hopefully provide a basic idea and a starting point for further research and analysis. The list is not presented in any particular order, as it is clear that every knowledge-holder and community is likely to have slightly different interests and priorities depending on their experiences. However, what is suggested is that such lists, when based on more thorough field work than this, are likely to be remarkably similar across communities and regions. Most of the elements in this list are fairly self-explanatory, although some become quite complex once any thought is given to realizing them. For instance, the wider application of traditional medicines involves a range of questions such as standardization, efficacy testing and registration that are highly controversial in most countries. Control is probably the most complex element of the list in that while some communities might perceive it in a sense understandable to Western thinking the majority do not seem to. They are normally concerned with abstract, often spiritual, issues related to their knowledge. This basically comes down to the fact that knowledge holders would like to see outsiders treat and respect their knowledge in the same manner as they treat and respect it. One Quechua representative has frequently stated this belief in the terms: we are prepared to share our knowledge with you, but only if you will also share; if all you wish to do is privatize our knowledge, we will not share with you.[12] Is this really the extremist position it is often presented to be?

The mention of privatization brings discussion to the question of financial returns, the last element on the list. This is the one element that is intentionally placed in rank order: last. Several of the initiatives discussed above, and many others, have focused on what has come to be known as 'benefit sharing' as the central raison d'être of debate on the protection of indigenous and local community knowledge. It would be naïve in the extreme to suggest that indigenous and local communities do not consider the option of financial returns on their knowledge. However, while most communities would not object to such returns, as with most people, it does not seem to be an overriding motive in most situations. Communities, and knowledge holders in particular, frequently seem to be interested in the other elements of the list whether there is a prospect of financial return or not. A good example of this is an initiative that has recently been developed in the Karamajong pastoral communities of Eastern Uganda.[13] This initiative involves the recording, validation and dissemination of ethno-veterinary knowledge. The principal motive for establishing the project, which is directed by Karamoja elders, was peacemaking among what are traditionally quite fierce communities. The idea is basically that by sharing information about what is most important to them, cows, communities will recognize the value of cooperation, rather than conflict. Quite late in the project activities it was recognized that the information collected had potential application throughout sub-Saharan Africa, and possibly elsewhere. It was decided that the community would be comfortable freely sharing its knowledge with other communities in Africa, seemingly in some ways as a method of reaching out to them. Financial returns only really became an issue when the question of possible application in developed countries is addressed. It was decided that on the one hand, such countries could afford to pay for useful knowledge and on the other that this was particularly true as they would be likely to privatize and generate profits from it anyway. The main point to be taken from this is that, yes, financial returns were considered, but they did not contribute to the creation of the project, have no impact on its continued vitality and, ultimately, are just an interesting potential spin-off that is still far from being realized.

---

[12] Alejandro Argumedo, Director, Kechua-Aymara Association for Sustainable Livelihoods (ANDES).
[13] KEVIN – the Karamoja Ethno Veterinary Information Network.

VI. What are the ways forward?

- Immediate
- Use the aspects of intellectual property rights that can be beneficial:
  - a combination of geographical indications, trademarks and maybe trade secrets

- Longer term
  - Must push for the expansion of intellectual property rights, or as 'sui generis' rights:
  - in intellectual property rights, this has been done for integrated circuits, databases, etc. so why not other knowledge?

This chapter is intended to have immediate and practical usefulness for both healers and policy-makers. As such, it is appropriate to conclude by proposing ways to move both the situation of indigenous and local community knowledge holders and the debate over their rights forward by at least a small amount. It is clear that knowledge holders, and particularly healers with possible means for mitigating the devastating effects of diseases such as malaria and HIV/AIDS, cannot afford to wait for policy changes and need some more immediate solutions to the problems facing them. The first, and by far the most important, thing that a knowledge holder can do as regards protecting rights to their knowledge is to conduct an honest assessment of their knowledge, a sort of personal intellectual property audit. The main elements of this assessment will be as follows:

1. What knowledge is actually possessed?
2. What elements of this knowledge are most reliable/efficacious?
3. What is the potential application of the knowledge in geographical terms?
4. How much of this knowledge is possessed by other healers?
5. With knowledge that is, or is likely to be, possessed by other healers how widespread is it?
6. What do I want to achieve by protecting my knowledge? (e.g. finance, respect, opportunities for collaboration, etc.)

At this point, one hopefully has a list of various remedies, at least in the case of healers. This list should be subdivided into knowledge that is likely to have relevance principally in developing countries and that which has potential in developed countries as well. It should also contain some consideration of the possible financial, or market, value of knowledge. Whether knowledge is widely held or exclusive to the particular healer will play a significant role in any assessment of value. Where knowledge is widely held, it will inevitably be less valuable to the individual as either one cannot seek monopoly profits or one has to enter into a joint agreement with all others who hold the knowledge and thus dilute the profits. Alongside this list (or lists) should be an accompanying list of goals listed in some order of priority.

Once one has completed an assessment along the above lines, decisions can be made as to how to proceed. For knowledge that one wishes to share but does not wish to see pirated a possible solution is publication. This is not as complicated as it sounds as one can submit the information to a national pharmacopoeia, such as those proposed by International Center for Ethnomedicine and Drug Development (InterCEDD), to a publication devoted to the dissemination of information on traditional medicines, of which there are several around the world, or one can use an approach developed by the International Plant Genetic Resources Institute (IPGRI) in China.[14] Is this really the extremist position it is often presented to be? The IPGRI initiative involves the documenting of information and then using a researcher who regularly publishes in the field to cite this document, thus placing it on record. As usual this is not an exhaustive list but is provided rather as a suggestion of some options. These options all have a reasonable, although not conclusive, chance of ensuring that information remains available to all while also including credit for the original provider of the knowledge. What none of them do is provide much control over what subsequently happens to that knowledge.

If one seeks greater control over one's knowledge, then current options are limited to adapting various aspects of the existing intellectual property rights system. This paper hopefully establishes that patents are not really a viable option for indigenous and local community knowledge holders, while it also hopefully illustrates that other mechanisms, such as trade secrets, are not sufficiently suited. However, there are aspects of trade secrets and other rights that are useful, and, since many of these rights are either relatively cheap, or even free, to maintain a combination approach can be adopted. For instance, with traditional medicines that are to be marketed as phytomedicines, rather than proceeding to become full-blown pharmaceutical products, it is often not simply the contents of the product that is the major selling point. Often, people are attracted by the fact that the product has a good reputation, or comes from a reputable source, or sometimes even simply that it is from a particular place or ethnic group that is famous for healing. In such cases, one can seek protection through rights such as trademarks and geographical indications. These mean that others cannot try to imitate the source of your product (trying to confuse customers that theirs is actually yours) or claim that the product originates somewhere that it does not (if you are from an ethnic group or region famous for healing, say so, and then you can prevent others not from that group or region from claiming the same thing). With such mechanisms, it is not possible to actually prevent somebody from copying the contents of a product, but if nobody will buy the copy, what is the point? These types of solution can be somewhat effective where one's goal involves financial returns and, to some degree, also where the concern is the integrity of the product. However, to have greater control over the integrity of a product, and thus to control its use, the best mechanism currently available is a trade secret. Basically do not tell anybody about it unless they agree that it is yours and

---

[14] See 'IK Journal' at http://www.ipgri.org/system/page.asp?frame = themes/human/home.htm

will treat it as a secret also, not substantially different from what many knowledge holders already do!

The fact that this paper recommends the piecing together of various elements of different rights to achieve a form of protection that may or may not be effective should be sufficient to indicate that a lot more needs to be done in this field. The intellectual property rights system has been constantly expanding and adapting over the last 200 years, the most active period of its history. There is no reason why it should not continue to do so, particularly in the case of indigenous and local community knowledge that plays so many important roles in the societies where it exists. The TRIPs Council and/or WIPO might be found to be unsuitable fora due to their nature and history, but there is no reason why new specifically tailored rights could not be overseen by another forum in which indigenous and local communities have more confidence. From discussions in a range of international and national fora, the development of new intellectual property or *sui generis* (unique) rights does seem to be on the table. What is currently missing is for policy-makers to focus on the fact that these rights have to be developed from the needs and aspirations of knowledge holders and their communities for them to be effective, or possibly even functional. A top-down approach is unlikely to ever succeed in resolving this question. If the development of these new rights (it probably will end up being right*s*, not right, as indigenous and local community knowledge and the goals of its holders are as varied as knowledge and goals found elsewhere) is to be done in a manner that is effective for, and appropriate to, those they are intended to serve, it must begin with an assessment of what knowledge holders actually want and then proceed to see how these needs and desires can be accommodated within a practical, unified framework.

Iwu and Wootton (eds.), Ethnomedicine and Drug Discovery

Bioprospecting: using Africa's genetic resources as a new basis for economic development and regional cooperation

ANTHONY ONUGU

Abstract

The rapid evolution of the life-science industry hand in hand with the globalization of the world economy is throwing up new challenges and opportunities in the use and management of natural resources. It is against this background that this paper seeks to identify the necessary and sufficient conditions for African countries to utilize their natural resource endowment as a basis for economic development and regional cooperation. It identifies stakeholders, the needs of each as well as their expectations, and concludes that broad stakeholder participation is a crucial basis for effective policy-making. With the coming into force of the Biosafety Protocol, additional challenges emerge for the international governance of biodiversity and the impact of biotechnology. The paper examines the implications of this for national policy and resource development in Africa.

Keywords: *bioprospecting, biodiversity, genetic resources, biotechnology, policy*

I. Introduction

Africa's place in the global economy and the debate on the legal regime for utilization of 'non-rival public goods' is defined by its natural resource endowment. The flora of Africa encompasses a wide array of potential resources, ranging from non-timber forest products such as cane, reeds, bamboos, etc., to commercial timber and new sources of value added products, food, cosmetics, drugs, pharmaceuticals and other compounds of chemical, economic and industrial importance. Advances in biotechnology and genomics have made it increasingly feasible to transform plant medicine from an almost invisible trade into a modern industrial enterprise capable of making a significant contribution to both healthcare delivery and the economic growth of developing countries. These developments are raising expectations in developing countries with a promise of potentials but increasingly a failure of expectation.

For many countries in the region, biodiversity is the most important natural resource available as the foundation for economic development. However, it remains a resource whose potential is a challenge yet to be fully addressed at the national and international level. On one hand, there are concerns about the viability of bioprospecting as a livelihood activity, providing little short-term financial benefits from sample fees.[1] Initial expectations about the ability of bioprospecting to have a significant impact on natural resource and economic development policy decisions have been said to be largely theoretical.[2] Yet, the fact remains that there is a large and growing market for products developed from genetic resources, and thus, bioprospecting activities, in one form or another, are likely to continue.[3]

II. Bioprospecting in the regions

II.A. Growing demand for genetic resources

The flora of Africa is particularly rich in plant materials with medicinal or other useful properties. As a result, several bioprospecting projects are currently conducted in the continent, and the contractual arrangements are such that crude plant materials are collected from the region and exported overseas for processing. As markets for genetic resources have grown and diversified over the past decade, a wide range of intermediary organizations have emerged to provide specialized services to commercial end-users of genetic resources.

It is reported that bioprospecting leading to export of medicinal plants rose steadily between 1986 and 1989. As in many African countries, marketing of medicinals is carried out in small quantities by a plethora of small businesses, based on the demand of exporters. In 1998/99, the Ghana/United States-based company Bioresources International (BRI) contacted potential partners within Cameroon to assist in the domestication of *Pentadiplandra brazzeana*. BRI is working in partnership with the US Company Ocean Spray to establish a partnership for sustainable bioprospecting for raw materials from farmers in West and Central Africa.[4]

II.B. Who are the stakeholders?

Although the Convention on Biological Diversity (CBD) places primacy on the sovereign rights of a country over its genetic resources, there are several players involved in this sector. Five broad categories of stakeholders exist in the mechanisms of access, use, conservation and commercial utilization of genetic resources. These are the state and its agencies, the private sector, intermediary organizations, non-governmental organizations. (NGOs) and local communities.

In Africa, legal rights over genetic resources largely reside with national governments, with a further devolution to state or provincial governments and local councils in federal systems. As stipulated by the CBD, governments are expected to make laws or legislation to regulate access, use and conservation of genetic resources. The dominance of the public sector stake in this is further underlined by the weakness of the private sector in Africa. Government institutions involved include the Ministries, and agencies concerned with natural resources, agriculture, economic issues including fisheries, forestry and national parks, customs, health, research, and justice.

The public-sector research culture has a long tradition of open sharing of genetic resources, germplasm and research findings between research centers, and especially local academics and their foreign counterparts. Universities and academics are dominant in the limited natural product research and development that is taking place. In many cases, they represent active players in local bioprospecting, and a source for unregulated transfer of genetic materials, often as part of informal research collaboration. This tradition of open sharing and exchange of genetic resources, however, is under challenge from recent developments in IPR coverage and implementation.

In several cases, governments have established research institutions in an attempt to capture the commercial benefits of biodiversity. In Nigeria, the National Institute for Pharmaceutical Research and Development (NIPRD) and the National Centre for Genetic Resources and Biotechnology (NAGRAB) were established to achieve this objective. In Rwanda, an institute for science and technology (IRST) was created along with a center for research on traditional medicine (CUPHARME-TRA).

Commercial use of genetic resources through the discovery, development and marketing of products is driven by advances in the life-science industry. Lacking the technological base for participation in this industry, foreign interests have largely dominated private sector involvement in bioprospecting in Africa. This they usually achieve through the use of intermediaries. The private industrial sector involvement in bioprospecting includes, in particular, pharmaceutical, plant-health, horticultural, personal care and cosmetics, flavoring and fragrance, food and beverage, and other biotechnological companies.

The role of intermediary organizations is aptly demonstrated by the case of Plantecam in Cameroon. Commercial sales of medicinal plants in Cameroon was dominated by Plantecam, a subsidiary of the French company Fournier Laboratories, under exclusivity trade and supply arrangements with European pharmaceutical and natural products companies. In addition, Plantecam operations in Cameroon were commissioned as a Free Trade Zone in April 1995, which precluded it from eligibility for national taxes or customs duty. Plantecam/Fournier Laboratories until its winding up, was wholly French-owned, and did not have Cameroonian shareholders. Annual turnover in 1997 was 1.5–2 billion CFA.[5] Other intermediaries include botanical gardens, foreign universities and research institutes, gene banks and commercial brokers.

NGO participation is often in the area of research and advocacy on behalf of local communities. The Bioresources Development and Conservation Program (BDCP) is involved in bioprospecting in West and Central Africa, under the International Cooperative Biodiversity Group. (ICBG) program. It involves a strategy for adding value to genetic resources locally, rather than export of crude extracts.

Local communities, including traditional healers and community groups, access and use community forests in food consumption and its importance in the diet, plant medicines and for house building, household and agricultural equipment, fire wood, fodder and in trade and processing activities is widely reported. By forging symbiotic relationships to nature, local communities have come to develop traditional knowledge and management systems for genetic resources. Indeed, a distinction has been made between the knowledge of the use of biogenetic resources, which resides with individuals and communities and ownership rights to genetic resources as vested in national governments, whose interest is not necessarily in harmony with the former.[6]

II.C. Stakeholder needs

These include the following:[7]

1. Ambiguities in the CBD and national access and benefit-sharing measures.
2. Difficulty in keeping up with the diversity of measures adopted by countries.
3. Bureaucracy and delay involved in following access procedures.
4. Unrealistic expectations on the part of Governments and provider-country institutions.
5. Perception that the Convention on Biological Diversity rejects scientific traditions of research collaboration and exchange of specimens (therefore damaging research on biodiversity and new product development).
6. Belief that the Convention on Biological Diversity and access legislation create a disincentive to conduct research on natural products.
7. Unreasonable transaction costs and cumbersome procedures.

Consequently, changes in business practices have included:

1. Increased reliance on material from ex-situ collections rather than samples acquired through collecting activities.
2. Reduction in corporate collecting activities, consolidation of collecting programs into fewer countries and in some cases concentration on domestic collections.
3. Increased use of intermediaries as brokers of access and benefit-sharing (for obtaining permits and negotiating access and benefit-sharing arrangements) and suppliers of samples.
4. Increasing use of material transfer agreements (MTA).

However, source countries want to receive benefits that are commensurate with the contribution of their genetic resources in product R&D, as well as the capacity to add value to their resources. Source countries are largely unconvinced of the willingness of companies to seek equitable relationships, rather than fast deals. The belief is that companies do not want to let go of the free rider privileges they enjoyed under the free access regime.

These divergent views and interests reflect the need for a participatory approach to policy-making on access and benefits sharing in the continent.

II.D. The market for genetic resources

The market for plant genetic resources, especially medicinal plants, is complex and varied. The domestic market comprises a hierarchical structure beginning with the rural markets and progressing to the urban market. At the same time, the role of middlemen is increasingly dominant as trade progresses from the rural market to the urban market. Growth in medicinal plants trade has also led to the establishment and licensing of collection companies. These companies vary in size and scope of activity. They include established companies who carry out direct sourcing for export to Europe or smaller-commodity companies who largely service the intermediary companies. Due largely to the contradictions in land and property rights, as well as the discordant nature of the market, most trade in plants is carried out in a parallel market predominated by individual harvesters and operating mostly on public domain forests. Whereas collections are by law made operational by licenses provided by the Forestry Department, the large majority of collections are undertaken without licenses by independent contractors and local communities, who are paid at the roadside parallel market for the delivery of plant materials.

Sourcing methods have changed drastically with the pressure of demand, as collectors largely fell trees to enable easier access for harvest. In each case, the plant material collected varies from one medicinal plant to the other.

II.E. Some emerging trends

In attempting to analyze user and provider experience, the benefit-sharing experience of different industry sectors provides the opportunity to identify their respective characteristics:[8]

The pharmaceutical sector is the largest industry sector using genetic resources in terms of market, research budgets and profit margins. With respect to the sharing of benefits, the situation has evolved over the past decade, and it has now become usual for pharmaceutical companies to pay royalties on net sales. A number of factors, reflected by specific clauses in supply contracts, generally have an impact on the magnitude of the royalties. These factors include:

1. The current market rate for royalties.
2. The likely market share of a given product.
3. The relative contribution of the partners to the discovery and its development.
4. The degree of derivation of the final product from the genetic resource supplied.
5. The provision of ethnobotanical data with the sample.

Benefit-sharing arrangements are often composed of a package of monetary and non-monetary measures delivered across time. With the increasing capacity of source countries to carry out quality research in a cost-effective manner, companies have become more open to collaborations with provider-country institutions.

In the area of *botanical medicines, personal care products and cosmetics*, benefit sharing has essentially been carried out through the supply of raw materials by provider countries for the manufacture of products. In certain cases, these benefit-sharing arrangements have taken place through partnerships that have included technology transfer and capacity building. Developments in scientific and technological capacity for the study and testing of natural products have encouraged the demand for access to new species. Closer relationships may therefore be envisaged between these companies and source countries. Benefit-sharing could then include commercial research and product development activities, thereby assisting in the development of local capacity and institution building.

In the *biotechnology sector*, outside health care and agriculture, companies are generally not familiar with the scope and coverage of the Convention on Biological Diversity. They often obtain samples free of charge through collaboration with academic researchers or obtain their materials from intermediary organizations, such as culture collections, in exchange for a license fee or a purchase price. In rare cases of benefit-sharing with provider countries, the company will generally collect the genetic resources itself or will establish an arrangement with an intermediary institution in the source country.

In the *agricultural seed industry* for major crops, an informal system of exchange is still generally in place, permitting reciprocal access to genetic resources. It is common for many seed companies to obtain genetic resources free of charge or in exchange of a small handling fee, particularly if the germplasm acquired is unimproved. A number of actors are involved from initial access, through pre-breeding and commercial development, to sale of the final product to the farmer or consumer. Agreements are more common towards the end of the chain, for example with the use of license agreements when seeds are patented. The increasing use of intellectual property rights is said to influence partnerships in the industry to a greater extent than the development of policies and legislation on access to genetic resources.

In the *horticultural sector* and, more specifically, in the field of commercial ornamental horticultural development, there exist commercial arrangements that involve:

1. Royalties.
2. Payment of fees.
3. In-kind benefits, e.g. reciprocal access to plant material between non-commercial organizations.
4. Acknowledgment of the name of the provider of the genetic resources in the name of the plant variety subsequently developed by a breeder.
5. Sponsorship of student placements within the company, fellowships or enrolment on courses of higher education.

II.F. What is the state of legislation in the continent?

Policy measures, including legislation addressing access to genetic resources and the equitable sharing of benefits, either have been adopted or are in the process of being developed in over 40 countries.[8] Based on an analysis of existing and draft legislation on access, the following typology for the classification of legislative frameworks has been suggested.[9] It illustrates the diversity of approaches taken to address access and benefit-sharing and the difficulty in drawing general conclusions:

1. Access provisions contained in general/framework environmental or sustainable development laws: The Gambia (1995), Malawi (1996), Cameroon (1996), Uganda (1995).
2. Access provisions in nature conservation or biodiversity laws.
3. Access provisions incorporated into existing laws through amendment: Nigeria (1998), Cameroon (1994).
4. Specific access and benefit-sharing laws.
5. Regional legal frameworks for access and benefit sharing: In March 1998, the Organization of African Unity adopted a Declaration and a Draft Model Law for the recognition and protection of the rights of local communities, farmers and breeders, and for the regulation of access to biological resources.

It has been noted that, so far, experience in creation of access and benefit-sharing legislation in Africa has been one of the 'missed opportunities'.[10]

In the first case, such legislation should be clear and simple to allow flexibility, transparency and reduce transaction costs, and will need to be tailored to the circumstances of individual countries. The development of international guidelines or principles for such measures could provide assurance to provider countries that their resources are used in accordance with the terms of the Convention.

Secondly, it is important that national legislation on access and benefit-sharing be made consistent with existing international obligations, and does not restrict or undermine the position of such countries in ongoing international negotiations, including adherence to future agreements such as the International Undertaking on Genetic Resources for Food and Agriculture being negotiated under the auspices of the FAO.

The OAU Model legislation and most of the access law and mechanisms in Africa reflect a predominant skew towards 'protective' legislation, as opposed to the 'promotional', one that creates an enabling environment for the sustainable and value-added use of genetic resources, and the clear emergence of a biological resource and biotechnology industry.

Generally speaking, attempts by African countries to regulate access to genetic resources reflect an orientation towards nationalism, and a design to control access and secure benefits, without first conducting strategic assessments of national biodiversity inventory, capacities and goals, in order to direct the development of their access legislation. As a result, existing national law on access and benefit sharing does not adequately support conservation and sustainable use. Furthermore, it often results in undesired consequences. Several national laws on access have, for example, unwittingly hindered domestic research and partnerships with foreign organizations, thus blocking the very capacity building that such laws may specify as an important objective.[10]

In large part, the processes leading to legislation on access and benefit sharing have lacked broad stakeholder participation, from local and indigenous communities, research institutions, natural product companies and foreign interests in this sector. Many countries are simply basing their legislation on generous adaptation from existing laws. The Kenyan case reflects considerable adaptation from the Philippine law's distinction between academic and commercial agreements, and on the 'intangible component' specified by the Andean Pact and Brazilian drafts. A key need in this regard is recognition of the difference between government policy and national policy, as well as the role of women and the poor, who, in most cases, are the custodians of biodiversity.

The general state of legislation in the continent has led to a vacuum in the implementation of bioprospecting agreements in the absence of a guiding legislative authority.

Controversy has recently arisen in Zimbabwe and Switzerland concerning the way by which the University of Lausanne gained access to genetic resources in Zimbabwe and how the benefit sharing has been negotiated. In the process, Zimbabwean organizations have rejected the US-patent 5'929'124 on antimicrobial diterpenes, which relies on traditional knowledge from Zimbabwe and on the root of the tree *Swartzia madagascariensis,* found throughout tropical Africa. In April 1997, an addendum to a material transfer and confidentiality agreement between the American pharmaceutical company Phytera and the University of Lausanne was signed, under which Phytera received an option for an exclusive world-wide license and in return agreed to pay royalties of 1.5% on the net sales of any product marketed under this license.

The University of Lausanne, however, is obliged to give 50% of any royalties received to the National Herbarium and the Botanical Garden of Zimbabwe and to the Department of Pharmacy at the University of Zimbabwe. Neither the state of Zimbabwe nor the traditional healers affected by the bioprospecting were correctly informed or gave their prior informed consent for the search of genetic resources in Zimbabwe.

In Zimbabwe, the mandate and the authority to allow access to genetic resources lies with the Ministry of Environment. However, no contract existed between the Ministry and the University of Lausanne, nor was there any contract that shifted the mandate from the Ministry to the University of Zimbabwe, the local intermediary through which the University of Lausanne gained access to the resources.

There were no mutually agreed terms for a fair and equitable benefit-sharing mechanism. An agreement signed between the University of Zimbabwe and the University of Lausanne stipulates that in the event of finding any product, which may require the application of intellectual property rights, this will be a subject of joint negotiation and application. Contrary to Article (F) of this agreement, as the Chairperson of the Pharmacy Department indicated, the University of Zimbabwe did not take part in the negotiation process between the University of Lausanne and Phytera. No one consulted stakeholders in Zimbabwe on whether they agreed with the amount of royalties they were to receive.

II.G. *Need for new policy directions*

Africa's rich endowment of genetic resources provides clear potentials for enhancing the economic development of the continent. However, natural resource endowment is not a sufficient basis for growth unless it is linked to the development of process technologies and their diffusion in the economy.[11] The rise and increasing convergence of modern biotechnology and information technology are establishing intellectual capital as the primary resource in the global economy. This has resulted in a greater reliance on secondary sources for bioprospecting, including published information, academic researchers and botanic gardens. In this regard, the role of institutions such as botanical gardens, field museums and herbariums as well as academic and research institutions will increasingly grow.

The field bioprospecting trips of old are increasingly proving to be unnecessary. The combined effect of this as well as the nature and rapidity of developments in these industries will more than likely render the provisions on access and benefit sharing a nullity and of little practical effect. For developing countries, the goal of equity and sustainability in the development process will increasingly need to address exclusion and not exploitation as the fundamental threat.

Increasing global focus on new approaches to the management of public goods and the need to activate a sense of common purpose led in large part to the conception, negotiation and enunciation of multilateral environmental agreements (MEAs), principal of which is the convention on biological diversity (CBD). The CBD sets the international framework for management of biological diversity. As part of the obligations on becoming signatories to the convention, African countries are required to initiate legislation on preservation and regulation of access to genetic resources. The convention equally exhorts countries to initiate action towards assessment of the national stock of genetic wealth. Buoyed by over-optimistic estimates of windfalls from bioprospecting and royalties from future drug discoveries, African countries have reached for the legislative option of nationalism. Frightened by negative perspectives of the role of biotechnology in sustainable use of genetic resources, African countries have enacted 'defensive' legislations.

It is a sad commentary on environmental advocacy that those who have the closest relationship to nature, and the most need for its resources and by-products, are most discouraged from positive action, by a skewed focus on negative statistics and impacts. The result, as shown by the protected area mechanism for instance, is that the people are seen as part of the problem in the conception of solutions. Why does the native Kikuyu not see enough beauty in a herd of elephants, or the Koma and Aka people in mountain gorillas of northwestern Nigeria and forests of the Congo basin to support a protected area for it? The truth is that their appreciation lies, not simply in the aesthetics (they see it everyday of their waking life), but on its use value. Vigorous use of natural resources is an integral part of indigenous conservation behavior. A recent study has indeed shown that areas of high biological diversity in Africa coexist with high population densities.[12]

There is a need for a new orientation of policy in Africa. There is a strong need for policy-making to resist the temptation to treat all external parties in Africa's genetic resources as being motivated by greed and piracy. There is a need for careful considerations of options to, for instance, avoid repeat of the single-minded focus on biosafety, important as it is. African countries need instead an approach that seeks a balance between risks and benefits associated with the use of biotechnology, and thus greater attention to the issue of technological cooperation.

Secondly, while it is important to address the relationship between Consultative Group on International Agricultural Research (CGIAR) and the CBD in terms of the legal status of the genetic resources held by the former, it is even more important to address policy (and international negotiations) to strategies for using the scientific and technological knowledge resident in these institutions for the benefit of local agro-biotechnology research and agriculture. As it is today, many of the CGIAR institutions in Africa remain islands unto themselves, with local scientists and government officials allowed the occasional privileged access, and visiting fellows and programs, of limited direct relevance to the local community and their priorities, except the intellectual progress of the individual scientists.

There is a wide gap between long-term regional strategic priorities in health and agriculture, and the negotiating position of Africa. Unfortunately current efforts are unlikely to achieve much in the absence of this linkage. While African countries identify agricultural productivity as a priority, there is a need for delimitation of specific long-term goals, and a linkage of this to the role that biotechnology will be expected to play. Similarly, a regional action framework must, as of necessity, identify existing local comparative advantages in agricultural and agro-biotechnology research, and use these as the basis for concerted action.

II.H. How can African countries use their genetic resource endowment as a basis for economic development?

Against the financial realities of drug development, the real benefit of genetic resources lies in its ability to provide a stimulus or seed money to establish or improve in-country capacity to conduct research on genetic resources and support the emergence of an indigenous biotechnology and pharmaceutical industry, not in windfall payments from royalties, as commercial end-products may never materialize. At the same time and on the basis of strict monetary calculations, pharmaceutical companies have insufficient motive to fund the search for new drugs to combat malaria and other tropical diseases, the so-called orphan drugs. Accordingly, the more important questions for concerted regional action are:

1. How to utilize the continents' biogenic resource endowment as a basis for tracking into the $US 230b LOHAS (life of health and sustainability) market (as India is currently pursuing).
2. Realignment of the relevant mechanisms of access and benefit sharing legislation in aid of biomedical research and public screening programs to meet Africa's health needs.

3. Utilization of current mechanisms of international governance of biotechnology, including the CBD, as instruments in the building of local capacity to enable Africa to participate in the biotechnology industry.
4. Use of these same mechanisms to build a regulatory framework that establishes a balance between the benefits and risks in this industry.

Given the commercial orientation of this industry, how can market instruments be adapted to such an initiative? What role can the private sector play in this?

II.I. Building relationships

Critical to sustainable use of genetic resources for economic development in Africa is the establishment of strategic partnerships with foreign companies and organizations, rather than a focus on quick deals. This becomes more important since what is being traded is not just a commodity but a priceless resource that embodies the cultural views, life style and even religion of a community.[13] It has been suggested that one approach to this is the identification of niche areas for specific focus. For this to be successful, the primary orientation should be to add value to local resources rather than export of raw extracts. However, changes in global patterns of funding for research and development[14] are leading to the emergence of private proprietary science, and a direct linkage between scientists, venture capital and corporate interests. At the same time, advances in information technology have led to wide diffusion of knowledge, with positive potentials for countries in Africa with a large population of qualified scientists, as is the case with Nigeria.

The demands of the biotechnology industry have often meant that developed country scientists relied on their counterparts from developing country for starting materials based on ethnobotanical or ethnomedical leads. Such collaborations, if properly structured, can lead to the emergence of new knowledge networks, which have the potential to benefit developing countries and facilitate their participation in the industry. There is a need for realignment of research in aid of public–private partnerships as the International Cooperative Biodiversity Group program on drug development and biodiversity conservation or the Human Genome Project in the United States appropriately demonstrate.

There is a need for greater focus on the role of the non-governmental sector as a new basis for development of internal capacities. This becomes especially important against the background of the 'brain drain' problem in Africa. Current information technologies make it possible to collaborate with widely scattered scientists, linking 'pockets of intellectual property' within the framework of the non-profit organization, to specific strategic goals and action-oriented research. By this dynamic collaborative approach, it becomes possible to optimize access to both technology and expertise, and by engaging in product R&D within a not-for-profit yet business-like framework, it links cutting-edge science and management, with a substantial advantage for lowering costs.

However, the quantum leap in the role of the private sector in a global economy and efforts by developed country governments to protect their competitive advantages has led to a weakening of civil society organizations and international bodies. There is therefore a need to build new relationships involving the public and private sector and drawing from the latter's social consciousness to negotiated policy-making.

II.J. What role for MEAs?

This will require a new approach to implementation of multilateral environmental agreements and institutional adjustments by NGOs to move beyond strategies of protest and advocacy to engage the private sector in new partnerships for development. The new challenge in the management of public goods is therefore moving towards the coexistence of negotiated policy and regulation from within, pari pasu with international legal regimes.

The articles of the CBD give recognition to the sovereign rights of African countries over their genetic resources, and their rights to exploit such resources in accordance with their environmental policies. In order to reinforce this, countries are encouraged to prioritize the issue of species identification, classification and monitoring. However, many of the commitments that African countries have undertaken have merely shifted responsibility to the national level without any corresponding commitments on the part of the international community. Assigning rights without concomitant empowerment amounts to very little in real terms.

In the first instance, there is a need to redesignate the issue of biodiversity assessment and monitoring as an 'enabling activity' under the Global Environment Facility (GEF) project pipeline. This will effectively place this as a building block on which other implementation mechanisms can be based. It would then be necessary to use the Global Taxonomy Initiative (GTI) to build the capacity of source countries to carry out taxonomic identification. In this regard, it is important that the implementation process be locally based. In many cases in Africa, what is lacking is not the human resources for taxonomic identification but the institutions. These institutional weaknesses are not addressed by direct co-option of international taxonomy organizations, as lead institutions in the process. Considerable capacity already exists from the work of the Organization of African Unity Science and Technology Research Committee (OAU/STRC), African Scientific Cooperation on Phytomedicine and Aromatic Plants (ASCOPAP), as well as SABONET and SAFRINET in southern Africa, and BOZONET in Eastern Africa.

The adoption of the Biosafety Protocol in concomitance with the globalization of the world economy and the evolution of the biotechnology industry in combination has opened new challenges and opportunities for the Convention and developing country parties. This opportunity will, however, only materialize if the Protocol enters into force and its institutions are effectively established and operated. With regard to Africa, the challenge lies in structuring national policy and international negotiating framework towards effective exploitation of Articles 12 and 16 of the Convention.[15]

In turn, there is an opportunity for the Convention to build local capacity in Africa. This includes assistance to developing country regulators to develop the following:

1. Various testing, monitoring, certification, biosafety requirements needed to aid adoption of new technologies and legal requirements.
2. Negotiating skills for access and IPR issues.
3. Anti-trust regulations.

Another particular need in this regard is to strengthen the role of national focal points to the extent that all Parties are able to participate fully in the processes of the Convention.

While these issues ultimately may require national or international regulation, regulatory policies may prove inadequate. It seems ironic that as the CBD locates the engine room of its implementation strategies in the public sector, most developing countries are repositioning the nexus of economic development in the private sector. In many countries, a philosophy of deregulation prevails. Besides, most of the forces, which drive innovation in the biotechnology and information technology industry, originate from the private sector. Ten years into implementation of the CBD, mechanisms of self-regulation are increasingly required to address the identified priorities.

Assigning regulatory functions to institutions outside government, provided there is a system to retain overall control, could thus become a powerful tool. Such strategies empower global players from the private sector of business and from the civil society sector of non-governmental organizations to assume greater responsibilities. Through this process of *controled self-regulation*, NGOs, CBOs, trade unions, organized industry and professional associations can become partners in the regulatory system. If properly instituted, this process could ensure not only that benefits are equitably distributed between foreign users of genetic materials and the source country institutions but that equity is established at local or national levels.[16,17]

References

1. Reid W, Laird S, Meyer R, Gamez R, Sittenfield A, Janzen DH, Gollin MA, Juma C. (1993) Biodiversity prospecting: using genetic resources for sustainable development. Washington, DC: World Resources Institute.
2. Simpson D. (1997) Biodiversity prospecting: shopping the wilds is not the key to conservation, resources. Washington, DC: RFF.
3. Downes D, Wold C. (1994) Biodiversity prospecting: rules of the game, Bioscience.
4. Odamtten GT, Laing E, Abbiw DK. (1996) The Economic value and potential for plant-derived pharmaceuticals from Ghana. In Ballick MJ, Elisabetsky E, Laird SA, editors. Medicinal resources of the tropical forest: biodiversity and its importance to human health. New York: Columbia University Press.
5. Sunderland T, Nkefor J. (1997) Conservation through cultivation: a case study. The propagation of Pygeum – *Prunus africana*. TAA Newsletter. December.

6. Iwu MM. (1996) Implementing the biodiversity treaty: how to make international cooperative agreements work. Trends Biotechnol 14:78–83.
7. ten Kate K, Laird SA. (1999) The commercial use of biodiversity – access to genetic resources and benefit sharing. London: Earthscan Publication.
8. K ten Kate K, Laird SA. (2000) Biodiversity and business: coming to terms with the 'grand bargain'. Int Affairs 76 I:241–264.
9. Glowka L. (1998) A guide to designing legal frameworks to determine access to genetic resources. IUCN Environmental Law Center.
10. ten Kate K. (1999) Proposal for a pilot project to integrate a component on access and benefit sharing into biodiversity strategies and action plans.
11. Juma C. (2000) Science, technology and economic growth: Africa's biopolicy agenda in the 21st century. http://www.cid.harvard.edu/cidbiotech/pp/policy.htm.
12. Balmford A, Moore JL, Brooks T, Burgess N, Hansen LA, Williams P, Rahbek C. (2001) Conservation conflicts across Africa. Science 291:2616–2619.
13. Iwu MM. (1996) Implementing the biodiversity treaty: how to make international cooperation work. Paper prepared for Trends Biotechnol.
14. Juma C. (1999) The limits to south–south cooperation. Nature (web), May 27 edition.
15. Convention on Biological Diversity (992) United Nations, New York.
16. Gupta A. (1999) Studies on the role of intellectual property rights in benefit sharing – case study one: the Nigerian case, a project report prepared for the World Intellectual Property Organization. (WIPO).
17. Moran K. (1998) Moving on: less description, more prescription for human health. EcoForum 21(4):5–9.

Iwu and Wootton (eds.), Ethnomedicine and Drug Discovery

Balancing conservation with utilization: restoring populations of commercially valuable medicinal herbs in forests and agroforests

RICHARD A CECH

Abstract

This paper discusses the medicinal herb industry in relation to conservation of medicinal plants, including an overview of sustainable wildcrafting and sustainable forest agriculture. Goldenseal (*Hydrastis canadensis*) and Black Cohosh (*Cimicifuga racemosa*) are given as examples of wild herbs, of economic importance, which may be readily cultivated. Also included are guidelines and techniques for forest farming of hardwood forest-dependent species.

Keywords: *medicinal plants, conservation, sustainable, farming, culitvation*

I. Introduction

Old Louie leaned on the fence, bending down the top wire, a wire already seriously stretched by the daily escape of our buck goat, who nobody except myself was willing to tackle and bring back to pasture. Not that the rest of my family was *afraid* of the goat, it was simply that he stank with an eye-smarting fragrance that only a doe goat in heat could admire, and only a staunch believer in social life after garlic could possibly withstand for long. However, Louie paid no attention to the vestigial aroma of goat-grease that exuded from the fence and (probably) from my soiled blue jeans. His good eye followed me from under the sun-scorched brow, in turns pleading and plotting, because there was something he wanted, and he was ready to use any persuasion available to get it (and given the daily marauding of my goats into his pasture, he figured he just might have some bargaining power). You see, there were chittam (otherwise known as cascara sagrada or *Rhamnus purshiana*) trees on my land, trees that yield a bark much in demand by pharmaceutical companies for the manufacture of a cathartic extract. It makes you go. In fact, the going joke among west-coast woodspeople, a joke that has been funny for several centuries, is to off-handedly mis-pronounce chittam as 'Shit 'em'. But I digress. The reason Louie wanted permission to cut the bark off my trees, was because all the trees on *his* side of the fence were already denuded of bark, from root to tip. I could see several white skeletons from my vantage point, and I was darned if the trees on my land would be similarly fated. So Louie went home disgruntled that evening, leaving me to reflect on how to keep a goat in, and when I grew tired of that, to continue to muse about conservation, and how to harvest without taking the life of the tree (or plant). I was somewhere in the middle of a long learning. The answer for chittam is that if you cut the tree off with a tall stump, it will coppice (i.e. re-sprout from the stump), producing multiple trunks that bear useable bark again in a couple of years. Louie did not know this, because he simply stripped the trees standing, and they died.

I have made it my life work to learn the reproductive habits of many medicinal plants of prairie, forest, mountain and swamp areas. Knowing how a plant reproduces is the first step to encouraging its growth in wild settings or in domestic culture. Every farm-grown medicinal plant has the potential of saving the life of a plant of the same species growing somewhere in the wild. Given that each ecosystem is defined in large by the plants that grow there (trees make the forest, and grasses make the prairie), the conservation of wild plants equates to the preservation of our wild places, a diminishing resource.

Among the major herb buyers, consisting of domestic and foreign herbal industry, the conversion from utilization of dwindling wild-harvested herbs to cultivated herbs is driven mainly by economic factors. If the wild crop becomes too rare and therefore unavailable or too expensive to buy, companies will seek to convert to a cultivated source. Open-market farm production of a number of significant herbs currently lags behind demand; hence companies are turning to growing their own medicinal plantations, or contracting directly with farmers who can grow crops for their products, and assure supply. Assisting this trend is a groundswell of consumer demand for sustainably cultivated medicinal plants. However, let us first examine the age-old occupation of wildcrafting (harvesting wild plants), in the light of our more recent concerns, which involve sustainability of wild medicinal plant populations.

II. What is sustainable wildcrafting?

The short answer is, that in *sustainable* wildcrafting, every act of taking is coordinated with an act of planting. Unfortunately, even the most conscientious wildcrafting may impact plant populations, given that human support of the natural process of regeneration is not always successful, and also taking into account the *age* of the plants involved. This lesson is nowhere more apparent than in the case of large-scale silviculture in the Pacific Northwest, where re-forestation is used as a justification for the cutting of old-growth forests. A lot of difference exists between a 'plug-one doug fir seedling' and a mature tree. In the same way, there is a lot of difference between digging a 30-year-old ginseng and pushing a ripe ginseng seed (or 12 of them, for that matter) into the ground.

We must encourage utilization of renewable portions of the plant. Digging roots generally kills the plant, but aerial portions of many plants contain the same active constituents as the roots. Of course, when delving into the possible substitution of a less used plant part for the traditional preparation, one must remain alert for any potential side-effects, toxicities or differences in potency that may bear on the advisability or pragmatics of utilizing a previously un-tested plant part. This is a wide-open field for applied scientific investigation into the active constituents of plants. An example of this kind of applied science is Herb Pharm's (unpublished) chromatographic comparison of Goldenseal root and rhizome to Goldenseal leaf and stem. This study demonstrated that the two main alkaloids (berberine and hydrastine) are present in significant concentration in the sustainably harvested leaf and stem as well as the root and rhizome (Personal Communication, Ambrose Amarquaye, Ph.D.).[1,2] If we can make medicine from the renewably harvested portions of the plant, our impact on wild plant populations will be minimized.

Traditionally, most herbal medicines are made from dried herbs. Dehydration is an effective method of preservation; it decreases transport costs, and often concentrates the relative weight of active constituents. In a few cases, the main active constituents or marker compounds are not even present in the fresh herb, but are formed as an artifact of the drying process. However, the potential for using fresh (undried) plant material is worth considering, since the process of dehydration does not always concentrate the weight of active constituents in relation to the total weight. In some cases, the process of dehydration actually diminishes relative constituent levels, through volatilization, chemical degradation or chemical transformation of the active constituents. To further complicate this issue, the breakdown products of original compounds may also demonstrate medicinal activity. Further study into the possible use of fresh plant material instead of dry will allow manufacturers to choose the method that gives the greatest yield of desirable constituents from each precious plant.

Seasonality of harvest bears strongly on the impact of harvest. In the case of Ginseng and Goldenseal, we have seen that harvest prior to the setting of seed is damaging to plant populations because the plant has no 'final chance' to reproduce itself through seed. Early harvest equates to bad herbalism, also, as a root is at its healthiest, heaviest and presumably highest potency in the late summer and fall, after recovery from dormancy and after accomplishing a new growth cycle. The common practice among states is to limit wild Ginseng harvest to a fall season. The main buyers of wild American Goldenseal have made a similar, unofficial agreement to buy Goldenseal in the fall, thereby discouraging wildcrafters from digging plants prior to maturation and dissemination of seed.

---

[1] Personal communication with Ambrose Amarquaye, Ph. D., of the Herb Pharm analytical laboratory. Organically grown goldenseal (*Hydrastis canadensis* leaf and stem obtained from Horizon Herbs in Williams, Oregon showed a concentration of 2.1% berberine and 1.7% hydrastine (W:W). Samples of organically grown goldenseal leaf and stem from Kentucky contained 1.2% berberine and 0.7% hydrastine (W:W).

[2] *Plant Drug Analysis, Vol II*, by Wagner and Bladt. The normal range for berberine in *H. canadensis* rhizome is 2.0% to 4.5%, and the normal range for hydrastine is 3.2 to 4% (W:W)

The bottom line is whether or not the act of wildcrafting has, on a season-to-season basis, reduced, maintained, or increased the mean age and population of desired plants. If economically significant species are maintained or increased, either through successful replanting or by removal of competitive weed species, the wildcrafting may be termed sustainable. Of course, this presupposes that wild places are being treated like wild gardens, which is rare, and that other wildcrafters are not impacting the same area, which is rarer still.

III. Forest farming

Plant ecologists argue the advisability of planting wild herbs back into the woods. It is difficult for humans to re-establish, in a few seasons, a balance broken by years of thoughtless resource utilization. Additionally, it is sometimes difficult to know which herbs belong where, or to obtain the appropriate landrace, which is adapted to a specific wild ecology. Luckily, when a forest is left alone, the earth tends to heal itself with time, and barring distinction of species, diverse native plants, which are best adapted, will eventually re-establish into their appropriate niches. Given that the forest is a whole system, dependent on the interraction of trees, insects, animals and non-woody plants, the practice of harvesting the wild plants as an alternative to cutting the trees is of short-lived viability. It is better to set aside areas of forest that can be farmed intensively for the production of desirable forest-dependent species. This ultimately improves the health of the forest, allows efficient production of large quantities of usable plant material, and diminishes traffic in wild areas.

Generally speaking, forested areas destined to provide cover for production of medicinal plants are climax hardwood forests. To start, it is necessary to remove and pile underbrush and weedy species. The ground is then worked up into beds, thereby defining production areas and access corridors. Tree roots, which get in the way in the beds, may have to be chopped out. Ground preparation may be done by hand, by rototiller or by tractor. As far as tractor implements go, a spader works better than a tractor rototiller or a plow, being better suited to moist conditions, less disruptive to soil strata, and less apt to be damaged by roots and rocks. In ideal situations, no-till agriculture is possible. Weedy species are hand-pulled, and the desired crop is planted without disturbing the natural soil structure, where rotting leaves disseminate nutrients down to the roots of the plants.

Maintaining soil fertility in agroforests is easier than maintaining the fertility of open fields. practices that encourage healthy growth of the trees, such as thinning, trimming dead wood, removal of underbrush and irrigation during dry spells, also serve to improve the health of the soil, which is further enhanced by leaf-fall. The herbal crop benefits are that mineral-rich soils (typical west of the Rockies) may be assisted by application of composts, while humus-rich soils (typical of the eastern hardwood forest biome) may be assisted by application of rock powders, such as limestone and rock phosphate. Compost may be added at any time during the growth cycle, but a note of caution is advised as composting can increase the danger of fungal diseases, which occasionally pose a challenge to forest farmers. Rock powders should be applied during bed preparation or during dormancy.

IV. Goldenseal (*Hydrastis canadensis*)

Goldenseal is following in the footsteps of American Ginseng (*Panax quinquefolius*) in terms of plant demographics (reduced wild populations), governmental status (both now regulated in Appendix II of the Convention for International Trade in Endangered Species of Wild Fauna and Flora (CITES)), and by way of rising popularity and escalating price. However, Ginseng is widely cultivated, while Goldenseal cultivation is poorly understood and lags far behind demand. Cultivation of Goldenseal is unquestionably the most significant current herbal agricultural opportunity, but is a path fraught with challenges. United Plant Savers (UpS) is sponsoring ongoing experimentation designed to determine sustainable wild-harvesting techniques and also to define preferred cultivation scenarios. Experimentation is taking place at the UpS Botanical Sanctuary in Rutland, Ohio. Results are reported on an ongoing basis in the 'United Plant Savers Newsletter'.[3]

As with cultivation of any forest-dependent species, successful cultivation of Goldenseal requires exact conditions of shade, soil and season. Choosing a site within the hardwood forest and within the original distribution range of the plant is more likely to prove successful than attempts to field cultivate the plant under shadecloth or lath, or to grow it in woods outside its native range. As one gets further away from native ecology, more problems tend to occur, and thus more time, money and energy go into adapting the environment to the plant. Ideally, one grows the plant where the right conditions already exist.

Monocropping in the forest may cause problems such as runaway disease and insect infestation. *Alternaria* fungus and Goldenseal mites are cases in point. Creating a diverse cropping system, where beds of the main crop are interspersed with plantings of their natural companions, helps the plants sequester soil bacteria and fungi, which assist in nutrient assimilation and protect the plant from insects, parasites and disease organisms. Good candidates for intercropping with Goldenseal are Jewel Weed (*Impatiens* spp.), Bloodroot (*Sanguinaria canadensis*), Black Cohosh (*Cimicifuga racemosa*) and Wild Yam (*Dioscorea villosa*.) The woody species Spicebush (*Lindera benzoin*) and Paw Paw (*Asimina triloba*) may prove useful in further shading the understory. Just as healthy natural ecologies are characterized by the diversity of their plant life, a diversity of plantings in the agroforest contribute to the rapid growth and disease resistance of crop plants.

Cultivation of goldenseal from seed is challenging, but this practice also yields several advantages. Genetic diversity is enhanced, and seedlings are more disease resistant than plants grown from cuttings. A single mature plant may be dug and the

---

[3] UpS Newsletter is available from United Plant Savers, PO Box 420, E. Barre, VT 05649

rhizome divided into three or four new plants, but that same plant, if allowed to go to seed, can produce approximately thirty new individuals on a yearly basis.

The seed is washed from the ripe fruit in the late summer and sown as soon as possible into prepared beds, under the forest canopy or in the shade garden. Initial germination occurs the following spring, with ongoing germination in the spring of the second year. Seedlings may be grown at close spacing for two years, then transplanted in the fall to a wider spacing, attaining maturity in seven years.

To propagate by root cutting, each mature rhizome is divided into three or more cuttings, which are planted at 12-inch centers in the fall or very early spring. The first year of growth from cuttings is usually a bit weak, typified by attrition of the rhizome, early dormancy and potential for fungal infection. This may be avoided, in part, by planting root cuttings in the fall, not the spring. Further protection from fungal disease is afforded by planting at a wider spacing, thereby improving air circulation around the plants.

The lateral rootlets bear reproductive nodules, which serve to spread the plant in nature. Masses of free rootlets found in Goldenseal shipments may be planted in flats, or the nodule-bearing rootlets may be removed from rhizome cuttings and planted separately in flats. These produce hardy plants that may be transplanted out after a year of containerized growth.

V. Black Cohosh (*Cimicifuga racemosa*)

Black Cohosh root is experiencing a current resurgence of popularity, due to its application in treating pre-menstrual syndrome, menopause, estrogen deficiency, dull pain and some kinds of depression. The widely distributed native populations of this herb have recently been challenged by wholesale harvest, which amounts to hundreds of thousands of plants (personal communication)[4] Unlike Goldenseal, which is sold mostly to domestic markets, foreign phytopharmaceutical companies use much of the Black Cohosh harvested in America. A CITES listing for Black Cohosh would definitely reduce the harvest of wild plants, more so than in the case of Goldenseal, which is consumed, for the most part, within the United States.

Harvest of Black Cohosh may be made more sustainable by using this technique: the plant is located in the fall by identifying the remaining blackened, characteristically pronged stem. The rhizome is uncovered, and the harvester breaks off the older portion (the 'back' of the rhizome, without bud) for use, leaving the remainder of the rhizome in place, with the nascent bud poised to continue the next year's growth. The plant is then covered up with soil, followed by leaf mulch. This aerates the plant, and may actually stimulate growth the following year.

Cultivation by dividing the rhizome is a dependable method. Divisions are made in the autumn, producing sturdy plants the following spring. A good division

---

[4] Personal communication in spring of 1998 with one mid-scale broker indicated a purchasing volume of 10,000 dry pounds per week, which equates to 140,000 plants at the average rate of 14 entire roots per dry pound.

contains a piece of rhizome with attached roots and a nascent bud. Older, woody portions of the rhizome, lacking a nascent bud, tend to rot instead of producing new growth.

Cultivation by seed is difficult, as the seed has a short viability, and stratification is required. The best approach is to use newly harvested seed, plant immediately in a moist, warm medium (70 degrees F) and keep it this way for 3 months, then overwinter (40 degrees F or less), followed by germination in the spring. Seedlings may be transplanted out after a year of containerized growth, or may be maintained in woodland nursery beds for a year, then transplanted.

Black Cohosh likes good soil, and the health of the crop as well as harvest weight may be increased by growing it in a deep, loamy soil, with the addition of composted hardwood leaves and/or composted manure. Irrigation during dry spells in the summer will prohibit early dormancy, thereby increasing yields. The plant tolerates much more sun than Goldenseal, and open field production with little or no shade is possible, but only when using very vigorous transplants.

VI. Conclusion

How much do we value wild ecologies? This is the practicality of medicinal plant conservation: plants will be taken from the wild until our understanding of cultivation improves to the point where it becomes more convenient and more profitable to domesticate them. Hopefully this will be a labor driven not only by economic gain, but also by a sincere interest in conservation. If we care about the wild places, then we will garden and farm in a way that improves the earth in the same way that the trees improve the soil of the forest.

Iwu and Wootton (eds.), Ethnomedicine and Drug Discovery

CHAPTER 10

Ethnomedicine of the Cherokee: historical and current applications

JODY E NOÉ

Abstract

Indigenous Americans have utilized their environment effectively throughout history in supplying their communities with what was needed. The use of very rich and diverse ethnomedicines by the Cherokee (*Tsa-la-gi*) Peoples, which include both the Eastern Cherokee (Qualla Boundary Reservation, Cherokee, North Carolina) and the Western Cherokee (Tahlequah, Oklahoma), has been historically documented. These ethnomedical practises are still used in today's Cherokee medicine. The Cherokee tribe was one of the only Native American indigenous cultures to create a written language or syllabary for their native tongue. This syllabry was created by a Sequoyah (whose name is George Guess) in the 1800s, despite the fact that their culture had already been exposed to European influences for over 200+ years. Therefore, you will find that some of the documented herbs used as medicines by the Cherokee are not actually native species but are naturalized species. In the Cherokee use of ethnomedicine, it is not just the plants that instilled the healing but rather the combination of plants, prayer, ritual, diet and mindset that all contributed to the methods of application of the medicines on each person. Despite the vastness of this subject, we will restrict our discussion to some commonly used herbal medicines used historically and currently by the Cherokee. We will also discuss harvesting and preservation practises used through the millennia by the Cherokee and how they are still working today to preserve the green pharmacy utilized by each Cherokee medicine person.

Keywords: *ethnomedicine, indigenous, language, culture, traditional medicine*

I. Cherokee past

The Cherokee (*Tsa-la-gi*) people are indigenous to the North American continent, with their nation once spanning over most of the mountainous southeast. Their nation stretched from the current day states of Virginia, North Carolina, South Carolina, Georgia, Alabama, Tennessee, Kentucky and West Virginia.[1–5] They were a farming and hunting community with a matrilineal clan system. By the time Hernando de Soto first entered their country in the mountains of Carolina and Georgia in 1540, they had a well-established eastern agricultural woodland culture. He called them Chalaques in his writings and never penetrated their interior communities.[2] During the continual European infiltration, the Cherokee became prosperous traders, merchants, farmers, teachers, writers and diplomats. The Cherokee trade route was well established amongst the Europeans, and many took Cherokee wives coveting their strength and autonomy as women equal to men amongst the tribe. During the American Revolution, the Cherokee sided with the British, but in 1785, they negotiated their first peace treaty with the United States, but by this time, their nation's land had shrunk to a few million acres.[1,2,4,5] Treaties continued to be made, all promising the Cherokee the preservation of their autonomy as a sovereign nation. Several of the stipulations in various treaty agreements were designed for the Cherokee to fail (for all Native Americans to fail!). George Guess, *Sequoyah*, invented an 85-character syllabary for the Cherokee language and published the first Native American newspaper, the *Cherokee Phoenix*, in 1828 as a response to one of the treaty stipulations. The Cherokee also established public schools during this period with widespread literacy following immediately with the invention of the syllabary. In 1827, the Cherokee drafted a constitution that incorporated the Cherokee Nation's hopes of remaining an individual sovereign Nation. By 1838, all their hopes and trust in the United States were destroyed during their removal, better known as *The Trail of Tears*. One more illegal treaty was made, called the New Echota Treaty of 1835, to cede all tribal lands and remove the Nation to Indian Territory (now Oklahoma). So, in 1838, when most of the tribe refused to leave, 7000 federal troops and 2000 state militiamen began forcibly evicting Cherokee.[3] Men, women and children were forced to walk across the Smoky Mountains to the Ozark Mountains during the winter, without proper clothing, bedding, food or supplies. This is known to the Cherokee as *Nunahi-Duna-Dio-Hilu-I*, or *Trail where they cried*, where some 4000–6000 Cherokee men, women and children died from hunger, disease and exposure.[1,3,4,5] The Cherokee Nation had members that remained behind in their homelands of the Great Smoky Mountains; there were also those who went ahead of the removal having been given a vision of what was to happen to the People. These fractions of Cherokee were known as the *Keetoowah* or 'Black Hawk Society'. *Keetoowah* is an ancient name of the Cherokee, known amongst other Native American tribes from the times when the Cherokee were a vast Nation and also known as the 'Mound Builders'. These traditionalists are those responsible for keeping the Cherokee cultural rituals, the medicine priests of the nation. Because of these traditionalists, the Cherokee sacred belts, ceremonial pipes and other sacred regalia were not lost and are still in possession by the tribe, to this day. Will Thomas, a white man who befriended the tribe, was systematically buying back the land in the North Carolina mountains from the United States government. When it became legal for the Cherokee to own land, he transferred ownership to the tribe, and the land eventually became the Qualla Boundary Reservation, in the North Carolina Smoky Mountain range, better known as the Eastern Band of the Cherokee. Cherokee Nation was established in Tahlequah, Oklahoma after the removal, and it is known as the home of the Western Band of the Cherokee.

II. Cherokee present

Acculturation took place amongst the Cherokee, and many sacred words, ceremonies, rituals and daily practices vanished amongst the Nation. The Cherokee also preserved many of their ancient rites and still practise them, as they have for thousands of years. The Seven Clans of the Cherokee remain, and even though some of the names are different between the East and West, the matrilineal society remains intact. The Cherokee consider themselves the *Ani-Yun-wi-ya*, the Principal People or Real People. *Tsa-la-gi* is the familiar name used amongst the people of the nation and where the English word Cherokee originated. Pronunciation of the Cherokee language differs in dialect amongst the East and West Bands of the Nation, just like the clan names. Currently, the seven-clan system consists of the following: *Ani-wa-yah*, the Wolf clan who were traditional hunters and the group that raised wolves in captivity; *Ani-ka-wi*, the Deer clan, traditional hunters and runners; *Ani-djisk-wa*, Bird clan members who were renown blowgun hunters; *Ani-wo-di*, the Paint clan (East), Red paint clan (West) known as great conjurers; *Ani-sa-honi*, the Blue clan known as the herb and root doctors; *Ani-go-ti-ge-wi*, the Wild Potato clan (East), Savannah clan (West) best known as farmers; *Ani-gi-lo-hi*, the Long Hair clan (East), and the Twister clan (West) known as dancers with elaborate hair styles.

The *Ki-tu-wa*, or *Keetoowah*, as noted above, were known as the traditionalists. The secret Keetoowah society is noted to have been formed by the full-blooded Cherokee to keep not only the cultural traditions but also the full blood line. This name, Keetoowah, is the oldest name by which the Cherokee are known. There are many reported stories about the Keetoowah, and the secret society still exists to this day in both Oklahoma and North Carolina nations. These traditionalists are those that have preserved several of the sacred objects, traditions, ceremonies and rites of the nation. The sacred ceremonial pipe that has been written about by historians, the sacred belts that tell the stories of the Cherokee, the Stomp dance, the Turtle shell dancers, the sacred rituals of healing with water, plants and prayer, have all been witnessed by the author as active Cherokee ethnomedicine practised today. A Keetoowah story told to the author by the grandson of Redbird Smith, Crosslin F. Smith, is that the father of Redbird had a vision, and the Keetoowah moved to Oklahoma before the removal. Here, they brought with them the sacred objects and rites to help in the preservation of the nation. My teacher, Crosslin F. Smith, is the high medicine priest of the Keetoowah in Oklahoma who keep the Stomp Dance. In Mr. Smith's recount of the story, the family and other Keetoowah made the trip earlier than the removal to establish a place for the people when they had to walk the Trail of Tears. Mr Smith to this day continues to practise traditional Cherokee medicine amongst the people not only of his tribe but also of neighboring tribes, and non-Native American people.

III. Ethnomedicine

The practice of traditional Cherokee medicine among the Eastern and Western Bands is still a viable, functioning alternative to allopathic medicine. There are established 'Indian Hospitals' that are placed in both the East and West by the Federal government. These allopathic establishments also honor the cultural traditions of the Cherokee, by allowing the medicine people to visit their patients in the hospital. Many Cherokees choose to practise traditional Cherokee medicine as their primary healthcare. This cultural tradition has many aspects; daily household medicine is practised by the women in the Cherokee household for colds and influenza, and to keep harmony in the family. The medicine people, or priests, function at a higher level in the society, and their practise is eclectic, including prayers, rituals, cleanings and ceremonies that address everything from chronic terminal physical illness, psychological or financial counseling, to whatever the mind, body or spirit needs. This practice of medicine is not just the use of plants as phytotherapeutic agents. In this cultural tradition, the plants are part of the healing, but also in the presence of the medicine priest, the prayers and delivery of the ritual all work to stimulate the person's whole healing, no matter where the etiology of the person's affliction may be. Thus, the people may come for a physical remedy, a spiritual remedy, a psychological remedy, or merely to seek counsel for finances or bless their tobacco to make it safe for smoking. All of these are important in the traditional Cherokee practise of medicine. So, when the word 'medicine' is used, it means the way you approach life, healing and the sacred in your daily life. These Cherokee traditional values have not been acculturated, and they remain as a backbone in the tribal culture.

IV. Plants used past and present

The Cherokee, both Eastern and Western Bands, continue to utilize their surrounding flora as primary medicines. For the sake of brevity, only a few plants will be discussed in this section, but an extensive collection can be found at Old Dominion University's department of Biological Sciences, Botany herbarium.[6] This herbarium is indexed under the International Index of Herbaria as ODU. Here, this author's ethnobotanical collection is housed with over 250 specimens preserved, along with audio and video media of Cherokee culture. The Cherokee, who continue to use the plants as medicines, continue also to practise the rites that preserve each member of the plant nation. The rule of fours, where one must pass a plant that she is looking for four times before gathering and then saying specific prayers of thanks for the medicine is a standard practise of species preservation created by the Cherokee. This starts with the gatherer consciously creating a 'good mind' prior to even embarking on the journey to seek the plants. This good mind allows the gatherer to listen to the plants and hear them speak, so even a plant that is not well known or used can call to the seeker for a specific use or person. Thus, Cherokee plant medicine is always dynamic and never stagnant, allowing the use of all plants as directed by the Great Spirit. In the continual preservation effort of the Cherokee to save some of their indigenous medicines, they have established remote patches of wild plants in both Eastern and Western Bands, where they can count on collecting their medicines.

IV.A. Urtica diocia, URTICACEAE

Nettles are traditionally used as a medicine tea for upset stomachs and for ague, and the twisted stems were used for bow strings.[7] Current clinical applications place the constituents of histamine, formic acid, chlorophyll, assorted minerals including iron and vitamin C as active. The plant acts as an astringent, potassium sparring diuretic and tonic in supporting the whole body and decreasing histaminic reactions. It is especially good to use it as a tea for eczema associated with anxiety as well as childhood eczema.

IV.B. Arisaema triphyllum, ARACEAE

Jack in the Pulpit or Indian turnip is named so by the tuberous root system. Traditionally, the Cherokee use was eclectic from a liniment of the beaten boiled roots that were also used as a poultice for boils and headaches. The ointment is also used topically for ringworm and tetterworm, and other topical fungal infections. In addition, the root can be used as a food and tea for stimulating expectorant properties, diaphoresis and as a carminative.[7] Current clinical applications are sparse, as this plant is very under-utilized by modern medical herbalism.

IV.C. Trillium erectum, LILIACEAE

Beth root or Red robin is traditionally used as a tea for profuse menstruation, hemorrhages and as a constitutional warm tea for the menopausal years. It is also used for coughs, asthma, bowel complaints as a tea and as a poultice for ulcers, tumors, and inflamed areas (personal communication, Mary Chiltoskey). Current clinical applications site the active constituents as steroidal saponins, steroidal glycosides and tannins. It acts as a uterine tonic, astringent and expectorant internally. It is indicated as a phyto-precursor for female sex hormones, acting as a uterine tonic. I use it successfully with post-partum hemorrhages, or in any styptic formula to stop any hemorrhaging. It is also a good plant to use in menorrhagia, or excessive bleeding associated with menses and menopause.

IV.D. Verbascum thapsus, SCROPHULARIACEAE

Mullein, Mule tail, or Kidney medicine is traditionally used as a root and flower tea for kidney dysfunction. The root and leaves are also made into a tea to help with female menstrual cycles. The flowers are used in an oil infusion as a topical vulnerary, while the leaves are wrapped around the neck for mumps and rubbed under the arms for prickly heat. The leaves are also used as bandages to hold in poultices applied to the site (personal communication, Mary Chiltoskey and G.B. Chiltoskey). Current clinical applications in modern herbalism use Mullein effectively as a demulcent for the mucous membranes of the lungs, to treat coughs, colds, bronchitis, and hoarseness while helping to stimulate the cough to be productive and expectorate.

IV.E. Castilleja coccinea, SCROPHULARIACEAE

Indian paintbrush is a plant still revered by both the Eastern and Western Cherokee. The Eastern Cherokee refer to it as the 'tea to destroy enemies' and also used it as a birth control medicine. The tea is taken for two weeks and synchronized with the menses, slowly titrated up and weaned off over the two-week period (personal communication, Gee George and Geneva Jackson). The Western Cherokee use the little 'tubers' of the underground part, which is actually a haustorium of an infective parasite. This haustorium is used as a tea for constipation and parasitic infections (personal communication, Crosslin Smith and Glenna Smith). Current clinical applications once again show negligible use in modern herbology.

IV.F. Sanguinaria canadensis, PAPAVERACEAE

Bloodroot, Red puccoon, Red turmeric, Red Indian paint, or Red root is used as a traditional dye for basket-makers to give the vibrant red hue used in the traditional art form. It is also used in small doses as a root decoction for coughs, croup and general lung inflammation. It can be used as a topical wash for ulcers and other open lesions. In addition, the root is powdered and sniffed for catarrh and nasal polyps.[7] Current clinical applications show some of the active constituents as the alkaloids sanguinarine, protopine, chelerythrine, homochelidonind. These give the phytotherapeutic actions as an emetic (hence the low doses!), expectorant, cathartic, antiseptic, cardioactive, spasmolytic and topical irritant.

IV.G. Datura stramonium, SOLANACEAE

Jimson weed, Thorn apple, or Devil's trumpet remains a very important Cherokee plant medicine used for people and live stock. When livestock eat the bountiful leaves, they are said to 'go crazy', and sometimes, the colloquial name of 'crazy weed' is given to the plant. The Eastern Cherokee use the leaves as a traditional asthma treatment. The leaves are dried and smoked for acute asthma attacks, or made into a hot poultice to draw out boils. The seeds are also eaten by the medicine priests to enhance their psychic abilities (personal communication with Gee George and Crosslin Smith). The Western Cherokee use the leaves in a cold-water extraction to be used topically as a poultice put directly on burns and open lesions. The seeds are eaten in small amounts to relieve severe pain (personal communication, Crosslin Smith). *Datura* has a long ethnomedicinal history in many cultures throughout the world. In current clinical practise, this plant once again is severely under-utilized, but with good indication, its constituents of scopalamine and other atropines make it a potentially toxic medicinal plant.

IV.H. Sassafras albidum, LAURACEAE

Sassafras or Leaves-that-look-like-a-shoe is a traditional Cherokee plant that was adopted by the European settlers, the Thompsonians, and the Eclectics used it as a tea for purifying the blood. The Eastern Cherokee use the tea today for the same reason, but it is indicated especially for skin diseases, venereal diseases, ague or rheumatism. The root bark is steeped and used for diarrhea, colds and as an appetite suppressant. It is used topically as a vulnerary wash and poultice.[7] The Western Cherokee use it in a tea form as an enhancer in any herbal formula and as a blood-purifier. The most important ethnological practise is in the collection of the sassafras. Only the young plants that have red stems are harvested (personal communication).

When in the field, it can be observed that the stems of the sassafras are differently colored, some with white stems and some with red. It is said by the traditional people that the red-stemmed plants are medicine, and the white-stemmed plants are poison. Current clinical applications have subsided in the later part of the 20th century due to the research that indicated that sassafras could possibly be carcinogenic. Could it be that the ancient knowledge of the Cherokee circumvented this problem by their traditional collection methods?

IV.I. Passiflora incarnata, PASSIFLORACEAE

Passion flower, Old field apricot, or May pops is another traditional eclectic medicinal plant. The ripe fruits are used as a kidney and bladder tonic tea or eaten as a fruit (personal communication with Gee George). The root tea is used as a beverage or taken as a tonic for the liver; it is also used internally to help with skin boils. Topically, the root is pounded and used as an anti-inflammatory. Current clinical applications are that the plant is an excellent nervine, antispasmodic and vitalistic spiritual awakener. I use the plant successfully in my clinical practise as a gentle nervine that is good for anxiety, stress as well as nervousness especially associated with menses in women.

References

1. Adair J. (1775) First publication, London. Cherokee Beliefs and Practises of the Ancients – Out of the flame, Reprint 1998. Nashville, TN: The National Society of Colonial Dames of America.
2. Mooney J. (1982) (reproduced) Myths of the Cherokee and sacred formulas of the Cherokees. Nashville, TN: Charles and Randy Elder Booksellers.
3. McClure TM. (1998) A guide for tracing and honoring your Cherokee ancestors. Somerville: Chuanennee Books.
4. Mails TE. (1992) The Cherokee people. New York: Marlowe and Company.
5. Starr E. (1993) Reprint. History of the cherokee Indians. Tulsa, OK: Oklahoma Yesterday Publications.
6. Noé, JE. (1991) Master's Thesis on the ethnobotany of the Cherokee. Old Dominion University Department of Biological Sciences, Dr. L.J. Musselman Department of Botany, Norflolk, VA.
7. Hamel PB, Chiltoskey MU. (1975) Cherokee plants – their uses. A 400 Year History. Hamel and Chiltoskey, Cherokee, NC.

Iwu and Wootton (eds.), Ethnomedicine and Drug Discovery

CHAPTER 11

Traditional medicines and the new paradigm of psychotropic drug action

ELAINE ELISABETSKY

Abstract

The pharmaceutical industry exploits traditional remedies for the development of new drugs, especially chasing for unusual molecules with interesting and innovative mechanisms of action. Even when traditional formulations – rather than just plant species – are taken into consideration, traditional medical systems are not regarded as potential sources of new paradigms for drug usage. Although the phytotherapy industry is more likely to innovate, and more easily accepts less conventional concepts of treatment and drug action, pharmacology as a discipline is refractory to the potential contribution of ethnopharmacology. These complex patterns suggest that the effects of plant drugs may often be based on a more diverse pharmacodynamic basis than the usual understanding of drug/effect relationships. In fact, the unusual pharmacological activity observed with plant formulations may result from effects of more than one active ingredient, from drug interactions among ingredients, from active ingredients possessing multiple mechanisms of action, or even from interference with targets not yet recognized by the current biomedical understanding of cell biology modulation. This paper explores the idea that the understanding of traditional medical concepts of health and disease in general, and traditional medical practices in particular, can lead to true innovation in paradigms of drug usage and development. Diet, prevention, well-being maintenance, low-dose/long-term posologies, and complex mixtures, often central to traditional medical treatments, are discussed. Traditionally used adaptogens, analgesics, antipsychotics, and anticonvulsants, discussed in the light of the current understanding of drug/receptor interactions and mental disorders, are remarkably well in line with newer paradigms of psychotropic drug action and therapy.

Keywords: *pharmaceutical industry, traditional remedies, CNS ethnopharmacology, psychopharmacology*

I. Introduction

Notwithstanding major discrepancies among traditional and modern (here used as Western biomedical) medical systems, peoples world-wide establish dialogues and make use of different systems to maintain or re-establish health.[1-4] Since traditional(s) and modern medical systems evolved from diverse definitions of reality and are based on different paradigms,[5] not surprisingly, confrontations between traditional and modern medical systems have existed throughout history.[6-8] Nevertheless, most systems share the practice of utilizing raw material from nature processed into drugs or food supplements with therapeutic goals.

Within the context of drug development, interest in traditional medical systems is mostly focused on the use of plants processed as traditional medicines. The underlying understanding is that at least some of the plant species that constitute the basis of such preparations may contain therapeutically useful compounds, to be further developed into Western-type drugs.[9,10] None the less, the fact that traditional medical systems are organized as cultural systems[11–13] allows for profound differences in meanings of health, disease, and disease etiologies.[14–16] Accordingly, such differences result in a variety of therapeutic practices not easily accommodated in the biomechanical paradigm of modern medicine.

The purpose of this paper is to review some prevailing features of traditional medicines in the context of the current understanding of mechanisms of action of psychoactive drugs. Comparisons of folk and contemporary healing concepts have been thoroughly explored by social scientists and medical anthropologists, in the light of ethnomedicine and of ethnopharmacology,[15,17] and are beyond the purpose of this paper. The scope of this discussion is limited to some of the features of traditional medicines that are of relevance for understanding the way in which traditional medical practices, including plant medicines, affect brain functions. Concepts such as diet, prevention, well-being maintenance, low-dose/long-term posologies, and complex mixtures, often central to traditional medical treatments will be discussed in this context. The analysis may be useful by revealing particularities from traditional treatments of relevance for developing central nervous system (CNS) drugs, and by re-examining the paradigm of psychoactive drugs.

II. Diet

Within the classical non-Western humoral systems (prevailing in Asian countries and several Central and South American ethnic groups), good health is perceived as a balance between physical elements and body humors.[7] Foods and disease conditions are assigned temperature and moisture qualities or attributes;* accordingly, hot illnesses are treated by consuming cold foods, and a wet disease is treated by dry foods.[18]

In line with the Western-like context of drug/body interaction, there is the commonly found concept of foods that contain substances that interact with body function. Observing a special diet, either preventing or increasing consumption of certain food-medicines, is frequently requested or recommended in association with other therapeutic measures in various societies.[19,20] Specifically for Kampö drugs, one advantage claimed by these drugs is that 'they contain the same stable components as the food we eat everyday'.[21]

In the context of CNS, persisting deficits in nutrition can lead to neurologic diseases; conversely, in several genetically determined neurologic diseases, treatments include either elimination of certain foods or supplementation with specific

---

* The Chinese system considers other dimensions of gender, light and time.[18]

vitamins.[22] It has been shown that alterations in tryptophan consumption significantly alter mood and behavior in selected groups of individuals,[23] and that phenolic antioxidants, such as those found in wine, have a protective effect against atherosclerosis.[24]

The increasing attention and comprehensive analysis of data relevant to the effects of ingestion of chemicals from plants in the maintenance/improvement of body functions and/or disease prevention[17,25–28] may reveal data to further substantiate specific interactions between dietary restrictions (or impositions) and functionality of given organs or tissues. It has been argued that the growing market, and the somewhat surprisingly quick acceptance, of nutraceuticals may be regarded as a return to earlier forms of behavior, when various societies did not make evident distinctions between food and medicine.[29]

III. Health maintenance and disease prevention

The concept of well-being varies among peoples. Besides the ingestion of curative or preventive foods, some types of remedies are used chronically 'because it is good for health', to prevent the appearance of certain diseases, and/or to easy or to slow the process of aging. For the Matsigenka Indians (Peru), well-being embraces physical and psychological health, successful gardening and hunting, and harmonious social interaction. Accordingly, Matsigenka medicine includes treatments for various culture-specific emotional and psychological syndromes, with plant species that apparently may be psychoactive.[30] For the Zoró Indians (Brazil), a healthy person is not only disease and pain-free, but also has to be 'big, strong, good looking and pleasant to meet'. Remedies are categorized as preventive, detoxifying (used along with diet or to facilitate poor digestion) and curative.[31] In Latin America, the use of 'tonics' or 'nerve tonics' is widespread, especially by the elders, by patients recovering from CNS illnesses, and as a 'general stimulant' to cope with highly stressful physical or psychological situations. The pharmacological meaning of such tonics has yet to be elucidated.[32]

In this context, it is illustrative the rekindled interest in adaptogens, a term coined in 1947 by the Russian scientist Lazarev, designating agents that allow organisms to counteract stressors, helping the organism to be 'adapted' to intense demands. It is beyond the scope of this paper to fully discuss this subject,[33–37] but it has been difficult to pigeonhole the therapeutic or pharmacological meaning of adaptogens in terms of Western concepts of drug action. One current view of adaptogenic effects is that they cause an increase in the basal level of the human dynamic equilibrium, thereby mediating the 'switch on' and 'switch off' systems. One can thus envisage an increased but balanced basal level of the most important mediators of stress system.[37] Relevant to this discussion is that the mode(s) of action of adaptogens have been associated with a variety of cellular and molecular mechanisms, including increased synthesis of proteins or nucleic acids, increased formation of glucose-6-phosphate, catecolaminergic synaptic modulation, antioxidant activity, immunosti-mulation, the prostaglandins, or even genetic expression.[33,36] Even recommending the substitution of the term adaptogen, Davydov and Krikorian[37] emphasize that the presence of various compounds in *Eleutherococus* spp., and the many pharmacological actions demonstrated in a wide range of tests, validates further research investments oriented to drug development.

We believe that adaptogens, rather than being dropped as a term or idea, have to be study in a comprehensive and open-minded manner, even if it takes time for the Western-oriented medical professions to eventually absorb new concepts of disease management (as has been the case with acupuncture and homeopathy). We suggest that thorough scientific studies of adaptogens are a domain where a non-Western category of drug action may reveal not only new prototypic drugs, but extremely useful new drugs and strategies to beneficially interfere with, minimize, or prevent multifactorial illnesses and age-associated decline.

IV. Low dose, long term

Posology is usually well defined by medicinal plant specialists.[38] Traditional formulations of medicinal plants processed for therapeutic purposes (such as teas, syrups, concoctions, beverages, etc.) are expected to be ingested over a given period of time, varying with the expected length of the condition to be treated, and the time required for the remedy to attain its curative goal. Often, remedies are recommended for a considerable length of time (weeks or months) or even maintained for several years; sometimes, it is expected that it actually takes weeks or months for the efficacy of the treatment (either cure or significant amelioration) to become apparent. Such long-term treatments are usually associated with diseases that are chronic and incurable, where the treatment aims to keep a given disorder under control (such as asthma, epilepsy, diabetes), or minimize diseases or processes (such as aging) over the course of several years. Considering the yield of active constituents usually obtained from the amount of plant material, and mode of preparation used to prepare home-made remedies,[39] it is arguable that more often than not, traditional therapies involve the repeated ingestion of low doses of active substance(s) over a significant period of time.

In pharmacodynamic terms, it is expected that this pattern of intervention with molecular targets may be profoundly different from an acute (single administration) or sub-chronic (few administrations) challenge to any given molecular target. Nevertheless, the traditional posology is rarely taken into consideration in evaluating medicinal plant extracts or substances in new drug screening/development programs. Failure in shaping the traditional uses associated with efficacy claims is in part related to practical matters: the research design needed to match the effects of such repeated interaction with tissues and/or molecular targets requires greater quantities of testing materials and poses several other obstacles that are difficult to manage. In fact, several in vitro methodologies are, unfortunately, inadequate for these purposes. Nevertheless, the consequences of constant and repeated challenges to molecular targets have to be taken into consideration at least in interpreting results, especially when effects are to be integrated with those evaluated through the use of in-vivo models.

Specifically referring to the CNS, attention has been called to the fact that 'it is the adaptive response of the nervous system to adequate repeated perturbations mediated through these initial targets that produces the therapeutic responses'.[40] The authors suggest that chronic perturbations can lead to different types of adaptations (quantitatively and qualitatively), eventually resulting in a new functional state. Traditional remedies, more often than not, consisting of long-term treatments with low to moderate doses of active ingredients (as those found in traditional preparations) are in line with the current paradigm of psychoactive drug action.

V. Complex mixtures

In discussing the basis for the effectiveness of Kampö medicines, Ishihara[21] calls particular attention to the pharmacological antagonism and synergism found among the chemical components of the crude extracts that form Kampö infusions. Such drug interactions have been further substantiated in Kampö formulations.[41] Complex chemical mixtures have the potential to bring about interactions, potentiations, and/or combinatorial responses, and the proportion of different components in the mixture may determine efficacy and safety.[37] It has been repeatedly demonstrated that a spectrum of substances are responsible for the polyvalent therapeutic action of the *Ginkgo biloba* extract Egb761, whose beneficial effects result from 'the combination of its various protective, curative and modulating properties against the pathological process'.[42]

In fact, effects of plant-based remedies may be due to one active compound with a single mechanism of action, to compound(s) that possess multiple mechanisms of action, to the combined activity of more than one active ingredient in a single species (e.g. flavonol glycosides, phenolic compounds, bilobalide, gingkolides in *Gingko biloba*),[42] or synergic interactions of different active ingredients from several plant species processed as a medicinal formula (e.g. harmane alkaloids and dimethyl-tryptamine (DMT) in the ayauhasca drink). Examples of investigations of medicinal plants with alleged psychopharmacological effects are useful to illustrate how these issues are of relevance in the context of CNS diseases and psychopharmacology.

VI. Analgesia from *Psychotria* alkaloids

In the Brazilian Amazon, *P. colorata* (Willd R & S) Muell. Arg is used to treat earache (a handful of flowers cut in pieces are packed in banana leaves and left over warm ashes; the warmed flowers are mixed with milk, preferably 'mother's milk' and filtered through a piece of cloth, and drops are topically applied to the ear), and abdominal pain (roots and fruits are mixed with water and left to boil; the decoction is taken orally).[43] We reported that alkaloids present in the leaves and flowers of *P. colorata* have a marked analgesic activity, as evaluated through various pain models.[42–44] Phytochemical analyzes of *P. colorata* flowers identified pyrrolidinoindoline alkaloids as major components, including hodgkinsine, psychotridine, chimonanthine.[44–46]

Pharmacological analyzes revealed that hodgkinsine produces a dose-dependent naloxone reversible analgesic effect in thermal models of nociception, and acts as a potent dose-dependent analgesic in the capsaicin-induced pain;[47] in vivo data complemented by binding studies demonstrate that the activation of opioid and blockade of glutamate NMDA receptors participate in the hodgkinsine mode of action. The analgesic properties of psychotridine turned out to be associated with the modulation of NMDA receptors.[47]

Because some NMDA-mediated painful events can be difficult to control with opioids alone (e.g. neuropathic pain states), it has been argued[49,50] that a combination of opioid and NMDA antagonism may be especially advantageous in specific clinical conditions. Interestingly enough, the combined (opioid and glutamatergic) mechanism of action of hodgkinsine alone or the alkaloid mixture in the traditional preparation is in line with current strategies for managing severe pain states that respond poorly to conventional analgesics.

VII. Antipsychotic properties of alstonine

It is estimated that two-thirds of healthcare practitioners in Nigeria are traditional healers of one sort or another.[51] A traditional psychiatrist in Nigeria, Dr C.O., introduced us to a plant medicine (*uhuma obi-nwoke*, 'heart of man' in Igbo), which he (and possibly others) uses to treat various kinds of 'madness' and some epilepsy fits, and as a sedative. To treat severe 'madness', the ground root is boiled in water. After an initial loading dose, patients may fall asleep for as long as 2–3 days; the dose is gradually tapered down, and patients will gradually be awake for longer periods of time each day, until the patient is completely off the medicine, and the symptoms of 'madness' have gone.[52] We reported that alstonine (an heteroyohimbine type alkaloid) is the major compound of Dr C.O.'s medicine, and that alstonine exhibits antipsychotic-like properties in several animal models.[53] In comparison to antipsychotic medications in clinics, alstonine behaves more like the newer atypical antipsychotic than the older neuroleptic drugs. Nevertheless, curious features of alstonine include its ability to diminish haloperidol-induced catalepsy, and the lack of direct interaction with D_1, D_2 and 5-HT_{2A} receptors. These data challenge any simplistic explanation regarding the mechanism of action of alstonine, still under scrutiny in our laboratory.

Rauwolfia alkaloids producing species usually produce several alkaloids, including reserpine – the first antipsychotic medication used in modern medicine (the major component of *Rauwolfia serpentina* used to treat mental illness in India for centuries). The various pharmacological effects of reserpine are so prominent that it is unsurprising that minor components likely to be present in crude extracts/ formulations with *Rauwolfia* spp. (including alkaloids like serpentine, ajmalicine and alstonine) have been neglected in earlier psychopharmacology evaluations. The ethnopharmacological study of traditional antipsychotics used by Nigerian traditional psychiatrists indicates that various alkaloids present in traditional preparations of species producing reserpine-like alkaloids (e.g. from genuses *Rauwolfia, Picralima Alstonia, Psychotria*) may have peculiar antipsychotic properties. These complex traditional formulations may consist of an unusual combination of compounds with diverse mechanisms of action, which are worth investigating in the light of the current understanding of schizophrenia and potential targets for medications.

VIII. Anticonvulsant properties of linalool

Linalool is a monoterpene compound, commonly found as major component of essential oils of several aromatic species, many of which are used traditionally as sedatives.[38,55] Psychopharmacological evaluation of linalool in mice revealed that this compound has dose-dependent marked sedative effects at the CNS, including protection against pentylenetetrazol, picrotoxin, quinolinic acid and electroshock-induced convulsions, hypnotic, and hypothermic properties.[55] Relevant to its mechanism of action, linalool behaves as a competitive antagonist of glutamate, and as a non-competitive antagonist of NMDA receptors in brain cortical membranes;[56,57] additionally, linalool reduces potassium-stimulated (but not basal) glutamate release.

Interfering with multiple mechanisms that underlie seizures may be necessary to effectively counteract epileptic phenomena,[58,59] and developing drugs that differentially affect normal and hyperexcitable neurons has been postulated to spare normal excitatory function.[58] Therefore, the fact that anticonvulsant effects of linalool can be attributed to both an inhibition of potassium-stimulated (but not basal) glutamate release in cortical synaptosomes, and antagonism of NMDA receptors, deserves further investigation as a strategy for antiepileptic-drug development.

IX. Ibogaine and drug addiction

Roots of *Tabernanthe iboga* are prepared as a narcotic of social importance, especially in Gabon and Congo, where it is used as a powerful stimulant and aphrodisiac, as well as employed ritualistically in hallucinogenic doses.[60] Ibogaine is an indole alkaloid extracted from *T. iboga* with reputed antiaddictive properties.[61] A remarkable aspect of this drug is its apparent ability to eliminate withdrawal symptoms and drug-craving for extended periods of time after a single dosage, as indicated by anecdotal reports and uncontroled clinical data, somewhat supported by animal studies.[62] Although the mechanism of action of ibogaine remains to be clarified, it has been shown that ibogaine acts upon various neurotransmitter systems.[63]

Glutamate, the main excitatory neurotransmitter in the mammalian CNS, is believed to play important roles in several physiological and pathological processes.

Specifically, the *N*-methyl-D-aspartate. (NMDA) ionotropic glutamate receptor subtype seems to be crucial in plastic changes associated with normal brain function (as in learning and memory), in neurodegenerative as well as drug addiction. In fact, NMDA antagonists have been explored as potential anticonvulsants, neuroprotective and antiaddictive drugs. Ibogaine acts as a non-competitive NMDA receptor antagonist, a property of relevance to its purported antiaddictive activity. Adding to the unique psychoactive profile of ibogaine in non-addictive and addictive human subjects,[64] in mice following a single administration of ibogaine, an unusual intermittent pattern of NMDA modulation has recently been identified (behaviorally and neurochemically).[65]

Although all of the various mechanisms of action so far proposed for ibogaine properties deserve further investigation, it is believed that its psychopharmacological profile (including antiaddictive properties) is likely to be the result of its interactions with various molecular sites.[63,64,66,67] Ibogaine has appeared in the drug-addiction therapy scenario as a truly innovative drug, arousing high expectations. Research with ibogaine, its main metabolite and derivatives, is bound to be fruitful, if not by ibogaine development as a drug, for the novelty in research avenues in the therapy of drug addiction.

X. Conclusion

Determination, rather than causality, is the current extended interpretation of cause–effect relationships in physics, 'cause' being substituted by 'determining conditions', where all the conditions of a process or state are equally important. It has been suggested that some (if not all) diseases should be understood as processes, where an interplay of multiple factors (genetics, environmental exposures, somatization targets, psychic conditions, etc.) have to be considered.[68] More in line with this multifactorial view of a disease as a process, in discussing *Ginko biloba*'s clinical effects, it has been concluded that a combined activity of several active principles is responsible for the overall effect (since no single constituent can mimic the effects of the total extract); moreover, 'a certain interdependency of action comes into play after administration of the complex extract'.[42]

Consideration of an interdependency of molecular and cellular functions, as well as among specific brain areas, is absolutely crucial for the understanding of physiological and pathophysiological processes that take place in the brain. The need for a few weeks to attain the full benefits of antidepressants and antipsychotics medications among others, as well as drug dependence and tolerance, is understood exactly as the result of long-term adaptations (plastic changes) consequent to a cascade of events triggered by the repetitive effects of these drugs.[40,69] According to this new paradigm of psychotropic drug action, effects of psychotropic drugs in the brain require adequate dosage, frequency and prolonged duration; these initial drug effects activate homeostatic mechanisms, leading to an adapted state, quantitatively or qualitatively different from the normal state. Unveiling the complex molecular pathways that underlie these processes is considered to be the current task of neuropsychopharmacology.[40]

Innumerous examples show that different medical systems are managed by a given culture in a selective and productive way, although embodying the concepts and ideology of the medical practice they represent.[70] Adding to Professor Prinzs'[71] observation that indigenous peoples do not use medicinal plants, but rather create relationships with them, Etkin and Ross[18] suggested that actually those people–plant relationships have physiologic implications. The analysis of treatments of mental illnesses in Nigeria provides evidence of treatment outcomes associated with religious–psychotherapeutic and religious–physiological elements.[72] Progress in neuropsychopharmacology and neurochemistry is constantly unveiling new cellular and molecular targets relevant to the drug's mode of action, and it can be expected that traditional treatments may act as disease modifiers by eliciting physiological responses through mechanisms yet to be elucidated.

Wing[14] suggests that probably the most significant aspect of traditional healing systems is that they pass the test of explanatory competence. As stated by Dr Ishihara,[20] 'the chief reason why Kampö traditional medicine survives today alongside Western medicine in Japan is simply that it is effective and practical'. As explored here within the context of the central nervous system, we suggest that a thorough understanding of traditional medical concepts of health and disease in general, and traditional medical practices in particular, can lead to true innovation in paradigms of drug action and development.

Acknowledgments

This study was supported by a CNPq fellowship. The author dedicates this paper to the late Professor Darrell A. Posey and wishes to thank him for passionate, pleasant and fruitful discussions on Science in general, and Ethnobiology in particular, over the last many and memorable years of friendship.

References

1. Young JC. (1980) A model of illness treatment decisions in a Tarascan Town. Am Ethnol 7(1):106–131.
2. Crandon L. (1986) Medical dialogue and the political economy of medical pluralism: a case from rural highland Bolivia. Am Ethnol 13(3):463–476.
3. Hardon AP. (1987) The use of modern pharmaceuticals in a Filipino village: doctors, prescriptions and self-care. Soc Sci Med 25:277–292.
4. Silva KT, de Silva MWA, Banda TM, Wijekoon. (1994) Access to Western drugs, medical pluralism, and choice of therapy in an urban low income community in Sri Lanka. In: Etkin NL, Tan ML, editors Medicines: meanings and contexts. Amsterdam: University of Amsterdam, pp. 183–206.
5. Staugard F. (1985) Introduction: modern and traditional medicine in a universal context. In: Traditional medicine in Botswana. Gaborone: Ipelegeng Publishers. Chapter 1, pp. 5–13.
6. Oppong ACK. (1989) Healers in transition. Soc Sci Med 28(9):917–924.
7. Logan M. (1973) Humoral medicine in Guatemala and peasant acceptance of modern medicine. Hum Org 32(4):385–395.

8. Landy D. (1974) Traditional curers under the impact of Western medicine. Am Ethnol 1:103–127.
9. Balandrin MF, Klocke JA, Wurtele ES, Bollinger WH. (1985) Natural plant chemicals: sources of industrial and medicinal materials. Science 228:1154–1160.
10. Abelson PH. (1990) Medicine from plants. Science 247.
11. Janzen J. (1978) The comparative study of medical systems as changing social systems. Soc Sci Med 12:121–129.
12. Kleinman A. (1978) Concepts and a model for comparison of medical systems as cultural systems. Soc Sci Med 12:85–93.
13. Sachs L, Tomson G. (1992) Medicines and culture. Soc Sci Med 34(3):307–315.
14. Bastien JW. (1985) Qollahuaya-Andean body concepts: a topographical–hydraulic model of physiology. Am Anthropol 87:595–611.
15. Wing DM. (1998) A comparison of traditional folk healing concepts with contemporary healing concepts. J Commun Health Nur 15(3):143–154.
16. Whyte SR. (1982) Penicillin, battery acid and sacrifice: cures and causes in Nyole medicine. Soc Sci Med 16(23):2055–2064.
17. Etkin N, Ross PJ. (1994) Pharmacological implications of 'wild' plants in Hausa diet. In: Etkin N, editor. Eating on the wild side: the pharmacologic, ecologic, and social implications of using noncultigens. London: University of Arizona Press, pp. 85–101.
18. Huard P, Wong M. (1969) La médecine Chinoise. Presses Universitaires de france, 2eme ed., 125 pp.
19. Etkin LN, Ross PJ. (1997) A discipline maturing: past trends and future direction in ethnopharmacology. In: Guerci A, editor. Salute e Malattia: Indirizzi e Prospecttive. Genoa: Erga Edizione, pp. 85–95.
20. Pieroni A. (2000) Medicinal plants and food medicines in the folk traditions of the Upper Lucca Province, Italy. J Ethnopharmacol 70:235–273.
21. Ishihara A. (1985) Traditional Kampö medicine of Japan. In: Herbal medicine: Kampö, past and present. Tokyo: Life Science Publishing, pp. 1–10.
22. Dreyfus PM, Seyal M. (1994) Diet and nutrition in neurologic disorders. In: Shils ME, Olson JA, Shike M, editors. Modern nutrition in health and disease. 8th edition. Philadelphia, PA: Lea & Febiger, Vol. 2, pp. 1349–1361.
23. Shansis FM, Busnello JV, Quevedo J, Forster L, Youg S, Izquierdo I, Kapczinski F. (2000) Behavioral effects of acute tryptophan depletion in healthy male volunteers. J Psychopharmacol 14(2):157–163.
24. Frankel EN, Meyer AS. (1998) Antioxidants in grape and grape juices and their potential health effects. Pharmaceut Biol 36:14–20.
25. Etkin N. (1991) Should we set a place for diet in ethnopharmacology? J Ethnopharmacol 32:25–36.
26. Johns T, Chapman L. (1995) Phytochemicals ingested in traditional diets and medicines as modulators of energy metabolism. In: Arnason JT, Mata R, Romeo JT, editors. Phytochemistry of medicinal plants. New York: Plenum Press, pp. 161–188.
27. Stavric B. (1997) Chemopreventive agents in food. In: Johns T, Romeo JT, editors. Functionality of food phytochemicals. Recent advances in phytochemistry. New York: Plenum Press, Vol. 31, pp. 53–88.
28. Wattenberg LW. (1998) Chemoprevention of carcinogenesis by minor dietary constituents: symposium introduction. Pharmaceut Biol 36:6–7.
29. Johns T. (1997) Behavioral determinants for the ingestion of food phytochemicals. In: Johns T, Romeo JT, editors. Functionality of food phytochemicals. Recent advances in phytochemistry. New York: Plenum Press, Vol. 31, pp. 133–154.
30. Shepard GH. (1998) Psychoactive plants and ethnopsychiatric medicines of the matsigenka. J Psychoactive Drugs 30(4):321–332.
31. Brunelli G. (1987) Des esprits aux microbes, santé et societé en transformation chez les Zoró de l'Amazonie Brésillliene. Memoire de Maîtrise, Université de Montréal, Montreal.
32. Elisabetsky E, Figueiredo W, Oliveira G. (1992) Traditional Amazonian 'nerve tonics' as sources of antidepressant agents. Food Products Press, an imprint of The Haworth Press Inc, Binghamton, NY, USA. Chaunochiton Kappleri, a case study. J Herbs Spices Med Plants 1(1/2):125–162.
33. Elisabetsky E, Siqueira IR. (1998) Is there a psychopharmacological meaning for traditional tonics? In: Prendergast HDV, Etkin NL, Harris DR, Houghton PJ, editors. Plants for food and medicine. Kew: Royal Botanical Gardens, pp. 373–385.
34. Rege NN, Thatte UM, Dahanukar SA. (1999) Adaptogenic properties of six *Rasayana* herbs used in Ayurvedic medicine. Phytother Res 13:275–291.
35. Panossian AG, Oganessian AS, Ambartsumian M, Gabrielian ES, Wagner H, Wikman G. (1999a) Effects of heavy physical exercise and adaptogens on nitric oxide content in human saliva. Phytomedicine 6(1):17–26.

36. Panossian AG, Gabrielian ES, Wagner H. (1999b) On the mechanism of action of plant adaptogens with particular reference to Curcubitacin R diglucoside. Phytomedicine 6(3):147–155.
37. Panossian AG, Wikman G, Wagner H. (1999c) Plant adaptagens III. Earlier and more recent aspects and concepts on their mode of action. Phytomedicine 6(4):287–300.
38. Davydov, M. and Krikorian, A.D. (2000) *Eleutherococcus senticosus* (Rupr. & Maxim.) Maxim (Araliaceae) and an adaptogen: a closer look. J Ethnopharmacol 72:345–393.
39. Elisabetsky E, Setzer R. (1985) Caboclo concepts of disease, diagnosis and therapy: implications for ethnopharmacology and health systems in Amazonia. In: Parker EP, editor. The Amazon Caboclo: historical and contemporary perspectives. Williamsburgh: Studies on Third World Societies Publication Series, Vol. 32, pp. 243–278.
40. Nunes DS. (1996) Chemical approaches to the study of ethnomedicines. In: Balick M, Elisabetsky E, Laird S, editors. Medicinal resources of the tropical forest: biodiversity and its importance to human health. New York: Columbia University Press, pp. 41–47.
41. Hyman SE, Nestler EJ. (1996) Initiation and adaptation: a paradigm for understanding psychotropic drug action. Am J Psychiatry 152:151–162.
42. Borchers AT, Sakao S, Henderson GL, Harkey MR, Keen CL, Stern JS, Terasawa K, Gershwin ME. (2000) Shosaiko-to and other Kampo (Japanese herbal) medicines: a review of their immunomodulatory activities. J Ethnopharmacol 73:1–13.
43. De Feudis FV. (1991) In vivo studies with Egb 761. In: Ginkgo biloba extract (EGb 761): pharmacological activities and clinical applications. Paris: Elsevier, pp. 61–96.
44. Elisabetsky E, Amador TA, Albuquerque RR, Nunes DS, Carvalho ACT. (1995b) Analgesic activity of *Psychotria colorata* (Will. Ex R. &S.) Muell. Arg. (Rubiaceae) alkaloids. J Ethnopharmacol 48:77.
45. Amador TA, Elisabetsky E, Souza DO. (1996) Effects of *Psychotria colorata* alkaloids in brain opioid system. Neurochem Res 21(1):97–102.
46. Elisabetsky E, Amador TA, Leal MB, Nunes DS, Carvalho ACT, Verotta L. (1997) Merging ethnopharmacology with chemotaxonomy: an approach to unveil bioactive natural products. The case of *Psychotria* alkaloids as potential analgesics. Ciência e Cultura 49:378–385.
47. Verotta L, Pilati T, Tatò M, Elisabetsky E, Amador TA, Nunes DS. (1998) Pyrrolidinoindoline alkaloids from *Psychotria colorata*. J Nat Prod 61:392–396.
48. Amador TA, Verotta L, Nunes DS, Elisabetsky E. (2000a) Antinociceptive profile of Hodgkinsine. Planta Medica 66:1–3.
49. Dickenson AH, Chapman V, Green GM. (1997) The pharmacology of excitatory and inhibitory amino acid-mediated events in the transmission and modulation of pain in the spinal cord. Gen Pharmacol 28(5):633–638.
50. Wiesenfeld-Hallin Z. (1998) Combined opioid-NMDA antagonist therapies – what advantages do they offer for the control of pain syndromes? Drugs 55(1):1–4.
51. Odebiyi AI. (1990) Western trained nurses' assessment of the different categories of traditional healers in Southwestern Nigeria. Int J Nurse Stud 25(4):333–342.
52. Costa-Campos L, Elisabetsky E, Lara DR, Nunes DS, Carlson TJ, King SR, Ubilas R, Iwu M. (1999) Antipsichotics profile of alstonine: ethnopharmacology of a traditional Nigerian botanical remedy. Acta Horticult 501:313–322.
53. Costa-Campos L, Lara DR, Nunes DS, Elisabetsky E. (1998) Antipsychotic-like profile of alstonine. Pharmacol Biochem Behav 60(1):133–141.
54. Elisabetsky E, Souza GPC, Santos MAC, Siqueira IR, Amador TA. (1995a) Sedative properties of linalool. Fitoterapia 15:407–414.
55. Elisabetsky E, Silva Brum LF, Souza DO. (1999) Anticonvulsant properties of linalool on glutamate related seizure models. Phytomedicine 6:113–119.
56. Silva Brum LF, Elisabetsky E, Souza DO. (2001) Effects of linalool on [^{3}H]MK801 and [^{3}H]muscimol binding in mice cortex membranes. Ohytother Res 15:422–425.
57. Dichter MA. (1997) Basic mechanisms of epilepsy: targets for therapeutic intervention. Epilepsia 38:S2–S6.
58. Löscher W. (1998) New visions in the pharmacology of anticonvulsion. Eur J Pharmacol 342:1–13.
59. Schultes RE, Hofmann A. (1980) In: The botany and chemistry of hallucinogens. Chapter IV: Plants of hallucinogenic use. USA: Charles C. Thomas, pp. 235–239.
60. Lotsof HS. (1995) Ibogaine in the treatment of chemical dependence disorders: clinical perspectives. MAPS 5:16–27.
61. Popik P, Layer RT, Skolnick P. (1995) 100 years of ibogaine: neurochemical and pharmacological actions of a putative anti-addictive drug. Pharmacol Rev 47:235–253.

62. Sweetnam PM, Lancaster J, Snowman A, Collins JL, Perschke S, Bauer C, Ferkany J. (1995) Receptor binding profile suggests multiple mechanisms of action are responsible for ibogaine's putative anti-addictive activity. Psychopharmacology 118:369–376.

63. Staley JK, Ouyang Q, Pablo J, Hearn WL, Flynn DD, Rothman RB, Rice KC, Mash DC. (1996) Pharmacological screen for activities of 12-hydroxyibogamine: a primary metabolite of the indole alkaloid ibogaine. Psychopharmacology 127:10–18.

64. Leal MB, Souza DG, Elisabetsky E. (2000) Long lasting ibogaine protection to NMDA-induced convulsions in mice. Neurochem Res 25(8):1083–1087.

65. Mash DC, Kovera CA, Buck BE, Norenberg MD, Shapshak P, Hearn WL, Sanchez-Ramos J. (1998) Medication development of ibogaine as a pharmacotherapy for drug dependence. Ann NY Acad Sci 844:274–292.

66. Glick SD, Maisonneuve IM. (1998) Mechanisms of antiaddictive actions of ibogaine. Ann NY Acad Sci 844: 214–226.

67. Vineis P, Porta M. (1996). Causal thinking, biomarkers, and mechanisms of carcinogenesis. J Clin Epidemiol 49(9):951–956.

68. Bisaga A, Popik P. (2000) In search of new pharmacological treatment for drug and alcohol addiction: *N*-methyl-D-aspartate. (NMDA) antagonists. Drug Alcohol Depend 59:1–15.

69. Nichter M, Vuckovic N. (1994) Understanding medication in the context of social transformation. In: Etkin N, Tan ML, organizers. Medicines: meanings and contexts. Amsterdam: Hain, pp. 285–305.

70. Prinz A. (1990) Misunderstanding between ethnologists, pharmacologists and physicians in the field of ethnopharmacology. In: Ethnopharmacologie: sources, méthodes, objectifs. Metz: Fleurentin Societé Française d'Éthnopharmacologie pp. 95–99.

71. Umoren UE. (1990) Religion and traditional medicine: an anthropological case study of a Nigerian treatment of mental illness. Med Anthropol 12:389–400.

CHAPTER 12

Drug discovery through ethnobotany in Nigeria: some results

JOSEPH I OKOGUN

Abstract

Ethnobotany and ethnomedicine are as old as man's history. The combination of ethnobotany and ethnomedicine is a popular research field in Nigeria that, unfortunately, has not led to the production of many market products. The plant *Zanthoxylum zanthoxyloides* (Lam). Zepernich & Timter (Rutaceae) is used in ethnomedicine for a number of ailments. The extracts of the plant contain a wide variety of secondary metabolites associated with a number of biological activities. The active principles in its extracts responsible for its antisickling effect have been identified, and the synthetic conversion of one of its metabolites yielded the antisickling acid, DBA. With assistance from various organizations, NIPRISANR, a herbal drug of ethnomedicine source that is used for sickle cell anemia disorder, is in its third clinical-trial development phase at the National Institute for Pharmaceutical Research and Development, Abuja, Nigeria.

Keywords: *ethnobotany, ethnomedicine, biological activities, antisickling, metabolites*

I. Introduction

Ethnobotany has a very long history dating back to the Biblical Old Testament times.[1] With the developments in modern science, a number of drugs owe their discovery and development to ethnobotany. Some examples are aspirin (acetylsalicylic acid) originally derived from the willow tree, *Salix* spp. (Salicaceae) used in Europe, reserpine from the Indian medicinal use of *Rauwolfia* spp., Afzel. (Apocynaceae), quinine from the South American *Cinchona* spp. (Rubiaceae), and eserine (Physostigmin) from the African use of the plant *Physostigma venenosum* Balfour (Pipilionaceae) in Nigeria. Just recently, artemisinin, an antimalarial, has been developed from the Chinese herbal medicine Quinghaosu.[2]

It has been established that up to 25% of the drugs prescribed in conventional medicine are related directly or indirectly to naturally occurring substances mostly of plant origin. This contribution is a credit to ethnobotany in drug discovery. Natural products from plants, microbes and animals contribute to about half of the pharmaceuticals in use today.[3,4] Farnsworth[5] has shown that 119 drugs of known chemicals in medical use arose from less than 90 plant species.

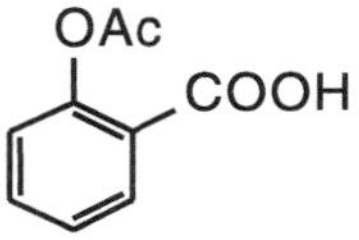

Aspirin

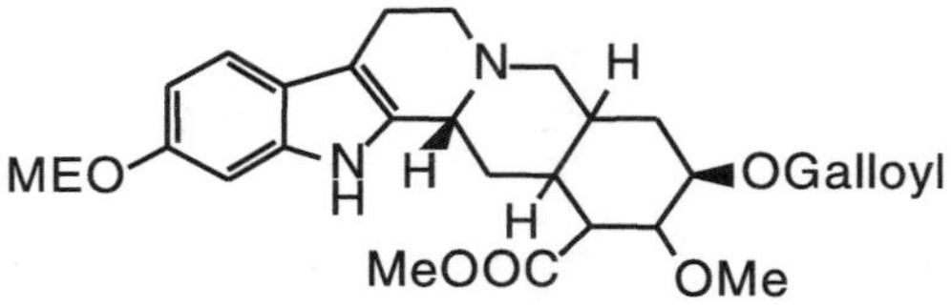

Reserpine

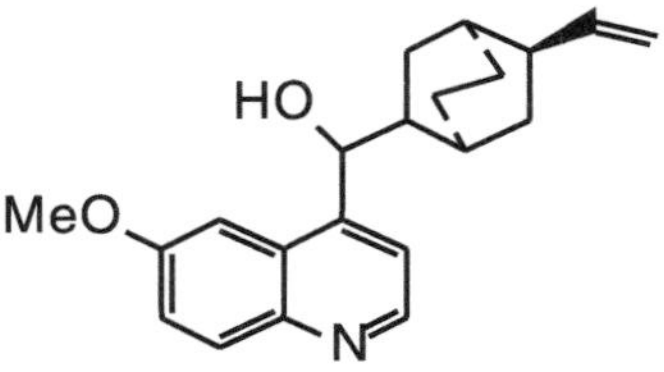

Quinine

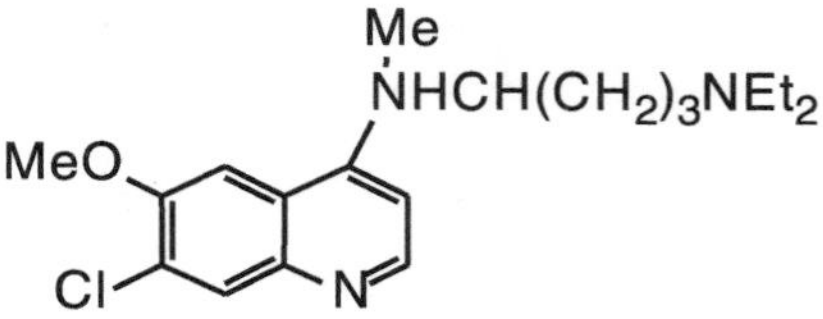

Chloroquine

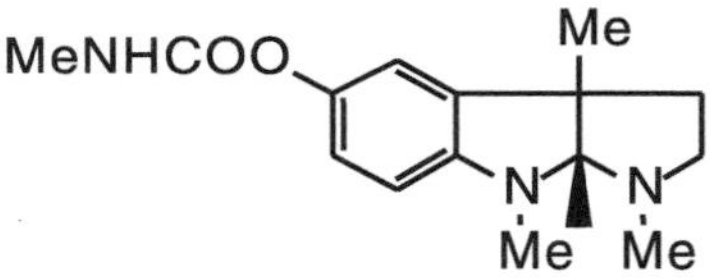

Physostigmine

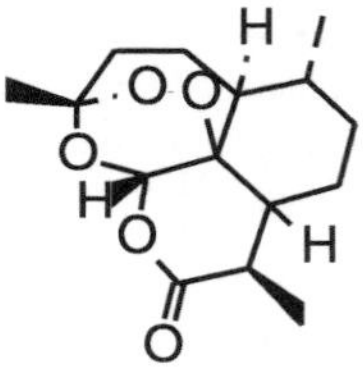

Artemisinin

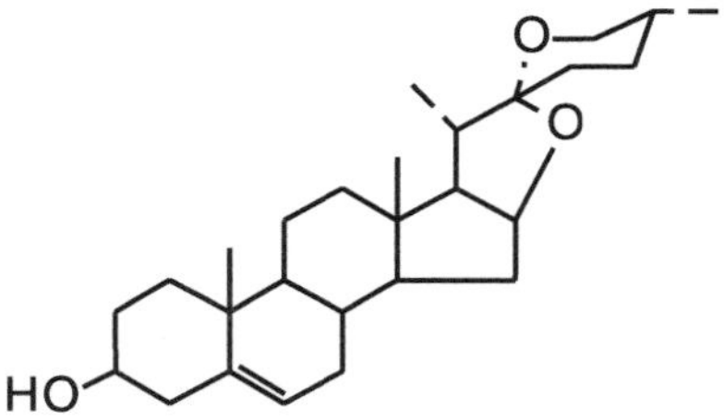

Diosgenin

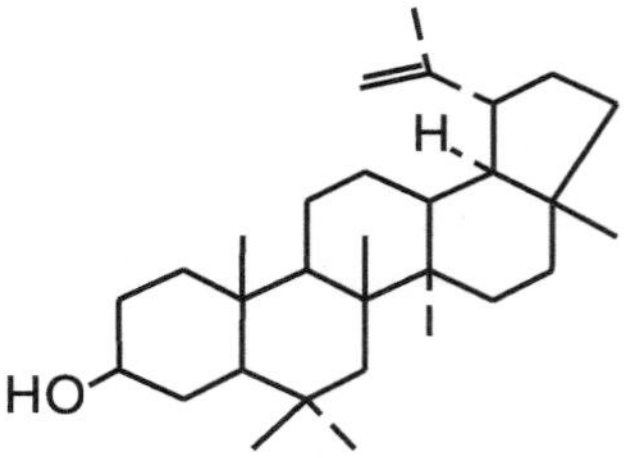

Lupeol

OMe
RO
OH
OH O
Hesperetin

CH₃(CH₂)₄ NHBu
Dieneamides

Fagarol

OMe
MeO
OMe N
Skimmianine

MeO
NMe⁺
OR
R = N Fagaridine
R = Me Chelerythrine

ON
MeO
OMe
MeO
HMe⁺
Fagaronine

OH
OH
Zanthoxyllol

OAc
COOH
Acetylphenylbutyric acid

OH
COOH
4-Hydroxyphenylbutyric acid

COOH
O
DBA

OH
HOOC OR
Hydroxybenzoic acids

Ethnobotany contributes to drug discovery by providing leads to:

1. Direct drug substances first isolated from nature as with reserpine[6] and eserine.[7,8]
2. Drug substances that have low desirable biological activities or have desired drug activities but with undesirable side-effects. Through modification of chemical structure by derivatization or synthesis of the same or similar chemical structures, drugs having the desired properties may be developed. Quinine[9] and chloroquine[10] illustrate this point.
3. Excipients in the formulation of drugs,[11] for example, gum arabic from the plant *Acacia verek* Guill. & Peir (Mimosaceae).

4. Raw materials for drug synthesis: Diosgenin from *Dioscorea composita* Hemsl. and *D. terpinapensis* Uline (Dioscoreaceae) serve as raw materials for the synthesis of steroidal drugs.[12]

These results have arisen in spite of the several known limitations of ethnobotany and the usually associated ethnomedicine. The limitations of herbal drugs derived from ethnobotany revolve around standardization, quality control, dosage and the common tendency to describe diseases and ailments vaguely. Standardization problems arise because constituents of the same plant may vary according to soil types, weather, time of the year and time of the day. Furthermore, plants may be wrongly identified, recipes may contain many components, and preparations may be unstable. Reasons such as these have necessitated the application of techniques in botany, chemistry, molecular biology along with pharmacology, toxicology and clinical medicine to drug development from ethnobotany.

II. Drug discovery and ethnobotany in Nigeria

There have been several relevant conference proceedings and publications[13–17] on this theme. Eserine, an alkaloid isolated from the Calabar bean *Physostigma venenosun* Balfour (Papilionaceae), called *esere* in the Efik language, has already been mentioned. It has been developed for use in the treatment of glaucoma.[7] The lead came from the ethnobotanical use of the plant as observed by early European explorers. In a review,[18] Akubue described pharmacological/toxicological studies on plants used in traditional medicine in Nigeria. Diseases or bioactivities listed in the review included malaria, antidiabetic, antisickling, antihepatotoxic, antibacterial, anticonvulsant, antisnake venom, molluscicidal, spasmolytic, anti-ulcer, analgesic, antipyretic, anti-inflammatory, and toxic agents. Mention was also made, in the review, of tinctures, infusions and extracts, some of which serve as excipients.

The review clearly showed that ethnobotany could serve as a drug source in Nigeria if certain actions were taken by the government, and if there were cooperation with the drug industries. It is the opinion of many that some overseas companies may be exploiting published results on Nigerian herbs by Nigerian scientists without the need for the original authors' accord or collaboration.

Ethnobotany continues to provide useful leads to drug-discovery efforts in Nigeria, yet, and unfortunately, to the best of our knowledge, such efforts have not surpassed the potential stages as far as entrepreneurial drug production is concerned. The results of work on a number of plants illustrate this point.

The plant *Garcinia kola* Heckel (Sterculiaceae) is a popular ethnobotany plant in Nigeria, and the title of my presentation demands that I mention it. It has been worked on extensively by Iwu, Igboko et al., among others[19–21]. It is a symposium sub-theme at this conference and so needs not be further discussed in this presentation.

III. Excipients from Nigerian plants

Irvingia gabonensis (Aubry-LeComte er O'Rorke) Bail (Irvingiaceae) that is cultivated in West Africa produces an edible fruit whose seed is used in the preparation of a delicious viscous soup for swallowing yam and cassava puddings. The fat is known as dika fat. A number of Nigerian scientists[22] have worked on the chemistry of the fat, which consists mainly of C12 and C14 fatty acids, along with smaller quantities of C10, C16 and C18 acids, glycerides and acids and protein.[23-26] Udeala et al.[27] have shown that dika fat could serve as a lubricant in tablet preparation, while the gum of *Prosopis africana* (Guill. & Perr.) Taub. (Mimosaceae) could be used[28,29] for the preparation of gel-forming sustained-release tablets.

The fruits. (Okra) of *Abelmoschus esculentus* (Linn.) Moench (Malvaceae) are edible and used for food in the same way as the seeds of *I. gabonensis*. They produce a viscous mucilage. Nasipuri et al.[30] have now studied and assessed the potential of this mucilage as an emulsifying agent. The emulsions prepared with various types of oils using the mucilage were compared with corresponding emulsions from acacia and tragacanth. It was found that okra mucilage emulsions are better than acacia emulsions with fixed oils, while the mucilage may be a useful substitute for tragacanth as an emulsion stabilizer.

Zanthoxylum zanthoxyloides was formerly known as *Fagara zanthoxyloides* (Lam) Zepernick & Timter (Rutaceae, Yoruba: Orin ata), and is a well-known local medicinal plant. It grows[31] in the semi-Savannah. Gbile[32] has carried out a detailed study of the plant genus to help in the identification of the various species, some of which have similar traits. This medicinal plant has been the subject of chemical investigation by various workers in Europe and Africa[33-48] since the beginning of this century.

The secondary metabolites isolated from the plant extracts included the ubiquitous plant steroid Beta-sitosterol, the triterpene lupeol, phenol zanthoxyllol, flavonoids diosmin and hesperetin, unsaturated amides, like *N*-isobutyl decadienamides and fagaramide, the lignan fagarol, the alkaloids skimmianine, chelerythrine isolated as artefact, fagaridin and fagaronine.

Thus, the plant provides a rich variety of isolatable secondary metabolites. Most of these investigations leading to the isolation of these compounds were carried out before 1970. Interest in the extracts of the plant was re-awakened in 1971 when Sofowora and Isaacs-Sodeye[39] announced the antisickling activity of the extracts of the plant and proceeded to demonstrate its efficacy as a drug to manage sickle cell anemia disorder.

However, after the announcement, the in vitro antisickling activity was reported not to have been observed by another bioassay,[41,42] and the antisickling activity of the plant extracts was questioned. Fortunately, chemical and pharmacognosy works were further carried out, and more secondary metabolites were isolated[43-49]. Hydroxybenzoic acids[46] and zanthoxyllol[47,48] have now been demonstrated in vitro to be the antisickling compounds in the extracts of this plant.

When zanthoxyllol was found to have little or no significant antisickling activity, it was reasoned that, in vivo, zanthoxyllol could be converted to a carboxylic acid similar to acetylsalicylic acid, which was also being used therapeutically to help sickle cell anemia disorder patients. The plan was to simply convert the propanol substituent on the benzene ring of zanthoxyllol to the corresponding butanoic acid and acetylation of the phenol group to give 4-[4-acetoxy-3-(3-methyl but-2-enyl) phenyl] butyric acid. This conversion could not be achieved in our hands. The acid was eventually made[50] starting with 4-hydroxyphenylbutyric acid.

IV. Antisickling, anti-inflammatory and analgesic acid, DBA

The 4-hydroxyphenylbutyric acid reacted with sodium hydride in toluene to give the di-anion, which reacted with 1-bromo-2-methylbut-2-ene to give the phenol of the required acid along with other compounds on acidification. The phenol was acetylated under mild conditions[51] to yield the required acetoxyphenylbutyric acid.

Attempts to synthesize this acid from zanthoxyllol via the phenylbutanonitrile led to the synthesis of 4-(2,2-dimethylchroman-6-yl) butyric acid also known as 3,4-dihydro-2,2-dimethyl-2H-1-benzopyran-6-butyric acid (DBA). DBA became the first synthetic compound to show potent antisickling activity[52] in vitro and has remained so. It also exhibited good anti-inflammatory and analgesic activities[53] that are more potent than those of aspirin.

V. Blood coagulation principles in extracts of *Z. zanthoxyloides*

Essien and co-workers[54] have investigated the effect on blood coagulation of the root extracts of *Z. zanthoxyloides*. It was demonstrated that the root aqueous extracts of the plant after lyophilization and removal of the ether, chloroform and methanol soluble components shortened the prothrombin (clotting) time of normal and factor viii-deficient plasma with no such effect on factor ix-deficient plasma. Zanthoxyllol and hesperidin[55], however, prolonged the clotting time of normal plasma. The observation was considered to be of medical significance. Hesperidin glycoside as citranin was once reported[56,57] to be an antifertility principle.

VI. Other ethnomedical uses of *Z. zanthoxyloides* and related species

Z. zanthoxyloides and related species are used in the treatment of tetanus (GA Iyoriobhe, personal communication, 1976) and toothache (NI Ologun, personal communication). We have successfully used the plant to manage toothaches, which manifested with swollen gums.

VII. Yield of ethnodrugs and synthesis of potential drugs from work on *Z. zanthoxyloides*

From these results, work on the extracts of *Z. zanthoxyloides* has identified ethnodrugs directly and, through chemical synthesis, yielded potential drugs.

VIII. The need for the conservation of *Z. zanthoxyloides*

Z. zanthoxyloides is very rich in secondary metabolites associated with a number of biological activities: anesthetic,[36] anti-inflammatory,[58] blood-clotting,[54,55] analgesic,[58] treatment of tetanus (GA Iyoriobhe), toothache treatment (NI Ologun), antifertility,[56,57] antitumor[39] and anti-HIV.

It is necessary to conserve and propagate the plant. We have successfully propagated the plant by transplanting a seedling collected in the wild. Attempts by experts to propagate the plant so far, through the usual simple methods, have not been successful (PRO Kio, DUU Okali, personal communications).

The reasons given for the failure were that the plant could not sprout from cuttings and that the plant is heterosexual and can only produce viable fruits if such fruits arise from ovary fertilization by pollen from a male plant growing in the neighborhood.

The *Z. zanthoxyloides* that we have grown successfully at our residential premises at the University of Ibadan, Nigeria campus is, as far as we know, the only plant of this species in the campus. It is unisexual and so produces non-virile fruits. Attention is hereby drawn to the need for a concerted effort to propagate this plant.

IX. NIPRISANR: a drug product at NIPRD, Nigeria from ethnomedicine for sickle cell anemia disorder

Through collaboration with a herbalist, the National Institute for Pharmaceutical Research and Development, in Abuja, Nigeria has produced NIPRISANR. NIPRISAN has five components apart from water and the usual excipients. Originally, it is a herbal preparation used by the herbalist for the management of sickle cell anemia disorder. NIPRD successfully extracted and formulated the herbal drug into standard forms including capsules, which have passed through the first and second phases of clinical trials and are now in the third and final clinical phase trial.

The success of NIPRISAN is due to the following:

1. The achievement of a cooperative arrangement between NIPRD's management headed by the Director-General, Professor CON Wambebe and the herbalist Reverend JO Ogunyale.
2. The commendable guidelines laid down by WHO, which allow the development of drugs from herbal preparations, which have been in use by indigenous peoples for generations.

3. The availability of in vitro models for anti-sickling activities and the established non-toxicity[59,60] of the drug along with its other beneficial effects.
4. The support for the work from the Federal Government of Nigeria, the Pharmacy Society of Nigeria and other Nigerian companies and learned societies.
5. The very important international support and grants from UNDP, which also assisted the patenting of the product.
6. NIPRD has all the relevant scientific and even marketing disciplines represented and active as staff at the Institute.
7. NIPRD engages in national and international collaboration and encourages staff development training.
8. The management in NIPRD cultures excellence and hard work.

X. Conclusion

As has been illustrated, there has been a good amount of exploitable scientific results on Nigerian medicinal plants, which can yield market products. The NIPRD model, when appropriately staffed and with collaboration, can collate available results for developmental purposes. The meaningful and fair collaboration among universities, research institutes and industry personnel is necessary. Such collaboration should be both intramural and extramural.

Acknowledgment

I thank Professor MM Iwu and his co-organizers of this conference for inviting and supporting me to attend the conference and the Director-General of NIPRD, Professor CON Wambebe for leave to attend the conference. I acknowledge the collaboration, discussions and contributions with or from my colleagues, which enabled me to present the work on *Fagara* and NIPRISAN. These include DEU Ekong, K Nakanishi, CON Wambebe, DT Okpako, JF Ayafor, PRO Kio, DU Okali, GA Iyoriobhe, NI Ologun, and a host of others.

References

1. Bitter water made sweet with a piece of wood inspired by Moses: Exodus 15:22–25; for a useful reference work on herbal drug substances, see Merck Index, 8th edition.
2. Klayman DL. (1985) Quinghaosu (Artemisinin): an antimalarial drug from China. Science 228:1049–1054.
3. Farnsworth NR, Bingel AS. (1977) Problems and prospects of discovering new drugs from higher plants by pharmacological screening. In Wagner H, Wolff P, editors. New natural products and plant drugs with pharmacological, biological or therapeutical activity. Berlin: Springer.
4. Clark AN. (1996) Natural products as a resource for new drugs. Pharmaceut Res 13(8):1133–1144; Wambebe C. (1998) Role of chemistry in the discovery of useful plant-derived drugs. Paper presented at the 21st Annual Conference of the Chemical Society of Nigeria, Ibadan, 20–24 September.
5. Farnsworth NR. (1990) The role of ethnobotany in drug development. Ciba Found Symp 154:2–11.
6. Schlitter. (1965) Rauwolfia alkaloids with special reference to the chemistry of reserpine. In: Manske RHF, editor. The alkaloids. Vol. 8 pp. 287–334.
7. Mann J. (1992) Murder, magic and medicine. Oxford: Oxford University Press, pp. 31–353.

8. Harley-Mason J, Jackson AH. (1954) Hydroxytryptamines. A new synthesis of phytostigmine. J Chem Soc 3651–3654.
9. Woodward RB, Doering WE. (1945) The total synthesis of quinine. J Am Soc 67:860–874.
10. Andersag H. (1948) Antimalariamittel aus der Gruppe halogensubstituierter Chinolinverbindungen. Chem Ber 81:499–507.
11. Handbook of pharmaceutical excipients. (1983) American Pharmaceutical Association and Pharmaceutical Society of Great Britain, pp. 1–2.
12. Coursey DG. (1967) Yams: an account of the nature, origins, cultivation, and utilisation of the useful members of Dioscoreaceae. London: Longmans.
13. Sofowora A. (1986) The state of medicinal plants research in Nigeria. Proceedings of a conference, co-organized with the Society of Pharmacology, 16, Ife, Nigeria.
14. Aladesanmi AJ, Elujoba AA, Adesanya SA, Igboechi CA, editors. (1990) Medicinal plants in a developing economy. Proceedings of the annual conference of the Nigerian Society of Pharmacognosy, January 9–13, Benin City, Nigeria.
15. Iwu MM. (1993) Handbook of African medicinal plants. Boca Raton, FL: CRC Press.
16. Sofowora A. (1992) Medicinal plants and traditional medicine in Africa. Chichester, UK: Wiley.
17. Okogun JI. (1985) Drug production efforts in Nigeria: medicinal chemistry research and a missing link. Proc Nigerian Acad Sci 29–52.
18. Akubue PI. (1986) Nigerian medicinal plants: pharmacology and toxicology. In: Sofowora A, editor. The state of medicinal plants research in Nigeria. Proceedings of a conference co-organized with the Society of Pharmacology, Ife, Nigeria, pp. 53–63.
19. Iwu MM, Igboko O. (1982) Epigemin, fisetin, and biflavonoid ametoflavone from *G. kola* seeds. J Nat Prod 45:650.
20. Iwu MM. (1985) Kolaviron and biflavanones from *G. kola* seeds. First symposium on plant flavonoids in biology and medicine.
21. Igboko OA, Iwu MM. (1986) Antihepatoxic biflavonoids and other constituents of *Garcinia kola*, Heckel fruit. In: Sofowora A, editor. Ref. 23 p. 364.
22. Okor RS. (1990) Applications of some selected natural products as excipients for dosage formulation. In: Aladesanmi AJ, Elujoba AA, Adesanya SA, Igboechi CA, editors. pp. 15–30.
23. Abdurahman EM, Rai PP, Shok M, Olurinola PF, Laakso I. (1996) Analysis of the fatty acid composition of the seed fat of two varieties of *Irvingia gabonennsis* by high resolution gas chromatography. J Pharm Res Dev 1:49–51.
24. Eka OU. (1980) Proximate composition of seeds of bush mango tree and some properties of dika fat. Nigerian J Nutr Sci 1:33–36.
25. Patel CB. (1950) Component acids and glycerides of dika fat. J Sci Fd Agr 1:45–51.
26. Abaelu AM, Akinrimisi EO. (1980) Amino acid composition of *Irvingia gabonensis* (Apon) oil seed proteins. Nigerian J Nutr Sci 1:133–135.
27. Udeala OK, Onyechi JO, Agu SI. (1980) Preliminary evaluation of dika fat, a new tablet lubricant. J Pharm Pharmacol 32:6–9.
28. Adikwu MU, Udeala OK, Ohiri FC. (1997) Evaluation of *Prosopis africana* for gel-forming sustained-release tablets. J Pharm Res Dev 2(2):42–46.
29. Ohiri FC, Udeala OK, Adikwu MU. (1997) Microbiological studies on tablets containing high proportions of *Prosopis africana* Gum, Ibid., 2(2):11–13.
30. Nasipuri RN, Igwilo CI, Brown SA, Kunle OO. (1997) Mucilage from *Abelmoschus esculentus* (okra fruits) a potential pharmaceutical raw material; Part II. J Pharm Res Dev 2(1):27–34.
31. Dalziel JM. (1955) The useful plants of West Tropical Africa, 2nd edition Crown Agents for Overseas.
32. Gbile ZO. (1975) Key to *Fagara* species in Nigeria. Nigerian J Sci 9:337–353.
33. Thoms H, Thumen F. (1911) Uber das Fagaramid einen nuen stickstoff haltigen Stoff aus der Wurzelrinde von Fagara xanthonyloides Lam. Chem Ber 44:3717–3730.
34. Paris MR, Mogse-Mignon L. (1947) Etude preliminaire du *Fagara xanthoxyloides* Lam. Ann Pharm France 5:412–420.
35. Carnmalm B, Erdtman H, Pelchowicz Z. (1955) Constitution of resin phenol and their biogenetic relationships XIX. The structure of sesamolin, the configuration of sesamin, and the nature of fagarol. Acta Chem Scand 9:1118.
36. Bowden K, Ross WJ. (1963) The local anaesthetic in *Fagara xanthoxyloides* Lam. Ann Pharm Franc 5:410–420.
37. Eshiet ITU. (1967) Extractives from the West African species of the genre *Fagara* with an appendix on a proposed diterpene synthesis. Ph.D. thesis, University of Ibadan, Nigeria.

38. Eshiet IT. (1968) The isolation and structure elucidation of some derivatives of dimethyl allyl coumarin, chromone, quinoline and phenol from *Fagara* species and from *Cedrelopsis gravei*. J Chem Soc 481–484.
39. Messmer WM, Fong HHS, Bevelle C, Farnsworth NR, Abraham DJ, Trojanek J. (1972) Fagaronin, a new tumor inhibitor isolated from *Fagara xanthoxyloides* Lam (Rutaceae) J Pharm Sci 61:1858–1859.
40. Sofowora EA, Isaacs WA. (1971) Reversal of sickling and crenation in erythrocytes by the root extract of *Fagara xanthoxyloides*. Lloydia 38:169.
41. Honig GR, Farnsworth NR, Terence C, Vida LN. (1975) Evaluation of *Fagara zanthoxyloides* root extract in sickle cell anemia blood in vitro. Lloydia 38:387–390.
42. Honig GR, Vida LN, Terence C. (1978) Effects in vitro of the proposed anti-sickling agent DBA. Nature 272:833–834.
43. Enyenihi VU. (1975) Synthesis and ultra violet spectra of 4-substituted coumarins: chemistry of the roots of *Fagara zanthoxyloides* Lam Ph.D. thesis. University of Ibadan, Nigeria.
44. Ayafor JA. (1978) Dictamine-type alkaloids from *Teclea verdoorniana* and extractives from *Zanthoxylum xanthoxyloides*. Ph.D. thesis, University of Ibadan, Nigeria.
45. Okogun JI, Ayafor JF, Ekong DEU, Enyenihi VU. (1978) Extracts from the Nigerian and Cameroon varieties of *Fagara zanthoxyloides*, Nigeria. J Sci 12:589–603.
46. Elujoba AA, Sofowora EA. (1977) Detection and estimation of total acid in the anti-sickling fraction of *Fagara* Linn. species (orin ata). Plant Medica 32:54–59; Sofowora EA, Isaacs-Sodeye WA, Ogunkoya LO. (1975) Lloydia 38:169.
47. Elujoba AA, Nagels L, Sofowora A, Dongen WU. (1984). Identification of phenolic acid from *Zanthoxylum zanthoxyloides* roots by GC – MS. Nigerian J Pharm 15(1):281–288.
48. Elujoba AA, Nigels L. (1985) Chromatographic isolation and estimation of zanthoxylol: an anti-sickling agent from the roots of *Zanthoxylum* species. J Pharm Biomed Anal 3(5):447–451.
49. Elujoba AA, Nagels L, Sofowora A, Dongen WV. (1989) Chroromatographic analysis of anti-sickling compounds in *Zanthoxylum* species. Int J Pharmaceut 53:1–3.
50. Fatope OM, Okogun JI. (1982) A convenient synthesis of 4–[4–acetoxy–3–(3-methylbut–2-enyl) phenyl butyric acid. J Chem Soc Perkin 1:1601–1603.
51. Bonner TG, McNamara PM. (1968) The Pyridine-catalyzed acetylation of phenols and alcohols by acetic anhydride. J Chem Soc B:795.
52. Ekong DEU, Okogun JI, Enyenihi VU, Balogh-Nair V, Nakanishi K, Natta C. (1975) New anti-sickling agent 3, 4–dihydro–2, 2–dimethyl–2H–1–benzopyran–6–butyric acid. Nature 258:743–746.
53. Okpako DT, Oriowo MA, Okogun JI, Ekong DEU, Enyenihi VU. (1983) 3, 4–Dihydro–2, 2–dimethyl–2H–1–benzopyran–6–butric acid. Its preparation and its anti-inflammatory and related pharmacological properties. Planta Medica 47:112–116.
54. Essien EM, Okogun JI. (1976) Effects of the root extract of *Fagara xanthoxyloides* or blood coagulation. Thrombosis and Haemostatis 26:525–531.
55. Essien EM, Okogun JI, Adelakun A. (1985) Blood coagulation of the root extracts of *Fagara xanthoxyloides* plants. African J Med Sci 14:83–88.
56. Sieve BF. (1952) A new anti-fertility factor (a preliminary report). Science 116:273–285.
57. Manwaring DO, Richards DW. (1968) The identity of cirantin, a reported anti-fertility agent with hesperidin. Phytochemistry 7:1881–1882.
58. Oriowo MA. (1979) Anti-inflammatory and related pharmacological properties of a *Fagara* extractive and DBA (a benzopyran butyric acid) derivative. Ph.D. thesis, University of Ibadan, Nigeria; Oriowo MA, Okpako DT in subsequent publication of results.
59. Awodogan A, Wambebe C, Gamaniel K, Okogun JI, Orisadipe A, Akah PA. (1966) Acute and short-term toxicity of NIPRISAN in rats I: biochemical study. J Pharm Res Drug 2:39–45.
60. Gamaniel K, Amos S, Akah PA, Samuel BB, Kapu S, Olusola A, Abayomi O, Okogun JI, Wambebe C. (1998) Pharmacological profile of NIPRD 94/002/1–0: a novel herbal anti-sickling agent. J Pharm Res Dev 3:89–94.

Iwu and Wootton (eds.), Ethnomedicine and Drug Discovery
© 2002 Elsevier Science B.V. All rights reserved.

CHAPTER 13

Plants, products and people: Southern African perspectives

NIGEL GERICKE

Abstract

This paper provides an overview of the expansive South African biodiversity, which contributes to the historical use of medicinal plants and the large existence of traditional healers. It also underscores the lack of acknowledgment that traditional medicines are given by local authorities and the flaws that exist within present validation systems. The author also briefly highlights a few indigenous South African medicinal plants to draw attention to some recent developments.

Keywords: *medicinal plants, traditional healers, bioassays, intellectual property, pharmacology*

I. Introduction

Southern Africa's remarkable diversity of climate, geology and soil is reflected in the region's biodiversity, and, combined with the rich cultural diversity, it is not surprising to find that about 3500 species of higher plants are used as medicines.

The sale of traditionally used minimally processed medicines is generally ignored by the authorities, but the lack of regulatory clarity for manufactured products has inhibited investment in this sector and delayed the development of a vibrant natural products industry.

Some medicinal plants have been selected to highlight recent developments, including *Harpagophytum procumbens* – Devil's Claw, *Pelargonium sidoides* – Umckaloabo, *Sutherlandia microphylla* – Cancer Bush, and *Sceletium tortuosum* – Kougoed.

Commercial opportunities in South Africa include:

- bioprospecting
- novel products
- new crop development
- raw material production
- extracts for clinical studies
- clinical studies.

Southern Africa's remarkable diversity of climate, rainfall, geology and soil types is reflected in the region's biodiversity. There are some 30,000 species of higher plants and a high degree of endemism, particularly in Cape Floral Kingdom and the succulent karroo vegetation types. The region also has a rich cultural diversity, so it is not surprising to find that about 3500 species of higher plants are used as medicines.

There are presently an estimated 200,000 indigenous healers in South Africa alone. About two-thirds of these practise full-time, and the remainder work as healers in their spare time, often working full time as subsistence farmers, domestic workers, and laborers. Healers rarely have a tertiary education such as a nursing qualification. Mr. Isaac Mayeng, a Tswana healer, is an exception indeed, since he has a B.Sc. degree in medicinal chemistry from New York State University in Buffalo, and is assisting the Ministry of Health develop regulations for the registration of indigenous medicines.

Plants are the main materia medica of local healers, and although there is no hard scientific evidence, I am quite convinced that many southern African medicinal plants have a history of use that dates back to the earliest ancestors of man. South Africa has a wealth of Australopithicene fossils dating back to some three million years, and is increasingly gaining recognition as a possible cradle of mankind. More recently, *Homo erectus* used hand-axes and other stone tools that are found throughout the region, and are between 1.2 million years to 200,000 years old, and these archaic hominids were likely to have been self-medicating with the local flora.

Somewhere in the region of 80% of people in southern Africa consume unprocessed or minimally processed indigenous medicines, usually in addition to using modern healthcare facilities and over-the-counter health products. In South Africa, some 20,000 tons of wild-harvested medicinal plants are traded in a year, worth approximately US$60 million. These figures are from a recent comprehensive study undertaken by Myles Mander on behalf of the Food and Agricultural Organization.[1]

Unfortunately there is still a great deal of regulatory uncertainty in South Africa, and it is not yet clear how the regulatory authority intends to deal with health products manufactured from indigenous plants. The sale of traditionally used minimally processed medicines is presently ignored by the authorities, but the lack of regulatory clarity for manufactured products has inhibited investment in this sector and delayed the development of a vibrant natural products industry in the country. Local pharmaceutical companies and natural products companies are largely importing products from Europe and the USA. It is presently easier to register products made from well-known European, North American, Indian and Chinese medicinal plants. This is a great tragedy from a cultural as well as an economic perspective, and I hope that a pragmatic regulatory framework can be developed to facilitate this fledgling industry.

The Medical Research Council of South Africa (MRC) is presently involved in supporting university programs that are collaborating with indigenous healers to find effective phytomedicines or new chemical entities for the major infective diseases in the region. Ongoing research into novel antimalarial drugs is taking place, which will most likely be extended to TB and HIV/AIDs, and AIDS-related opportunistic infections. One of the researchers at the Pharmacology Department of the University of Cape Town, Mr Gilbert Matsibisa (personal communication), has confirmed the antimalarial activity of a South African plant traditionally used to treat malaria, and has further isolated the pure compound responsible for this activity. It is highly active *in vitro* in exceedingly low doses against both chloroquine-sensitive and chloroquin-resistant malaria. The identity of the plant is being kept confidential as intellectual property is still being generated. Unfortunately, many of these positive research results simply earn the researcher a post-graduate degree, and do not end up being studied clinically. Partnerships that may be forged with the WHO Tropical Diseases Research Program and improved culture of university–industry collaboration could change this.

One of the major flaws in research being done to validate traditional uses is that this validation is primarily being done using a limited range of *in vitro* bioassays. I am struggling to convince research organizations of the value of direct observation of traditional treatments and outcomes in the field by MDs, and the need to develop ethical guidelines and formal ethics-committee approval for this type of activity. On a recent visit to an area on the Zimbabwe border, where endemic malaria is a major problem, I found that a highly respected Venda healer, Mr Joseph Tshikovha, has been successfully treating malaria in adults and children for many years with a decoction of two plants: *Sclerocarya birrea* and *Siphonochilus aethiopicus*. Knowing the healer, and the context in which he practices, I think that his information is highly credible, and should be thoroughly researched. An added advantage is that the two plants are widely used for other medicinal purposes, and are regarded by southern African healers as being entirely safe.

The following descriptions of a few indigenous South African medicinal plants highlight some recent developments and issues.

II. *Harpagophytum procumbens* – Devil's Claw

This is a perennial plant with stems spreading from the top of a central tap-root that grows throughout the Kalahari Sands area of southern Africa. The common name is derived from the claw-like fruit. The thick, fleshy secondary tubers are the parts used medicinally in manufactured products, and there is a world-wide trade of about 500 tons of wild-crafted material annually. In European phytomedicine, it has a reputation for efficacy in osteoarthritis, fibrositis and rheumatism, and is also taken as a bitters to stimulate the appetite, and for indigestion. In clinical practice, I have found that small daily doses of 200 mg of dry tuber, or 1 ml of 1:5 tincture are highly effective in small joint disease, and also have a subtle laxative effect that can more properly be described as regularizing bowel function.

In indigenous medicine, *Harpagophytum* is taken in the form of infusions and decoctions. It is regarded as an important tonic for all health conditions by the San community at Molapo in the Central Kalahari in Botswana and is used to treat fever and all infectious diseases including venereal disease and tuberculosis (Personal

Communication, Matambo). Interestingly, it is taken in small doses for diarrhea and in higher doses to treat constipation. The women of this community report that it is abortifacient in higher doses, and should not be taken during pregnancy. In South Africa, it is also used for diabetes (personal communication, Isaac Mayeng), hypertension, gout, and peptic ulcer.

The plant chemistry is well described, and the iridoids harpagoside, harpagide and procumbide have some documented analgesic and anti-inflammatory activities, and these, together with beta-sitosterols, may be responsible for some of the efficacy. Clinical studies support the use of Devil's Claw in painful joint conditions,[2] and low back ache,[3] and serum cholesterol and uric acid levels were reportedly reduced in a study.[4] It is now fashionable for companies to tout the high levels of harpagoside that their particular extracts of secondary tuber possess. The Kalahari San (or Bushmen) use any underground part of the plant, and usually in far lower doses (about 200 mg three times daily) than are recommended by the German Kommission E Monographs (1.5–4.5 g daily).[5] It would seem that the plant has such general utility that it should be considered an ideal health-maintaining and convalescent tonic. The very broad spectrum of use suggests that it may have immunomodulatory activity.

With the increasing demand in Devils Claw globally, there are now successful efforts underway in South Africa and Namibia to mass-propagate the plant, as it is likely that demand will outstrip supply.

III. *Pelargonium sidoides* – Umckaloabo

This plant is a perennial herb that is traditionally used in South Africa for diarrhea and non-specific chest conditions. It is very similar to *P. reniforme*. *P. sidoides*, which has purple–black flowers while *P. sidoides* has pink flowers. The German company Schwabe has produced a tincture of the dried roots of the plant for use in treating upper-respiratory tract infections and bronchitis in children, and its efficacy in children has been confirmed in some clinical studies (personal communication, Ulrich Feiter).

The activity of this phytomedicine may be due to the presence of umckalin and structurally related coumarins,[6] although it is not yet clear whether Schwabe's product is standardized to a defined active.

I have used a simple 1:5 tincture of the fresh roots in clinical practice to treat strep throat and bronchitis in adults and children to good effect, and a colleague, Dr Greg Cleveland (personal communication), has found the same tincture to be very effective in treating upper-respiratory tract infections and otitis media in children. The plant is not abundant in the wild, and it is being farmed successfully on a commercial scale in South Africa on contract to Schwabe.

I anticipate that with time, this modest plant will have a great future as an important product on the global market.

IV. *Sutherlandia microphylla* – Cancer Bush

This plant, and its close relative, *S. frutescens*, are among the most multi-purpose and useful of the medicinal plants in southern Africa. Tinctures, infusions and decoctions of the leaves and young stems of the two species have been used in the Cape from early times, the uses learned originally from the Khoi khoi and San.

Conditions that have been treated with *Sutherlandia* spp. include fever, poor appetite, indigestion, gastritis, esophagtis, peptic ulcer, dysentary, cancer (prevention and treatment), diabetes, colds and flu, cough, chronic bronchitis, kidney and liver conditions, rheumatism, heart failure, urinary tract infections, stress and anxiety. *Sutherlandia* preparations can be taken on a chronic basis when necessary, but should not be taken during pregnancy as teratogenicity and abortions are known to have occurred (personal communication, Dawid Bester).

My colleague, Dr Carl Albrecht, and I have examined a few strong, recent, anecdotes and believe that Cancer Bush can be of significant benefit in treating pancreatic and other cancers, and can improve the quality of life in terminal cancer patients. It has also been of great benefit in reducing the pain and inflammation in some patients with rheumatoid arthritis (personal communication, Dr Greg Cleveland).

A number of highly active compounds, including canavanine, pinitol and GABA, occur in quantity in both species of *Sutherlandia* (personal communication, Ben-Erik van Wyk) and suggest that there is indeed a sound scientific basis to some of the folk-uses for serious medical conditions. L-Canavanine is a potent L-arginine antagonist that has documented anticancer[7,8] and antiviral activity, including activity against the influenza virus and other retroviruses.[9] Pinitol is a known anti-diabetic agent,[10] that also may have an application in treating wasting in cancer and AIDS,[11] and GABA is an inhibitory neurotransmitter that could account for the use of the plant for anxiety and stress.

There is a great deal of variability in wild material, and carefully selected chemotypes are being successfully propagated for use as a general tonic and for clinical research.

V. *Sceletium tortuosum* – Kougoed

This plant was likely to have been used by pastoralists and hunter-gatherers as a mood-altering substance from prehistoric times. The earliest written records of the use of *Sceletium tortuosum* date back to 1662 and 1685. The traditionally prepared dried plant material is chewed, smoked, or powdered and inhaled as a snuff. *Sceletium* elevates mood and decreases anxiety, stress and tension. In intoxicating doses, it can cause a euphoria, initially with stimulation and later with sedation. The plant is not hallucinogenic, and no severe adverse effects have been documented.

Sceletium was originally also used to decrease thirst and hunger, and as a local anesthetic and analgesic for extracting teeth from the lower jaw. The plant is used to this day in minute quantities for colic in infants, added to a teaspoon of breast milk.

A few drops of sheep-tail fat in which *Sceletium* has been fried is also given to an infant with colic. *Sceletium* is used as a sedative in the form of a tea, decoction or a tincture, and it is used effectively by indigenous healers in Namaqualand to wean alcoholics off alcohol. To this day, the plant is sometimes called 'onse droe drank' (our dry liquor). Chronic use does not appear to result in a withdrawal state.

Plants from particular areas are believed to be more potent, and these wild plants are harvested at the end of the growing season. The succulent plant material is crushed with a rock, and then put into closed plastic bags and left in the sun to 'sweat' for a period of eight days. In earlier times, leather bags were used. The macerated material is then spread out in the sun to dry, and stored for later use. *Sceletium tortuosum* was once widely traded in the Cape, and some trading stores in Namaqualand used to stock it until about 10 years ago. Alcohol, tobacco, and probably also *Cannabis* have displaced its use, and stocks of wild plants have dwindled from over-harvesting, from habitat destruction, and possibly also from infection by diseases of introduced crops.

The active constituents of the plant are alkaloids, including mesembrine, mesembrenone, mesembrenol and tortuosamine. The alkaloid concentration in the dry material ranges from 0.05 to 2.3%. Mesembrine is usually the major alkaloid present and has been demonstrated in laboratory studies (sponsored by the National Institute of Mental Health in the United States, and conducted by the company NovaScreen) to be a very potent serotonin-uptake inhibitor. This receptor-specific activity and some receptor activities found on nicotinic, dopamine and nor-adrenaline sites certainly validate the traditional uses, and suggest additional therapeutic potential.

In clinical practice I have used low doses of tinctures of selected chemotypes of *Sceletium*, standardized to the total alkaloid content, and found them to be extremely useful for anxiety, stress, and depression. A South African company is presently in the advanced stages of development of a more sophisticated phytopharmaceutical that will be taken into clinical studies, which I hope will establish the true therapeutic value of this important plant.

Carefully selected plant material has been successfully cultivated on a small commercial scale as an essential prerequisite to ongoing research and development, as the plant is not common in the wild, and is probably endangered.

Southern Africa clearly has a cornucopia of indigenous plants that have been consumed by humans for a very long time, and the commercial and health potentials of these plants are only just starting to be explored in any depth. The region offers significant opportunities for companies and research organizations interested in harnessing the synergies between indigenous knowledge systems, scientific research and modern technologies. Equitable partnerships and ecologically sustainable practices are essential and could have a significant impact on development in the region over the next two decades.

Commercial opportunities include:

- *Biodiversity screening*. Screening programs by a number of companies are already underway in the southern African region, which has perhaps the best-developed plant taxonomy on the continent. Apart from simply using biodiversity as a source of chemical diversity for high-throughput screens, the application of ethnobotanically directed and chemotaxonomically directed plant selection, combined with a de-replication step is a powerful drug discovery tool. It is clear that all screening activities have to be in the spirit of the Rio Convention, and in keeping with evolving national laws and regulations.

- *Novel products.* Local indigenous knowledge systems are rapidly being eroded by access to modern healthcare facilities, schooling, and urbanization. None the less, extant plant use and a solid literature base provide a firm foundation for the development of products across the full spectrum of foods, nutraceuticals, cosmeceuticals, and phytomedicines.
- *New crop development.* We have excellent experimental farmers who have already succeeded in domesticating a variety of indigenous medicinal plants, including Aloe (*Aloe ferox*), Buchu (*Agathosma betulina*), Honeybush (*Cyclopia intermedia*), Uzara (*Xysmalobium undulatum*), Devils Claw (*Harpagophytum procumbens*), African wormwood (*Artemisia afra*), Umckaloabo (*Pelargonium sidoides*), Kougoed (*Sceletium tortuosum*), Sutherlandia (*Sutherlandia microphylla* and *S. frutescens*), Wild ginger (*Siphonochilus aethiopicus*), and Pepperbark (*Warburgia salutaris*).
- *Raw material production.* Apart from indigenous medicinal plants, there are a growing number of farmers supplying high-quality raw materials of well-known botanicals on contract to European phytomedicine companies, including Echinacea and Hypericum. Some of these farmers are in the process of implementing the EU Good Agricultural Practice (GAP) guidelines. The Agricultural Research Council of South Africa (ARC) is offering a new crop development service if a market can be demonstrated, and importantly, the ARC undertakes to manage technology transfer to small-scale farmers.
- *Extracts.* The Biochemtech divison of the CSIR in South Africa has recently commisioned an extraction facility that can make standardized plant extracts in pilot amounts for use in clinical trials. This facility operates at FDA-GMP.
- *Clinical studies.* South Africa has excellent university and private clinical research capabilities for undertaking efficacy studies on phytomedicines, and these are already being utilized to study the efficacy of products from well-known plants such as *Hypericum* for European companies. The studies are conducted according to EMEA and FDA–GCP. Depending on the plant and the needs of the study, pharmacokinetic studies can also be undertaken.

Acknowledgments

The following people are thanked for providing information on *Sceletium*: Lodewyk Mories, Gert Dirks, Jap-Jap Klaase, Dr. Greg McCarthy, Earle Graven, Scott Perschke of NovaScreen, and Linda Brady of National Institute of Mental Health.

References

1. Mander M. (1998) Marketing of indigenous medicinal plants in South Africa. A case study in KwaZulu-Natal. Food and Agricultural Organization of the United Nations.
2. Lecomte A, Costa JP. (1992) Harpagophytum dans l'arthrose: Etude en double insu contre placebo. Le Magazine 15:27–30.
3. Chrubasik S, *et al.* (1996) Effectiveness of *Harpagophytum procumbens* in the treatment of low back pain. Phytomedicine 3:1–10.
4. Brady LR, *et al.* (1981) Pharmacognosy. 8th edition. Philadelphia, PA: Lea & Febiger, p. 480.
5. Blumenthal M, *et al.* (1998) The complete German commission E monographs. Therapeutic Guide to Herbal Medicines. Austin, TX: American Botanical Council.
6. Wagner H, *et al.* (1975) Cumarine aus Sudafrikanischen *Pelargonium*-arten.
7. Swaffar DS, *et al.* (August 1995) Combination therapy with 5-fluorouracil and L-canavanine: in vitro and in-vivo studies. Anticancer Drugs 6(4):586–93.
8. Crooks PA, Rosenthal GA. (Filed Dec 5, 1994) Use of L-canavanine as a chemotherapeutic agent for the treatment of pancreatic cancer. United States Patent 5,552,440.
9. Green MH. (Filed Jan 25, 1988) Method of treating viral infections with amino acid analogs. United States Patent 5,110,600.
10. Narayanan, *et al.* (1987) Pinitol – a new anti-diabetic compound from the leaves of *Bougainvillea spectabilis*. Current Science 56(3):139–141.
11. Ostlund RE, Sherman WR. (Filed March 4, 1996) Pinitol and derivatives thereof for the treatment of metabolic disorders. US Patent 5,8827,896.

Iwu and Wootton (eds.), Ethnomedicine and Drug Discovery

Development of antimalarial agents and drugs for parasitic infections based on leads from traditional medicine: the Walter Reed experience

BRIAN G SCHUSTER

Abstract

Tropical plants have provided mankind with a dynamic natural laboratory not only as sources of medicine, cosmetics and food but also as an essential element in the stabilization of the ecosystem. Pharmaceutical development from botanical resources in the forested areas offers great potential for contributing to sustained growth. Consequently, there is an urgent need to conserve tropical forests as biological resources in order to ensure the future availability of known and yet undiscovered medicinal substances for future generations. This can only be achieved through community participation and return of benefits to local communities as economic incentive for conservation. The International Cooperative Biodiversity Group. (ICBG) program is an unparalleled effort to integrate improvement of human health through drug discovery, conservation of biodiversity and development of sustainable economic activities that focus on environment, health and population. The Walter Reed Army Institute of Research is collaborating with the Bioresources Development and Conservation Program, Smithsonian Institution and universities in Nigeria, Cameroon and USA in this initiative to demonstrate that sustainable drug development is a viable alternative to destructive activities such as logging as a source of income to local communities. The key concept is to increase the net worth of tropical forests as a resource base while using drug development as a catalyst for the conservation of biological diversity. The African ICBG emphasizes the evaluation of plants in Nigeria and Cameroon as cures for parasitic diseases such as malaria, leishmaniasis, and trypanosomiasis as well as a source of phytomedicines, which are affordable to local people. It also focuses on forest dynamics research to understand the effect of sustainable harvest and cultivation of medicinal plants, training of African scientists and capacity building.

Keywords: *tropical plants, pharmaceutical development, biodiversity, conservation, parasitic diseases, capacity building*

The Walter Reed Army Institute of Research (WRAIR) has a long and successful history of developing new drugs for indications that pose serious disease threats to the US military. This mainly encompasses parasitic diseases (primarily malaria), although other threats such as viral diseases, and diarrheal diseases are also important. The WRAIR drug development program operates as a 'virtual' drug company. It has, in house or on contract, all the disciplines required to develop a new drug from drug discovery through clinical trials and registration. Drug development at WRAIR follows the guidelines of the United States Food and Drug Administration.

If we look at the process of discovering and developing new drugs in the United States, we can see that it is a heavily regulated process that takes a long time (8–15 years) and consumes millions of dollars ($150–$500 million). Drug development by this process is a high-risk endeavor (see Figures 1 and 2).

For every 5000–10,000 compounds that enter a screening program, only about 2.5% make it to preclinical evaluation, and only about 2% of those proceed to clinical trials. Of those, perhaps only one of every five will get all the way through the process and be approved as a new drug by the Food and Drug Administration. This common or 'classic' drug-development paradigm progresses from the test tube, through animal testing, and then into humans before a New Drug Application is filed. It becomes very clear when we study this process that it is inherently high-risk, time-consuming and expensive – for a general description of the processes involved in drug development, see Ref. 1.

At the WRAIR, we asked ourselves, 'How can we do this better? How can we reduce the risks inherent in this "classic" approach to new drug development? How can we cut the costs and speed up the timelines for developing important new therapeutic compounds?' We decided to explore natural products and traditional medicine. It is for good reason that natural products used as traditional medicines have been gaining widespread attention. Their use is fast becoming a multibillion-dollar business. Why would we, as 'classic' drug developers, be interested in developing natural products? What would traditional medicines have that we want?

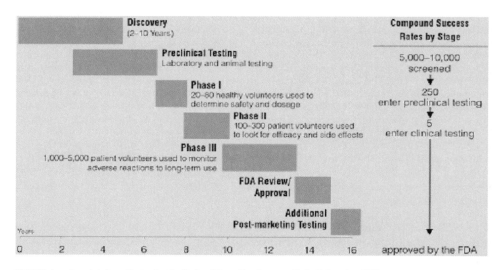

PhRMA, based on data from Center for the Study of Drug Development, Tufts University, 1995

13Apr 2000   WRAIR

Fig. 1. Compound success rate by stages.

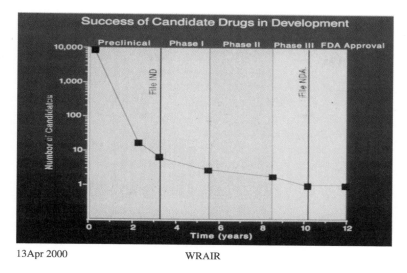

13Apr 2000   WRAIR

Fig. 2. Risk involved in drug development.

We believe that the potential to reduce the risk, time and expense of the 'classic' development paradigm rests with evaluating these traditional medicines.

Natural products themselves are known for their wide diversity, of both chemical structure and biological functionality. This makes them attractive, at least as lead compounds, for new drug development. Many technological advances have occurred within the last few years in chemistry, molecular biology and data processing, which now allow the isolation and identification of complex chemical structures from natural products. The process of evaluating natural products from the traditional medicines of the developing world is facilitated by the development of reliable communications. We, as scientists, must work very closely with the practitioners who are the source of most of the ethnomedical knowledge and natural products from the developing world. Success depends on our ability to communicate readily, build extensive data bases, and exchange these data with these collaborators.

By evaluating traditional medicines as potential new drugs, we are actually reversing the 'classic' drug development paradigm that proceeds from test tubes through animals and eventually to humans. Instead, we introduce a new drug-development paradigm whose process starts with human data. We select those traditional medicines that are successful phytomedicines, and this means that overall, we are beginning with a compound that is less risky in terms of its human-safety profile and therapeutic activity. Since we are beginning with a compound already being used in humans, we expect to decrease the high risk of failure and some of the costs that are inherent in the classical drug-development paradigm (see Figure 3).

Natural products have proven to be important as potential drugs. Consider past success in the pharmaceutical industry where over 50% of the best selling pharmaceuticals in use today are derived from natural products. It is incredible to

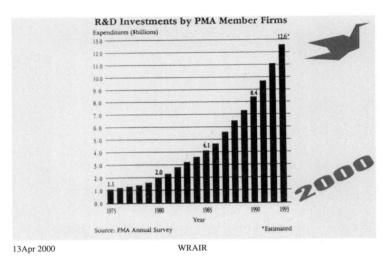

13Apr 2000   WRAIR

Fig. 3. Costs involved in R&D investments.

note that less than 10% of the estimated 300,000 flowering plant species have been examined scientifically for their potential as candidate drugs.

During the evolution of the natural products drug-development program at WRAIR, we began looking for ways to integrate our search for natural products with preservation of the biodiversity that these products represent. Precious biodiversity was being threatened by exploitation of forest products. In the late 1980s and early 1990s, articles published about the threat of extinction of species of higher plants estimated that 60,000 species of higher plants, that is one in four, would likely become extinct by 2050. We asked ourselves if our bioprospecting might become a tool for biodiversity conservation. We believed that drug discovery from natural products could provide an economic incentive for the required conservation of the world's biodiversity. In June 1992, the US National Institutes of Health (NIH), National Science Foundation (NSF), and USAID published a request for proposal (RFP) soliciting applicants who could put together programs to address the inter-dependent issues of biodiversity conservation, sustained economic development, and improved human health. It was hoped that these international programs would: improve health through the discovery of new therapeutic agents; conserve biodiversity by adding value to diverse biological organisms and developing local capacity to manage them; and promote sustainable economic activity in less developed countries by sharing the benefits of the drug discovery and conservation research processes. These three major goals of conservation, economic development and drug development would come together in organizations called International Cooperative Biodiversity Groups (ICBGs). Since the Walter Reed Army Institute of Research had already recognized the importance of natural products and their potential as a source of new drugs, we had already given thought to the potential loss of this biodiversity. Putting together a group from Nigeria and Cameroon, we formed the West and Central Africa ICBG. We have as our objectives: to establish and maintain an inventory of species that are used in traditional medicine to:

- identify lead compounds for the treatment of human diseases;
- establish and maintain study plots for long-term assessment of rain-forest ecological dynamics;
- conduct economic value assessment of major species in our host countries; and
- train scientists and technicians from participating countries in various aspects of drug development, plant research and biodiversity conservation.

Our African ICBG has a number of collaborating institutions. In the United States, the WRAIR collaborates with Pace University, the University of Pittsburgh, the Southern Research Institute, the Smithsonian Institution (the Tropical Research Institute and the MAB Program). Overseas, we collaborate with the University of Ibadan in Nigeria, the University of Dschang in Cameroon, the Center for Medicinal and Aromatic Plants at the University of Nigeria, the Biodiversity Support Program c/o the World Wildlife Federation (WWF), and have as our major partner the Bioresources Development and Conservation Program. Our ICBG activities include:

- ethnobiological surveys;
- plant collections;
- phytochemistry;
- isolation and structure elucidation;
- conservation;
- species catalogs;
- community extractive reserves;
- biodiversity plots;
- forest dynamic plots;
- economic value assessment studies; and
- antiparasitic drug screening (e.g. malaria, leishmaniasis, trypanosomiasis).

We also conduct screening programs against fungal diseases, viral diseases, and opportunistic infections. We have training programs in ethnobiology, ecological evaluation techniques, taxonomy, phytochemistry, conduct of bioassays and parasitological lab techniques. Thus, the WRAIR West and Central Africa ICBG is a large complex organization consisting of fifteen partner institutions in three countries organized into six associate programs.[2] This organization is needed to accomplish our major objectives of developing new drugs against parasitic and tropical diseases and using the drug development process to add value to the biodiversity in Nigeria and Cameroon and using the processes of drug-development and biodiversity conservation in a way that will result in sustainable economic development.

West and Central Africa ICBG objectives:

1. Establish and maintain inventory of species used in traditional medicine.
2. Identify lead compounds for the treatment of human diseases.
3. Establish and maintain study plots for long-term assessment of rain-forest ecological dynamics.
4. Conduct economic value assessment of major species in the host country and study area.
5. Train scientists and technicians from participating countries in various aspects of drug development, plant research and biodiversity conservation.

Several collection methods are used to search for potential new drugs from natural products. The methods that are most often used are random collections, chemotaxonomic collections, and ethnomedical collections. We at WRAIR decided to emphasize the ethnomedical method in our natural product drug-discovery program. This is a selection method based on traditional medical use and requires a multidisciplinary team and detailed data collection. We feel that this approach based on an approach pioneered by Shaman Pharmaceuticals has the potential to expedite our drug-development process.

Ethnomedical approach:

- Multidisciplinary approach
  - Ethnobotanist, physician, ethnopharmacologist, natural product chemist, anthropologist, traditional healer.
- Requires detailed information
  - The therapeutic value of the medicinal plant is rated by the evaluation of the ethnomedical information.

Ethnomedical approach:

- A Field Research Program
  - Healer
  - Disease prevalence interviews
  - Ethnomedical interviews
  - Direct patient evaluation and case discussions with local healers.

The ethnomedical approach requires an ethnobotanist, a modern physician, an ethnopharmacologist, a natural product chemist, an anthropologist, and a traditional healer. It requires the collection of detailed information. The therapeutic value of a medicinal plant is gauged by the evaluation of specific ethnomedical and botanical information. It becomes obvious that the ethnomedical approach is a field research approach, i.e. it must be conducted on site where the healer operates. It depends on healer interviews, disease prevalence interviews, ethnomedical interviews, observed patient responses, and case discussions with the local practitioners. It is this information, which allows us to evaluate the natural product and its effect on a selected disease. The botanical data include: all the names of the plant; voucher numbers and specimen locations; the collection site coordinates, latitude, longitude, and altitude; the description of the habitat that is associated with the collected species; the type of terrain and soil; and physical descriptions of the life cycle of the plant. The ethnomedical data include the actual sources of the ethnobotanical data, i.e. the interviewer and the interviewee, as well as: the number of plants in the medicine; how and when it is collected; the plant part/s used in the preparation of the medicine; a description of the disease (i.e. signs and symptoms); patient demographics; the therapeutic activity of the medicine; the preparation of the medicine; any additives that might be used; the dosing (amount, route of administration, duration); and the side-effects. We believe that our ethnomedical approach results in a deeper and more comprehensive understanding of the local diseases and the phytomedicines that are used to treat them. Our ethnomedical database has all the information that is required to prioritize and select potential lead natural products for screening and also makes the recollections, if necessary, easy to accomplish.

The value of the EM approach lies in a deeper and more comprehensive understanding of:

- local disease
- botanical medicines used to treat them
- plant availability
- local cultivation and harvesting
- medicine extraction and preparation
- dose size, interval, and duration of treatment
- onset of action and side-effects.

There are limitations to an ethnomedical approach. An obvious limitation occurs when natural products are sought for diseases that are not familiar to the traditional healer, either because they are primarily internal, such as cancer, or because they do not occur in the geographic area where the traditional healer practises. Due to the

Table 1

Botanical collection form	Ethnomedical data form
All names of plant, voucher numbers and specimen locations	Sources of ethnobotanical data, interviewer
Collection site coordinates (Lat, Long, Alt)	Number of plants in the medicine, how and when collected, plant part
Habitat-assoc. species; terrain; soil	Disease description: S & Sxs; pt. demographics; therapeutic activity of medicine
Physical description and life-cycle of the plant	Disease de preparation of medicine additives, dosing route and duration and side-effects

intensive process of on site data collection and heavy reliance on data review, the ethonomedical approach can be slow to generate high numbers of lead compounds. It is also important to have specialized screening systems that can properly corroborate disease efficacy. Crude natural product medicines or phytomedicines pose many technical difficulties for most screening systems that are designed for pure chemicals. The possibility always exists that there is a placebo effect inherent in the administration of a traditional medicine, which could lead to a false conclusion about activity.

The limitations are as follows:

- diseases that are familiar to the traditional healer
- slow to generate high numbers of compounds
- need good screens to corroborate disease efficacy
- beware the placebo effect.

If we look at the leads from our ICBG program, i.e. leads that are ethnomedical in origin, we can see that the WRAIR program has been very successful in identifying active natural products. For example, in malaria, 70% of 500 samples we have tested from traditional medicine leads had activity in an in vitro malaria screen. We are further evaluating 23 lead extracts. Forty per cent of another 130 samples tested positive for antileishmanial activity and have resulted in six lead compounds. Sixteen of 20 samples tested for cytotoxicity, i.e. a cancer screen located at the University of Utah, have provided five lead compounds. We are also having comparable success with screens against several viruses, trypanosomiasis, trichomonas, cryptosporidium, and toxoplasmosis.

The WRAIR ICBG program has selected 18 plants for secondary malaria studies. These plants will undergo further bioassay-guided fractionation. Already, 23 compounds with 12 chemotypes have been isolated and characterized as having very potent antimalarial activity. Two plant families, Annonaceae and Apocynaceae, have been noted because they appear to be common ingredients in the preparation of traditional malaria remedies in West and Central Africa. WRAIR/ICBG is especially confident that the new chemotypes such as picralima alkaloids and related betacaboline indoles, cryptolepine analogs and enantin and related isoquinolines, labdane diterpines, and morindone and other anthraquinones will provide lead compounds or templates for semi-synthesis of new antimalarials. The WRAIR malarial program will now begin an extensive program of lead optimization, bulk synthesis for in vivo testing, sample generation for high-throughput screens, and molecular probes for new biochemical targets, using these newly discovered natural products.

WRAIR ICBG – malaria:

- 500 samples (extracts) from plants used in traditional medicine for the treatment of malaria screened in vitro against *P. falciparum*.
- 70% showed activity.
- 18 plants selected for further studies (bioassay-guided fractionation).

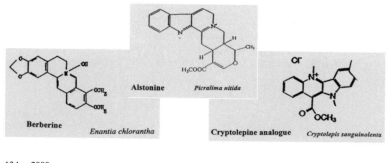

Fig. 4. New malaria chemotypes.

- 23 compounds (12 chemotypes) isolated and characterized with very potent antimalarial activity.
- Two plant families, Annonaceae and Apocynaceae, appear to be common ingredients in the preparation of traditional malaria remedies in West and Central Africa.

The WRAIR West and Central African ICBG program has added many dimensions to the drug-discovery program at our institute. In addition to the discovery of important new chemotypes (Figure 4), WRAIR ICBG have seen a gratifying infrastructure growth within our collaborating partner countries of Nigeria and Cameroon.

References

1. Hamner CE. (1982) Drug development. Boca Raton, FL: CRC Press.
2. Schuster BG, Jackson JE, Obijiofor CN, Okunji CO, Milhous W, Losos E, Ayafor JF, Iwu MM. (1999) Drug development and conservation of biodiversity in West and Central Africa: a model for collaboration with indigenous people. Pharmaceut Bio 37(Suppl):1–16.

Iwu and Wootton (eds.), Ethnomedicine and Drug Discovery

Health foods in anti-aging therapy: reducers of physiological decline and degenerative diseases

LISA MESEROLE

Abstract

Dietary plants based on whole foods with minimal processing offer enormous advantages over modern processed foods, as do many non-standardized herbs, tonics and teas over their fractionated modern progeny. Some of the primary nutrients and constituents of plant foods, spices and beverages undergo biochemical transformation immediately or shortly after harvest as well as from processing techniques and shelf life deterioration. Dramatic biochemical differences exist between 'locally produced and consumed foods' and their industrially mass produced counterparts, transported great distances and stored prior to consumption; thus impacting the quality and quantity of what is delivered to target markets and the consumer. Health foods as medicinal plant preparations may display the greatest health benefits when (1) minor constituents may be lost because incomplete understanding forces erroneous conclusions about the identity of the active and inert ingredients, (2) inert and unidentified compounds mitigate unfavorable side-effects or enhance therapeutic actions, (3) active compounds become unstable or less potent when isolated or manipulated in relation to their environment, and (4) unknown pharmacokinetics are altered by increasing active compound ratios. Standardized plant products are preferable when analytical, pharmacological, biological, etc data are adequate or complete and standardized products that are truly superior in efficacy and safety can be successfully produced.

Keywords: *whole foods, non standardized, nutrients, plant constituents, food plants, antioxidants, biological response modifiers, adaptogens, heterogeneous herbs, dietary quality*

I. Introduction

Traditional diets were based on whole, minimally processed foods and beverages. Modern diets, in many cases, have lost the many naturally occurring vitamins, minerals, antioxidants, flavonoids, tannins, biosaponins, volatile oils, retinoids and other constituent groups now known to offer wide ranging benefits that reduce the effects of cellular and physiological aging. These trace compounds exert their effects in very subtle but significant ways, and their absence from regular diets often leads to many degenerative diseases. As has been noted by Iwu[1] and others,[3] the presence of beneficial nutrients and active constituents in foods in the diet may provide the ultimate prophylactic approach in disease risk reduction and 'longevity medicine' based on newly understood biochemical mechanisms underlying optimal health and diet. In addition to concern about the loss of beneficial food constituents is the contemporary problem of the introduction of harmful new chemicals into the food chain. Emerging evidence for some chemicals' deleterious effects on fauna, flora and humans in aquatic and land food supplies is a public health concern and an important topic for sustainable food systems, agriculture and food security. The unpredictable consequences of such sophisticated manipulation (intentional and unintentional) of the food supply is born out by health care professionals' observations of their patients and traceable links between physiological chemistry alterations, biological exposure and tissue load of these new chemicals, and disease incidence.[3,4]

II. Traditional vs. modern diets

Dietary staples such as rice, yam, potato, maize, wheat, cassava were once the foundations of traditional diets; delivering fibers, lectins, trace minerals, amino acids, tannins, polysaccharides, phenolics, etc. to the ancient peoples who consumed them. In the past era, most 'food processing' was done shortly before the meal – at the hearth – or at least within a short distance of the household. Once, vast varieties of many staple crops (30,000 types of rice, over 100 varieties of yam, over 4000 varieties of potato) assured not only whole food nutrition but also a diversity of compounds among the varieties consumed. Advances in agriculture and food processing have supplied homemakers with convenience and hygienic food, but only a few hybrid crop types (there are now 10 types of rice grown commercially) that presently dominate the global cuisine. Milling of rice and corn has improved shelf life but causes the vitamin B deficiency diseases of pellagra and beriberi. Nutritional deficiency diseases are an age-old problem, but their presence in the face of abundant food and clean water is novel and was unknown in earlier centuries. The volatile oils and bitter principles present in herbs and spices (such as pepper, kola, mint, lemongrass, basil, ginger, turmeric, burdock root, horseradish, wasaabe, dandelion) act (among other things) as digestive stimulants and antimicrobials in the digestive tract. These are often removed or spontaneously degenerate in highly processed foods (air, light, heat, and other processing reduces their content).

The result of these advances and trends is that populations are consuming foods with reduced levels of what the same foods once contributed (antioxidants, flavonoids, etc.); and people may stand to benefit greatly by the restoration of some of these compounds (via nutraceuticals, special 'health foods', medicinal plant supplementation, etc.). Although standardizing a supplement or herb for content of key active compounds is desirable when the unique activity of those compounds in vivo is known to function independently of cofactors and coexisting inert ingredients, it may be undesirable when the full mechanism and science of biological activity remain partially unidentified.

Unstudied or incomplete analysis and understanding of mechanism of action are the rule rather than the exception for foods and medicinal plants. This is partially due to the simple fact that exhaustive constituent analysis is exorbitant in cost and therefore impractical. No foodstuff or medicinal plant has ever been fully analyzed for all constituents nor for full ranges of biological activities in vivo. This creates a dilemma where there is a lack of complete data, and this data deficit is likely to persist in the future and on many fronts. Nevertheless, voluminous and emerging data that are being generated through research and market development seem useful in validating and refining many of the traditional uses of plant foods, and also in identifying new potential applications for them.

Non-standardized foods and medical plant preparations may display the greatest health benefits when (1) minor constituents may be lost because incomplete understanding forces erroneous conclusions about the identity of the active and inert ingredients; (2) inert and unidentified compounds mitigate unfavorable side-effects or enhance therapeutic actions; (3) active compounds become unstable or less potent when isolated or manipulated in relation to their environment; (4) unknown pharmacokinetics are altered by increasing active compound ratios. Standardized products are preferable when analytical, pharmacological, biological, etc. data are adequate or complete, and standardized products that are truly superior in efficacy and safety can be successfully produced.

Although, a few centuries ago, Captain Cook observed limes' (and other foods') ability to prevent and treat scurvy, the vitamin C in limes was considered the curative agent until 1938, when Szebtgyorgyi reported that citrus peel pectin prevented scurvy-induced capillary bleeding and fragility. Limes display three features that maximize their antiscorbutic effects: their content of vitamin C and of specific flavonoids, and their acidic pH, which stabilizes C once exposed to air and light.

Unrefined plants contain many components recently being identified as biologically active and ubiquitous rather than unique in the plant world. Examples include the flavonoids (kampferol and quercetin), retinoids/phenylpropanoids, polysaccharides and sugars that act as biological response modifiers and immunomodulators, non-essential amino acids and small proteins, organic acids, and biosaponins – as well as vitamins, minerals and trace elements.[4]

Some of the rare nutrients that survive the harsh agricultural treatment and storage conditions are sometimes lost during food processing. Eating healthy food was extremely important to people of an earlier era and constituted a major aspect of healthy living. It has been observed that in order to avoid resorting to unpleasant therapies such as purging or bloodletting, doctors carefully monitored their wealthy patients' daily habits, including the composition of their diet to ensure that food substances with the required nutrients were consumed.[5]

III. Health foods and adaptogens

Taken as 'health foods', dietary supplements, and used as topical creams and washes, medicinal plants in a relatively crude form have been part of self-prescribed and prescribed treatment in every culture for millennia. Many of these non-standardized plant preparations protect the mucous membranes of the upper and lower respiratory tract, the reproductive tract, and the digestive tract. (Examples include

Hydrastis canadensis, Berberis vulgaris, Plantago major, Ulmus fulva, Alchemilla millefolium, Althea officinale.) Others (such as *Crataegus* sp., *Ginkgo biloba, Centella asiatica*) support cardiovascular function and optimize peripheral and brain blood flow. Many people seek antianxiety and sedative plants (*Valerian officinale, Melissa officinalis, Nepeta cataria, Piper methysticum*) to calm themselves; others seek stimulant (*cola nitida, Coffea arabica*) and adaptogen plants (*Panax ginseng, Eleutherococcus senticosis, Withania somnifera* 'Ashwaganda', *Pfaffia paniculata, Garcinia cola*) to boost work, stamina and well-being.

Topical fruit acids have healing and rejuvenating effects on the skin and hair, hence the 'strawberry or papaya face mask'; certain amino acids, minerals, essential oils, etc. in various plants explain the modern products of the nettle, ginkgo and rosemary hair rinses. Moisturizing oils and fatty acids from coconut, jojoba, shea, almond, and castor bean have long histories of use in skin care and are found in today's cosmetic lotions and creams.

Health foods and adaptogens categories:

- adaptogens
- anti-inflammatories
- bioflavonoids
- immunomodulators
- digestives
- smart drugs
- function enhancers
- beauty enhancers.

IV. Crude plants and their properties

Traditional foods were less refined than is typical today, and even maintaining uniformity from one harvest to the next was less regulated. Now – with irrigation and other agricultural and environmental controls – fruits, vegetables and crops need not shrink in dry years or in poor soils. Yet, even today, there is attention to standardized size and cosmetic appearance of crops, but little to standardized nutrient content or biological activity. One orange may contain less than 30 mg of vitamin C, another over 60 mg.

Flavonoids as examples of some of the challenges for natural product science

Flavonoids are now known to be ubiquitous in many plants. Quercetin, which occurs in many plants, including the onion (*Allium cepa*) skin is used as a hay-fever remedy and a digestive anti-inflammatory, and is the active ingredient in a popular OTC nasal spray for sinusitis and allergies. Classified as phenols, 80% of the known flavonoids occur in ferns and higher plants. Dominant among flavonoids are the flavones, flavonols, anthocyanidines, flavanones and isoflavonoids. Human hormone activity has been demonstrated by soy isoflavones, which can eliminate menopausal hot flushes in some women and reduce the need for hormone replacement therapy dosage in others. The bilberry (*Vacinium myrtillis*) and hawthorn (*Crataegus* species) anthocyanidines are known to support vision, eye health and the cardiovascular system. Flavonoids have been shown by Yin to inhibit thyroid cancer cell lines.[8]

IV.A. Biological activity

The biological activities of flavonoids have been demonstrated as anti-inflammatory, antihepatotoxic, antitumor, antimicrobial, antiviral, enzyme inhibitors, and antioxidant, and as having central vascular effects. Investigations into the structure-function relationships of 54 flavonoids by Lasure revealed a dose-dependent response on the inhibition of complement mediated hemolysis and also demonstrated quercitin and to be among the most potent inhibitors for the classical complement pathway while baicalein and myricetin were among the most potent inhibitors for the alternate complement pathway.[9] Such observations merely argue for retaining a chemically complex rather than reduced therapeutic and preventative arsenal. Dietary flavonoids are also inversely related to cardiovascular disease, according to some studies.

The difficulty in studying flavonoids discussed in Samuelsson's *Natural Products* demonstrates the difficulties in studying many of the natural product constituents in plants:

1. There are unknown effects of flavonoids' structural changes in the gastrointestinal tract.
2. Many other cofactors that may modify flavonoids' actions and pharmacology exist in flavonoid-high foods.
3. The main effect of many flavonoids is more preventative than therapeutic; this is much harder to study than purely therapeutic effects.
4. Flavonoids have a very broad spectrum of action and seem to have few or no unique effects (solely attributable to flavonoids themselves).
5. Human metabolism and pharmacokinetics of flavonoids are still poorly understood.

Examples of specific flavonoids are rutin, kavalactone from kava (*Piper methysticum*), and silybin of the silimarin complex in milk thistle (*Cardus marianum*).

Physiological actions of minor food nutrients and examples (Table 1).

V. Traditional tonics

Traditional tonics and adaptogens have been used for thousands of years. These include *Panax ginseng* (the longevity and a male virility herb, it also 'moves and warms stagnant blood'), *Eleutherococcus senticosis* (Siberian ginseng) extensively studied and shown to reduce the negative effects of physical, emotional or

Table 1

Physiological action	Example
Anti-inflammatory Digestives	flavonoids, lignans tannins, bitters, alkaloids, essential oils
CNS stimulants/sedators	alkaloids, flavonoids (kavalactone)
Anti-infectious	flavonoids, alkaloids, essential oils
Anti-autoimmune	flavonoids
Anti-allergic	flavonoids, tannins
Anti-ischemic	flavonoids
Anti-arrhythmic	flavonoids, alkaloids, glycosides

environmental stress, *Withania somnifera* – Ashwaganda – studied in India to minimize negative effects of chemotherapy and enhance vitality; *Ginkgo biloba* a mental-function enhancer; *Cordyceps sinensis*, the Chinese royalty longevity herb and 'Chinese viagra'; and *Garcinia kola*, bitter cola used by young and old to increase energy and disease resistance.

VI. Anti-aging medicine

Anti-aging is a contemporary marketing term that refers to preventative and longevity medicine for body, mind and spirit. New research is confirming and discarding some of the age old techniques. Empirical and research support exists for the following mechanisms of food and medicinal plans' effects on human physiology

- detoxification (of heavy metals, drug or hormone overload, etc.)
- biochemical support for aging organ systems
- capacity to balance catabolic with anabolic processes
- support aging organ systems
- balance catabolic with anabolic processes
- hormone support
- circulatory support
- CNS support
- cell membrane protection.

A generally accepted theory is that a significant reduction of the aging process, at least at the cellular level, can be achieved by increasing the intake of antioxidants. Several investigations have shown that cellular damage can be minimized by using antioxidants to reduce the damaging effects of free radicals.[7] In atherosclerosis, for example, the disease is believed to be the outcome of a cascade of biological reactions involving the oxidative modification of low-density lipoprotein.[8] Active oxygen also plays major roles in other diseases, including cancer, liver damage and heart disease.[8] Foods rich in anti-oxidants include fruits (cherries, blueberries, kiwis, pink grapefruits, raisins, prunes, raspberries, strawberries, plums, red grapes and oranges), vegetables (onions, spinach, red bell peppers, alfalfa, sprouts, beets, brussels sprouts, corn, kale, egg plant, broccoli flowers) and several flavonoid- or anthocyanadin-containing food substances. Some plants may serve as antidotes or modifier to toxin absorption or effects. Flavonoids (silimarin complex) from *Cardus marianum* has been shown to reduce death rates from *Amanita* mushroom poisoning via hepatocellular protection. Adequate dietary calcium intake is known to reduce environmental lead and cadmium absorption.

VII. Conclusion

There has been a drop in the nutrient and health value of foods and medicinal plants in modern times – just when citizens of the world are becoming increasingly afflicted with diseases of today: stress, anxiety, depression, fatigue, hormone and fertility imbalances, inflammatory and immune diseases, environmental toxin exposures, and coronary vascular disease. Emerging science that elucidates the compound analysis and biological activities is growing. However, little is known about the physiological, pharmacological, pharmacokinetics efficacy and safety of many medicinal foods, tonics and herbs. Even the science of stabilizing antioxidants and other unstable compounds when they are removed from their source is in its infancy. Many traditionally used foods, tonics and traditional plants have been used safely and effectively despite a lack of presence of or identification of standardized constituents or marker compounds. The goal of 'longevity' or 'anti-aging' medicine is to enhance the quality of life through health and well being, including cognitive function and physical appearance (weight/height proportion, glowing skin and hair, clear eyes, etc.). Anti-aging medicine is now a medical subspecialty and annual congresses include protocols for cardiovascular, brain and immune protection and optimization during aging.

Traditional tonics and folk medicines have been used throughout centuries and many are safely and effectively used today. Epidemiological studies, health surveillance data, research in molecular biology as it relates to human physiology, environmental toxicology and pharmacology have demonstrated some of the cause and effect relationships between diet composition, pollution, and morbidity and mortality statistics. There is growing evidence for the health benefits of fresh, whole foods and medicinal plants vs. those that have been hybridized, processed or stored for long periods and therefore have less complex or altered chemical and nutrient profiles. Public Health and individual quality of life and longevity depend upon the gene pool biodiversity of the food supply and medicinal flora. The 'health foods' market is ever popular and anecdotal reports are now sometimes substantiated by discoveries in medical physiology and pharmacology.

Since modern clinical data will remain incomplete regarding most herbs and food plants for the foreseeable future, relying on traditional usage and dosing is especially important. In the best of worlds, traditional use teamed with modern medicine and research can be blended to serve populations and the individual with enhancing his/her quality of life. Non standardized and whole plant foods and some of their newly recognized constituents left in sito have a critical role to contribute as biological response modifiers that enhance vitality and function during aging.

References

1. Iwu MM. (1986) Empirical investigation of dietary plants used in Igbo ethnomedicine. In: Etkin N, editor. Plants in indigenous medicine and diet – biobehavioral approaches. Bedford Hills: Redgrave, pp. 131–150.
2. Berg PR (1988) 'Effects of Flavonoid compounds on the immune response'. In Biology and Medicine II: Biochemical, Cellular and Medicinal Properties, pp. 147–171.
3. Winter CK, Seiber JN, Nuckto CF. (1990) Chemicals in the human food chain. New York: Van Nostrand Reinhold, p. 275.
4. Samuelsson G. (1999) Drugs of natural origin: a textbook of pharmacognosy, 4th edition. Swedish Pharmaceutical Society, Swedish Pharmaceutical Press, Stockholm, Sweden, pp. 226–230.
5. Goldborn T, Dumanoski D, Myers JP. Our stolen future. London: Abacus, p. 249.
6. Laudon R. (2000) Birth of the modern diet. Scientific American, August 76–81.
7. Beckman KB, Ames BN. (1998) The free radical theory of aging matures. Physiol Rev 78:547–581.
8. Yin C. (1999) Growth inhibitory effects of Flavonoids in human thyroid cancer cell line. Thyroid, pp. 369–376.
9. Pieters L. (1999) Low molecular weight compounds with complement activity. In Wagner H, editor, Immunomodulatory Agents from Plants, Progress in Inflammation Research, Birkhauser Verlag, Switzerland, pp. 142–143.

Iwu and Wootton (eds.), Ethnomedicine and Drug Discovery
© 2002 Elsevier Science B.V. All rights reserved.

CHAPTER 16

Indigenous peoples and local communities embodying traditional lifestyles: definitions under Article 8(j) of the Convention on Biological Diversity

KATY MORAN

Abstract

The UN International Convention on Biological Diversity (CBD) establishes the sovereignty of nations over their bioresources, enabling signatory nations to proceed with access and benefit-sharing processes. Who should benefit, and how, at the local level, however, is a confusing and contentious issue, due partly to the ambiguous CBD language lumping 'indigenous and local communities embodying traditional lifestyles'. This paper attempts to separate them into discrete groups of people, processes and products, comparing their differences. It is hoped that clearer definitions under the CBD will better enable access and benefit-sharing mechanisms to proceed.

Keywords: *access and benefit-sharing, agriculture, bioproducts, Convention on Biodiversity, indigenous peoples, pharmaceuticals, traditional knowledge*

I. Introduction

Sovereignty over bioresources by nation states is perhaps the critical component of the Convention on Biological Diversity (CBD) for biodiversity-rich countries. No longer are biotic resources considered 'the common heritage of mankind', the pre-CBD paradigm that provided open access to bioresources. Today, the CBD enshrines the goals of (1) the conservation of biodiversity, (2) the sustainable use of its components and (3) the equitable sharing of benefits that arise from its commercialized products. The 'grand bargain' of the CBD is to balance how all interest groups involved can gain from the use of biodiversity by recognizing the economic, social and environmental values of bioresources and the costs of preserving them. Since the CBD was introduced nine years ago, some of the 178 signatory nations have introduced legislation requiring benefit-sharing for access to their national bioresources by outsiders with commercial interests.[1]

What has not yet been adequately addressed by the CBD, however, is how the on-the-ground stewards of biodiversity, 'indigenous and local communities embodying traditional lifestyles', can share in product benefits. If they contribute their 'knowledge, innovations and practises' to the use of bioresources in such sectors as healthcare and agriculture, they also should receive benefits from the new medicines and improved crops they help to produce. One reason why this critical issue remains unresolved is that the wording of the CBD is ambiguous. It lumps indigenous peoples and local communities together throughout the document, without separating them into the discrete groups, which they are. Benefits from agricultural and medicinal products are, consequently, also bound together, which has polarized both positions on how to share them. The lack of clear separation of these people and products allows those with vested interests to exploit the very concerns that the CBD was created to resolve.

What are these differences? According to most anthropologists, the best single indicator of distinct cultures is spoken language, because it explicitly encapsulates the unique view of the universe held by the group that speaks the language.[2] Of the 6000 languages spoken today, close to 5000 are languages of indigenous peoples. Some 300 million indigenous peoples in about 6000 groups are spread around the world in more than 70 countries.[3] Characteristics that identify indigenous peoples, though elusive, have been defined in many fora.[4] In the New World, the term 'indigenous' is appropriate, due to its colonial history and indigenous societies' relative isolation from the national culture today. But in parts of Asia and Africa, terms such as tribal, native, aboriginal or ethnic minority are better used.

Indigenous use of plants as medicines spans time, from ancient to modern.[5-7] It is estimated that 74% of plant-based prescription drugs on the market today have the same or related use in Western medicine as originally practised by native healers.[8] A few examples are the anti-malarial quinine found in *Cinchona*, the tranquilizer reserpine from the East Indian snakeroot, cardiac glycosides from *Digitalis*, and the analgesics codeine and morphine. Schultes[9] stated in 1988:

> The accomplishments of aboriginal people in learning plant properties must be a result of a long and intimate association with, and utter dependence on, their ambient vegetation. This native knowledge warrants careful and critical attention on the part of modern scientific methods. If phytochemists must randomly investigate the constituents of biological effects of 80,000 species of Amazon plants, the task may never be finished. Concentrating first on those species that people have lived and experimented with for millennia offers a short-cut to the discovery of new medically or industrially useful compounds.

This knowledge is embedded in indigenous peoples' cultural systems and intimately connected to their religious beliefs and cosmology.[10] Often, a group's biological knowledge of their natural environment reaches a high level of abstraction, far from biological facts, and symbolically represented in their myths, rituals and religion. Accumulated over millennia and passed down generationally within communities, traditional knowledge of medicinal plants is as rich and diverse as biotic resources, and just as threatened with extinction.[11]

The people the CBD ambiguously describes as 'local communities embodying traditional lifestyles', are much more difficult, if not impossible, to identify discretely, let alone enumerate, world-wide. Farmers, for example, can be described as such, and are included under agricultural biodiversity sections of the CBD. But farmers are lumped with indigenous peoples throughout the document, without respecting both groups' differences. The folk varieties grown by small farmers are critical to world agriculture. As opposed to large-scale monocultures, the genetic diversity found in small farmers' fields may harbor traits that can adapt more readily to changing climates, more marginal growing conditions and evolving plant pests.[12]

Using 'traditional knowledge, innovations and practises', both groups have made valuable contributions to the discovery of new seeds and new medicines, and both deserve benefits proportional to their input. Their agendas overlap with respect to compensation, so both have a stake in implementation of the CBD. But first, serious consideration must be given to each group's unique circumstances. For the people who are the innovators, the processes by which they innovate and the new products, seeds and drugs, produced through their contributions, are quite distinct. Differences must be examined and definitions refined. Only then can benefit-sharing mechanisms for both farmers and indigenous peoples be accomplished.[13,14]

II. Methods

Although there are many bioproducts, the focus here is confined to new medicines and improved seeds. For the purposes of this paper, differences, though by degree, between indigenous peoples and farmers are listed under three tables that list components of bringing bioresources to commercialization. It is recognized, however, that some of the following categories are not mutually exclusive. An indigenous person, for example, can also be a farmer, but not all farmers are indigenous. Likewise, distinguishing between food and medicines is often blurred among both farmers and indigenous peoples.[15] For benefit-sharing purposes under the CBD, the following three tables will attempt to compare these differences by separating them into people (Table 1), processes (Table 2) and products (Table 3).

There are distinct differences between indigenous peoples and local communities of farmers embodying traditional lifestyles (Table 1). Generally, indigenous social organization is described as tribal, referring in this paper to specific systems of governance and to collective resource management. Farmers, however, are most often organized into peasant societies, constituting innumerable local communities that practise traditional lifestyles. Most peasant societies farm individual plots and practise some sort of market agriculture.

Indigenous peoples can be defined as culturally and linguistically isolated groups, which generally have minority status within modern nation states. They primarily seek collective rights and self-determination for use of their biological and cultural resources. The 's' on the end of peoples is important for their purposes. Under the

Table 1
The people

Farmers	Indigenous peoples
Peasant social organization	Tribal-like social organization
Land tenure and usufruct right sought	Territorial rights, self-determination sought
Identify with nation state	Peoples, distinct from nation state
National language and religion practised	Specific language and religion practised
Manipulation of germplasm used for improved crops	Traditional generational knowledge used for new medicines

Table 2
The process

Improved crops	New medicines
Use of biological resources	Use of cultural resources
Product input is secular	Product input is often sacred
Germplasm creates new varieties	Natural compounds valued for medicines
Change by human manipulation	Use of natural compounds
Putting genetic material into plants	Taking biological material out of plants

Table 3
The product

Seeds	Medicines
The customer is not the ultimate end-user	The customer is the consumer
Annual market size – $30 billion	Annual market size – $190 billion

United Nations' definitions, only 'peoples' have the right to self-determination, sovereignty and territorial rights.[4] Most farmers, however, identify with the nation states in which they live. They typically speak the national language and practise the religion of the majority. Farmers primarily want individual economic rights such as land tenure and usufruct rights. Both groups under Table 1 seek recognition and benefits from their input into improved crops and new medicines (refer to Table 2).

As seen in Table 2, farmers choose plant characteristics from the diversity of germplasm brought about by biological evolution to create improved crops. But cultural knowledge, passed down through generations of indigenous healers, is the primary source of traditional medicines. To transform a plant into a medicine, the healer must know the correct species or varieties, their location, the proper time of collection – both seasonal and circadian, which solvent to use and how to prepare it, the route of administration and the dosage. This biological plant knowledge is inseparable from the healer's knowledge of the spiritual properties of plants in the healing process. The preparation of medicines is steeped in religious ritual, mediated by the healer, reflecting indigenous beliefs in the spiritual force of all living things.

Farmers' manipulation of germplasm for improved crops is typically individual, secular and not so grounded within a cultural context. Traditional knowledge of the use of plants for medicines is, however, generational, often sacred and always deeply imbedded in indigenous culture.

For farmers, domesticated germplasm and its naturally occurring relatives create varieties valuable to agriculture. Farmers select and replant varieties among their harvest that are, over time, likely to improve yields. This change for improved crops is brought about by human manipulation, whereas the chemicals within wild plants are valuable by themselves in the use of natural compounds for medicines. Basically, the processes are quite different when developing new products. To improve crops, useful genes are transferred or put into plants to create new varieties. The discovery of new medicines, however, particularly pharmaceuticals, comes about by a series of steps to take useful chemical compounds out of a medicinal plant. These steps include extraction, isolation, characterization, improvement and efficacy testing.

As can be seen in Table 3, both farmers' and indigenous healers' products have the potential for high technological added value. Indigenous knowledge of the use of plants for medicines can lead researchers to bioactive compounds that may be developed into new pharmaceuticals and phytomedicines. But further technological investments in expensive, time-consuming research and development programs are needed. The final product is a pill, ointment or other type of therapeutic used at the patient's convenience. It is purchased by the consumer as a preventive medicine or as the immediate solution to a health problem.

The seed purchaser must contend with external conditions such as weather, pests, and diseases for future value. Other than highly heritable disease resistance, the usefulness of germplasm in improved seeds is difficult to assess fully except over a multi-year time frame. Perhaps the greatest non-monetary benefit for agricultural biodiversity is the reciprocal access to informal systems of germplasm exchange, which often includes use in breeding programs. Databases on plant medicines, both public and private, also exist, but only non-proprietary information is exchanged.[16]

An enormous difference exists in the annual size of both markets world-wide. The market size of the commercial seed industry is about $30 billion dollars annually, but the final value of agricultural produce reaching the consumer can be as high as $450 billion, some 15 times the sales of seeds to farmers. During this process, germplasm passes through many hands, being improved at each stage by different actors. Assessing the value added at each stage is difficult. The increased value, created by innovation, is distributed to those who enhance the value of the product during the lengthy chain of access. Most of the value goes to the technology provider, with somewhat less going to the farmer. The least value, to date, goes to the consumer.

The combined market for pharmaceuticals derived from natural products and botanical medicines is estimated at $190 billion, but development costs are equally staggering and high-risk ventures. Pharmaceutical research and development require expensive and time-consuming studies to secure government regulatory approval before any drug can be marketed. In the US, a pharmaceutical product typically takes from 10 to 15 years to materialize, after an investment of over $300 million by the company and investors who take the financial risk to develop, test and market a new drug. Costs for development of botanical medicinal products are much lower, but so are profits. For information on traditional use to guide the discovery of new medicines, benefits are paid through contractual agreements by many companies at different stages of the drug-discovery process.[17] Advance payments can be fees for samples or a lump payment to designated recipients such as individual healers, healers' associations, research institutions and/or culture groups.[18] Milestone payments are paid during research and development, and royalties, or a percentage of product profits, are paid after a product is commercialized.[19]

The above market differences mean that farmers have a much smaller benefit-sharing pie to be divided compared to indigenous peoples, in a more complex manner and among a much larger number of people.

III. Discussion

It must be stated again that the above differences are by degree and that there are also similarities between the systems of farmers and indigenous healers. Both are traditional, but dynamic rather than static. Both incorporate new ideas into their practices when there is a cultural and biological 'fit' with prevailing systems. During research by commercial interests, non-monetary benefits are similar for both, particularly local employment opportunities for sourcing raw materials. Training for sourcing at the local level, as well as more technical training in research and development for staffs of source country institutions, is common.

Both seed and drug companies provide research grants to public institutions to conduct specific research, the results of which are available to the company. Participation in research also leads to exchanges of non-proprietary technical information such as data on biodiversity management. Technology transfer of field or lab equipment is also part of current contractual agreements.[17]

Monetary benefits are quite different. Benefit-sharing for the use of indigenous knowledge for new medicines, as opposed to compensating farmers for their contribution to improved crops, has progressed far faster. The development and commercialization of new medicines are more dependent on the private sector, as compared with crop breeding, which has primarily relied on public sector institutions. In the last 20 years or so, however, governments have reduced research funding, and the private sector has picked it up.

Case studies where private companies have already initiated benefit-sharing mechanisms for drugs and botanical medicines are well documented, even though no pharmaceutical product using traditional knowledge has been commercialized, nor commercial profits realized, since the CBD was introduced.[20,21] Millions of dollars, as well as non-monetary benefits such as training and technology transfer have already benefited many source country governments and research institutions, healers and their associations and contributing culture groups. Principles such as advance payments and milestone payments for the use of traditional knowledge and bioresources are well reported. Mechanisms such as trust funds run by healers' associations and community representatives who control the distribution of the benefits are established mechanisms. Joint ventures between companies and in-country research institutions are additional components.[17,19,22–26]

Natural compounds for new medicines are valuable by themselves and easily identified. But crop germplasm from one landrace, or genotype, is combined into hybrids with other landraces that often are held in public sector research institutions. It is difficult, if not impossible, to designate the contribution of a single source to the final variety or to gauge its commercial value. Many plants were collected so long ago that their specific origins are no longer known.[27]

In 1987, before genetic resources became defined as national patrimony in the CBD, an International Fund for plant genetic resources was established at the UN Food and Agriculture Organization (FAO). The FAO Conference Resolution 4/89 tied the fund to recognizing farmers, primarily in developing countries, for their efforts over the millennia towards conservation and development of bioresources, a concept embodied in the new term called Farmers' Rights. However, no benefits were designated to return to farm communities, only to genetic conservation. Presumably, industrial countries and seed companies could contribute to the fund, but to date, only minimal contributions have been made.[28]

In 1994, the FAO and the international agricultural research centers of the Consultative Group on International Agricultural Research (CGIAR) signed a memorandum of understanding. The Understanding placed the genetic resources donated by developing countries to CGIAR centers under the auspices of the FAO (Commission on Genetic Resources for Food and Agriculture). Recipients of the genetic resources in these collections are not permitted to establish intellectual property rights over the materials, thus protecting them for continued use and development by farmers without impediments that are raised by someone attempting to prevent them from using their own varieties. In this way, nation states and the FAO claim to be trustees for farmers' intellectual property in folk varieties.[29]

IV. Conclusion

In the case for compensation, indigenous peoples and farming communities are not polarized or even conflicted under the CBD. But the lack of a clear understanding of the differences constructs unnecessary impediments in resolving who gets which benefits for what, particularly among farmers. Just the difference in market size, for example, complicates the process. Transaction costs for compensation may vary significantly, and the difference in annual market sizes requires very different compensation processes for both groups.

It seems in the best interest of indigenous groups for they, themselves, to offer a clear definition of who they are for use under the CBD, rather than allow others to make such distinctions. At the local level, the principle of self-identification is effective, but outside the community, at national and international levels, tighter definitions are required. At national and international levels, using 'indigenous peoples and local communities embodying traditional lifestyles' interchangeably is dangerous. A tighter definition, particularly at international fora, of who are, and who are not, indigenous avoids the concern that they will be immersed under farmers' movements and their specific concerns subordinated to those of the more powerful and numerous farmers' groups.

Finally, the lack of clear definitions allows appropriation of conservationists' language by those with opposing interests, enabling them to manipulate the CBD for their own purposes. In the United States, 'local communities embodying traditional life styles' characterizes logging, mining and other extractive communities. Many of these 'local communities' utilize powerful archetypes such as the sanctity of the community, but sometimes operate out of nothing but self-interest. 'Local communities' are buzzwords for extractive interests with anti federal government rhetoric, whose goals are often directly opposite those of the CBD.

Acknowledgments

I gratefully acknowledge the critical input to this article from Val Giddings, Tom Greaves, Len Hirsch, Steve King and Henry Shands. Data from the excellent compilation of information in Ref. 16 were used extensively. Generous funding from ShamanBotanicals.com, the Jocarno Fund, The Nelson Talbott Foundation and numerous individual contributors to the Healing Forest Conservancy made this publication possible.

References

1. UNEP. (1992) UN Convention on Biological Diversity (CBD) Nairobi: UNEP.
2. Berlin B. (1992) Ethnobiological classification: principles of categorization of plants and animals in traditional societies. Princeton, NJ: Princeton Press.
3. Clay JW. (1993) Looking back to go forward: predicting and preventing human rights violations. In: Miller MS, editor. State of the peoples: a global human rights report on societies in danger. Boston, MA: Beacon Press.
4. UN Working Group on Indigenous Peoples. (1992) Report on the intellectual property rights of indigenous peoples. July 6, 1992. New York: United Nations.
5. Balick M. (1990) Ethnobotany and the identification of therapeutic agents from the rainforest. In: Chadwick DJ, Marsh J, editors. Bioactive compounds from plants. New York: Wiley, pp. 22–39.
6. King SR, Carlson TJ. (1995) Biocultural diversity, biomedicine and ethnobotany: the experience of Shaman Pharmaceuticals. Interciencia 20:134–139.
7. Schultes RE, Raffauf RF. (1990) The healing forest: medicinal and toxic plants of the Northwest Amazonia. Portland, OR: Dioscorides Press.
8. Farnsworth NR. (1988) Screening plants for new medicines. In: Wilson EO, Peters FM, editors. Biodiversity. Washington, DC: National Academy Press, pp. 83–97.
9. Schultes RE. (1988) Primitive plant lore and modern conservation. Orion Nature Quarterly (3) 8–15.
10. Reichel-Dolmatoff G. (1976) Cosmology as ecological analysis: a view from the rainforest. Man J R Anthropol Inst 11:307–318.
11. Moran K. (1997) Returning benefits from ethnobiological drug discovery to native communities. In: Grifo F, Rosenthal J, editors. Biodiversity and human health. Washington, DC: Island Press, pp. 243–263.
12. Cleveland DA, Murray SC. (1997) The world's crop genetic resources and the rights of farmers. Curr Anthropol 38:477–516.
13. Moran K. (1998) Moving on: less description, more prescription for human health. EcoForum 21:5–9.

14. Greaves T. (1994) Intellectual property rights for indigenous peoples. Oklahoma City, OK: Society for Applied Anthropology.
15. Moerman DE. (1996) An analysis of the food plants and drug plants of native North America. J Ethnopharmacol 52:1–22.
16. ten Kate K, Laird SA. (1999) The commercial use of biodiversity: access to genetic resources and benefit-sharing. London: Earthscan Publications.
17. Mays T, Duffy-Mazan K, Cragg G, Boyd M. (1997) A paradigm for the equitable sharing of benefits resulting from biodiversity research and development. In: Grifo F, Rosenthal J, editors. Biodiversity and human health. Washington, DC: Island Press, pp. 267–280.
18. Carlson TJ, Iwi MM, King S, Obialor C, Ozioko A. (1997) Medicinal plant research in Nigeria: an approach for compliance with the convention on biological diversity. Diversity 13:29–33.
19. Iwu MM. (1996) Biodiversity prospecting in Nigeria: seeking equity and reciprocity in intellectual property rights through partnership arrangements and capacity building. J Ethnopharmacol 5:209–219.
20. Laird SA (editor). (2000) Equitable partnerships in practise: the tools of the trade in biodiversity and traditional knowledge, a people and plants program conservation manual. London: Earthscan (forthcoming).
21. King S, Carlson TJ, Moran K. (1996) Biological diversity, indigenous knowledge, drug discovery and intellectual property rights: creating reciprocity and maintaining relationships. J Ethnopharmacol 51:45–57.
22. Moran K. (1988) Mechanisms for benefit-sharing: Nigerian case study. In: Case studies on benefit sharing arrangements. Conference of the parties to the Convention on Biological Diversity, 4th Meeting, Bratislava, May.
23. Reid WV, Laird SA, Meyer CA, Gamez R, Sittenfeld A, Janzen DH, Gollin MA, Juma C (editors). (1993) Biodiversity prospecting: using genetic resources for sustainable development. Washington, DC: WRI.
24. Rosenthal J. (1988) The International Cooperative Biodiversity Groups (ICBG) program. In: Case studies on benefit sharing arrangements. Conference of the Parties to the Convention on Biological Diversity, 4th Meeting, Bratislava, May.
25. Guerin-McManus M, Famolare L, Bowles I, Stanley AJ, Mittermeier R, Rosenfeld AB. (1998) Bioprospecting in practise: a case study of the Suriname ICBG project and benefit-sharing under the Convention on Biological Diversity. In: Case studies on benefit sharing arrangements. Conference of the Parties to the Convention on Biological Diversity, 4th Meeting, Bratislava, May.
26. Brush S, Stabinsky D (editors). (1996) Valuing local knowledge: indigenous people and intellectual property rights. Washington, DC: Island Press.
27. Cleveland DA, Murray SC. (1997) The world's crop genetic resources and the rights of farmers. Curr Anthropol 38:477–516.
28. Twenty-fifth Session of the FAO Conference. (1989) Rome, 11–29 November.
29. Commission on Genetic Resources for Food and Agriculture (CGRFA-8/99/7 E). (1999) Progress Report on the International Network of Ex Situ Collections under the Auspices of FAO, Rome, 19–23 April.

Iwu and Wootton (eds.), Ethnomedicine and Drug Discovery

CHAPTER 17

Garcinia kola: a new look at an old adaptogenic agent

Maurice M Iwu, Angela Duncan Diop, Lisa Meserole, Chris O Okunji

Abstract

Garcinia kola Heckel (family, Guttifereae), known in commerce as 'bitter cola', is a highly valued ingredient in African traditional medicine. The plant is cultivated throughout West Africa for its edible fruit and seeds. Seeds of *G. kola* have been employed in folk medicine as rejuvenating agents and general antidotes. Bitter cola seeds have been shown to contain a complex mixture of biflavonoids, prenylated benzophenones and xanthones. Many pharmacological effects have been demonstrated for *Garcinia* biflavonoids, among them antiviral, anti-inflammatory, antidiabetic, bronchodilator, and antihepatotoxic properties. Other studies show that its antimicrobial activity is due to polyisoprenylated benzophenone. Some proprietary dietary supplements containing *G. kola* extractives already exist in US and African markets. This chapter focuses on *G. kola*, its constituents and application in medicine. The ethnobotany, clinical uses, chemistry, pharmacology and commercialization of *Garcinia* extracts and compounds, as well as analytical methods for quantification of marker compounds in these preparations will be discussed. Additionally, the proprietary products are discussed and evaluated as they relate to efficacy and human safety.

Keywords: Garcinia kola, *adaptogenic, nutraceutical, antiviral, anti-inflammatory*

I. Introduction

Garcinia kola is an edible seed, which belongs to a unique group of plants that help organisms to adapt to stress by influencing multiple regulatory systems responsible for stimulus–response coupling such as the immune system and act also as a general anti-infective agent. *G. kola*, commonly known as bitter kola or male kola, is used in traditional African medicine for the treatment of various infectious diseases, such as hepatitis, infections due to the influenza virus and other viral diseases. The plant elaborates a complex mixture of biflavonoids, prenylated benzophenones, xanthones and calanolide-type coumarines. Recently, Axxon Biopharm Inc. has presented dietary supplements containing *G. kola* extractives as the main constituent. These products were standardized using the biflavonoids, GB1, GB2 and kolaflavonone as the marker compounds.

II. Traditional uses of *G. kola*

Garcinia kola, sometimes confused with the west African cola nut, *Cola nitida/C. acuminata*, enjoys a folk reputation as a general antidote, tonic and antitussive agent, and has been incorporated into many traditional formulations for the treatment of various diseases. The fresh fruit has been used as a food and a tonic in most households where its cultural use is preserved. It was taken to fight off incipient flu and sore throat, and as a poison antidote. *Garcinia* has also been used to treat gastroenteritis, rheumatism, asthma, menstrual cramps and to improve the voice. As a medicine, its traditional uses include a wide range of ailments facing the healer: tablets, vials, lozenges, poultices and washes are delivery forms used. The remedy is used across all age groups. In the southern part of Nigeria where it is cultivated, the younger trees, providing chew-sticks and the yellow pulp of the fruit and seeds, are eaten. The seeds, which are known as 'bitter kola', and the roots and twigs, used as chew sticks, are alleged to aid oral hygiene, presumably by exerting an antibacterial action on the mouth. The seeds of *G. kola* are traditionally used in several West African countries for the treatment of head and chest colds, dysentery, diarrhea,[1–3] and urinary tract infections, and as an antidote for poisons.

The stem bark has a purgative action, and the powdered bark was used topically for tumors. The sap is used as a topical 'unguent' against parasitic skin diseases. The latex (gum) is used as a topical treatment on fresh wounds, and was taken internally for gonorrhea. The fresh seeds are chewed as an energy stimulant, used to prevent or treat colic in babies, and taken to treat upper-respiratory tract infections, bronchitis and throat infections. The seeds are also used as an aphrodisiac, and the dried seeds are used in dysentery and infectious diarrhea, and liver problems. Other uses of the seeds include treatment of liver disorders and coughs.

III. Chemical constituents

Phytochemical studies on the fruits have resulted in the isolation and characterization of kolanone, a novel polyisoprenylated benzophenone with antimicrobial properties and a biflavonone designated kolaflavonone. The seeds of *G. kola* contain several known simple flavanoids together with biflavonoids, amentoflavanone, kolaflavanone, GB-1, GB-2, and GB-1a. In high-speed counter-current chromatography separation, the biflavonoids are isolatable in very high yields. Two novel arylbenzofurans, garcifuran-A and garcifuran-B, were from the roots. Chemical studies of several other *Garcinia* species have provided triterpenes, bioactive xanthones, benzophenones, polyoxygenated phloroglucinols, and xanthones–benzophenone dimmers (Figure 1).

IV. Biological activity

Kolanone, one of the major constituents of the petroleum spirit extract of *G. kola*, possesses significant antimicrobial properties against both Gram-positive and Gram-

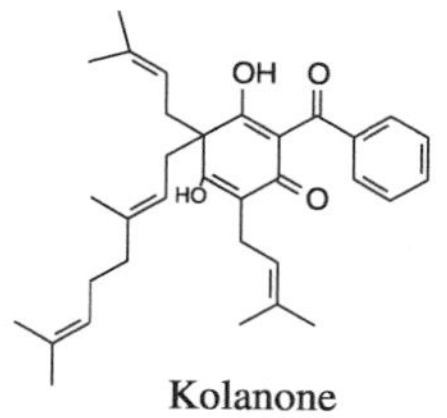

Kolanone

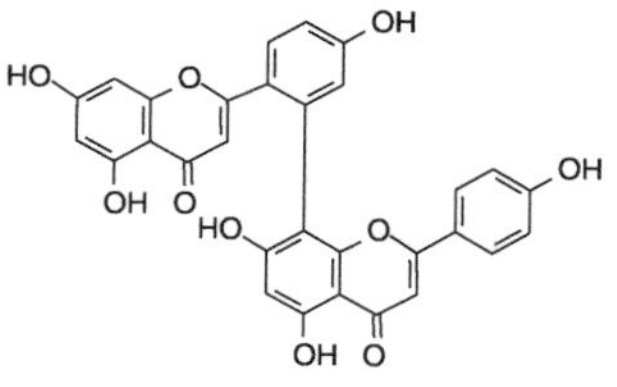

Amentoflavone

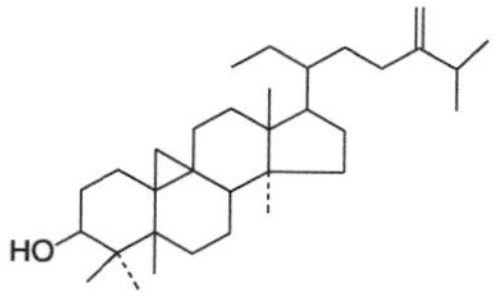

24-methylenecyclartenol

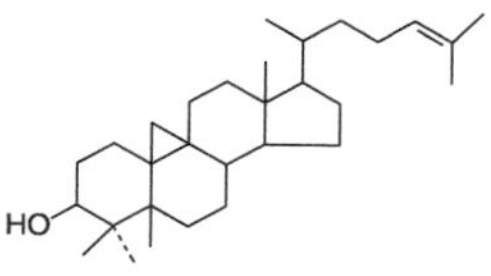

Cyclartenol

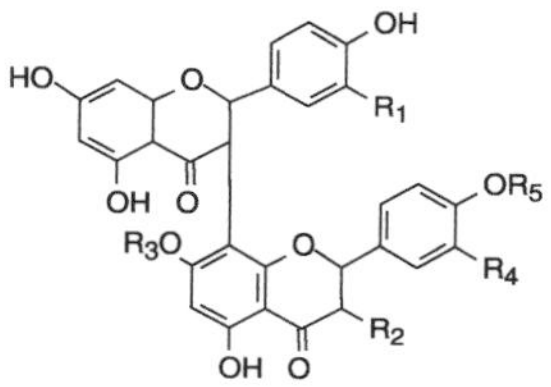

	R_1	R_2	R_3	R_4	R_5
GB1	H	OH	H	H	H
GB1a	H	H	H	H	H
GB1a-7"-O-glucoside	H	H	Glc	H	H
GB2	H	OH	H	OH	H
Kolaflavonone	H	OH	H	H	CH_3

Fig. 1. Some chemical constituents of *Garcinia kola*.

negative organisms. Its immune-boosting activity is believed to be due to the presence of biflavonoids, GB1, GB2, kolaflavonone, and garciniflavonone. Other activities attributed to the biflavonoids include anti-inflammatory, antimicrobial, antiviral and antidiabetic effects. They have shown a remarkable activity against a variety of viruses including Punta Toro, Pichinde, Sandfly fever, Influenza A, Venezuelan Equine Encephalomyelitis and Ebola. The IC_{50} values are in the range of 7.2–32 μg/ml with a MTC of more than 320 μg/ml.

V. Plaque-inhibiting activity of biflavonoids of *G. kola*

The mechanism of plaque formation is believed to involve the attachment of *Streptococcus mutans* to the surfaces of teeth and their subsequent colonization. One per cent (w/v) solutions of the organic fraction of methanol extract of *G. kola* fruit, kolaviron, or GB-1 were found to be effective in inhibiting the attachment of clinical isolate of *S. mutans* to glass or saliva-coated hydroxy-apatite beads. There was a marked inhibition (40%) of the synthesis of water-insoluble polysaccharides.

The three solutions also demonstrated in-vitro inhibitory activity against the growth of *S. mutans, S. mitis, Bacteroides gingivalis, B. melaninogenicus* and *Lactobacillus acidophilus*, with minimum inhibitory concentrations (MIC) of 0.02–25 $\mu g/ml$. The extracts were also found to be inhibitory at various concentrations against 36 other partially characterized oral isolates at various concentrations. In another study, *G. kola* extract showed potent activities against methicillin-resistant *Staphylococcus aureus*, vancomycin-resistant *Enterococcus*, and multidrug-resistant *Burkholderia cepacia* and *Pseudomonas aeruginosa*.[4]

VI. Clinical applications

The complex mixture of compounds found in *G. kola* makes it a useful medicine, not only as an adaptogen and stimulant, but also for many other complaints and illnesses seen in the clinical setting. Its documented and suggested clinical uses include: bronchitis and asthma, liver diseases, gall-bladder problems, intestinal problems, systemic inflammation, digestive inflammation, drug or environmental detoxification, blood-sugar regulation, male virility, lipid disorders, weight loss, and infectious diseases. The whole crude nut can be safely self-prescribed as a dietary tonic by individuals, as well as being taken in crude or extracted form when prescribed for illnesses.

Its bronchodilatory effect has been evaluated with a clinical study on 19 male adults. The only respiratory parameter change by 15 g of *G. kola* was a peak expiratory flow rate, indicating a mild brochodilatory effect.[5] Observational studies at the International Center for Ethnomedicine and Drug Development (InterCEDD) Nsukka indicate potential application as a remedy for colic in dysentery, liver disorders, upper respiratory infections, asthma, cough, sore throat, laryngitis, arthritis, menstrual and intestinal cramps, and headache. Topical uses awaiting clinical validation include its use as a wound dressing, as an anti-parasitic lotion, and in oral hygiene (as a chewing stick). Among all the possible clinical uses for *Garcinia*, probably the most relevant in terms of public health need is the potential for its use as an antiviral, anti-hepatotoxic and immune booster in cases of flu, liver and respiratory disease.

There are some potential clinical uses consistent with the activities of some of *Garcinia*'s constituent groups that are not directly supported by traditional use or as yet by clinical research. These include its potential as an agent to strengthen weak digestion due to a strong bitter principle. It is a reasonable hypothesis, based on its constituents and other traditional indications, that *Garcinia* may stimulate bile flow, thereby acting as a gall-stone preventative and as an aperient (mild laxative). The flavonoid profile of *Garcinia* suggests a possible role in treating gastric ulcers – but clinical evidence to support any relationship between its flavonoid content and anti-ulcer properties has not been reported. Other potential uses awaiting confirmation or rejection by research include: antidysentery, anti-infectious diarrhea, antiperiodontal disease, antihepatitis, antibronchitis, antisinusitis, anti-allergic, dysmenoreah and blood-sugar instability. Its flavonoids may also support anticlotting, and anti-inflammatory effects (as displayed by certain related flavonoids) in patients with cardiovascular problems, but this has not been substantiated by research.

VII. Toxicity and safety

G. kola appears relatively safe when used in traditional forms and dosages for traditional indications in traditional patient groups. *In-vitro* and animal studies on *G. kola's* effects suggest safety and benefit, although one animal study identified organ changes in kidney, liver and intestine[6] – though these were not linked to morbidity in that study. The few human studies performed have not identified toxicity or adverse events. More *in vitro*, *in vivo* and clinical studies and observations are needed to help define the best risk/benefit parameters for self-prescription and therapeutic usage when *Garcinia* is used for new therapeutic indications or new patient groups.

VIII. Standardization of *G. kola* formulations

Given the extent of use of herbal remedies and food supplements world-wide, a comprehensive quantitative analysis for monitoring the quality of these products is essential. The seeds of *G. kola* have been used in many herbal preparations either singly or in combination with other plants. Among these are Hepa-Vital Tea (a blend of *G. kola* and *Combretum micranthum* and Hangover Tonic (a blend of kolaviron and *Cola nitida*). Most of *Garcinia's* pharmacological activities are believed to be related to its anti-oxidative activity and to its ability to increase immunity in general. A number of products derived from *G. kola* have been manufactured and marketed as phytomedicines.

Like many dietary supplements and phytomedicines, *G. kola* products are susceptible to chemical variability due to growth. Plant collection through production of finished phytomedicines is required to ensure safety and protect the consumers. Quality specifications for *G. kola* have been established, based on classical pharmacopoeial methods. These methods include: plant morphology, chromatography and examination of pesticide residues and heavy metals, etc.

Capillary electrophoresis (CE) has proved to be very useful in a quantitative analysis of proprietary dietary supplement products containing *G. kola*, biflavonoids as the major biologically active compounds. This technique has been successfully applied in the analysis of Hepa-Vital Tea and Hangover Tonic. These two formulae contain *G. kola* as the major ingredient. The dietary supplement Hepa-Vital Tea, containing *G. kola* and *Combretum micranthum*, is used to maintain a healthy functioning liver, providing botanical support for eliminating undesirable substances common to today's environmentally stressed lifestyle. Hangover Tonic, containing *G. kola*, *C. nitida* and kolaviron, a unique blend of plants traditionally used by African cultures, is known for its rejuvenative properties and its ability to aid the body in handling non-specific stress and other toxic conditions.

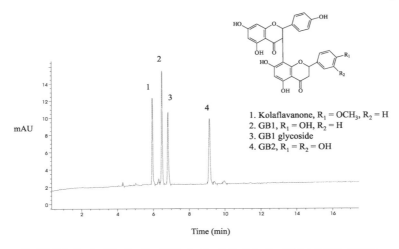

mAU

Time (min)

1. Kolaflavanone, $R_1 = OCH_3$, $R_2 = H$
2. GB1, $R_1 = OH$, $R_2 = H$
3. GB1 glycoside
4. GB2, $R_1 = R_2 = OH$

Okunji O Chris. Skanchy J. David, and Iwu M. Maurice, Quantitative Estimation of Kolaviron in *Garcinia kola*
Preparations Using Capillary Electrophoresis, paper presented at the American Society of Pharmacognosy
Interim Meeting , April 29-May 1, 1999 Grand Casino Convention Center and Veranda Hotel Tunica, Mississippi

Fig. 2. Electropherograms of biflavonones from *Garcinia kola*.

A CE method has been developed to estimate the pharmacologically active biflavonoids in the Hepa-Vital Tea and Hangover Tonic formulations. This method allows for quantitative determination of independent biflavonoids in traditional formulations. Quantitative analysis of the total content of *Garcinia* biflavonoids was established by using capillary electrophoresis (CE) performed on fused silica. This use of CE provides a sensitive, simple, and quantitative method that simultaneously quantifies the four major biologically active biflavonoids of *G. kola* (GB1, GB2, GB-glycoside and kolaflavonone) This standardization method utilizes chemical markers for quality control of various *G. kola* preparations. Calibration curves have been established by spiking authentic compounds into herbal samples. The biflavonones completely resolve in less than 10 min using a 100 mM borate buffer, pH 9.5. The limits of quantification for the three flavonones are 2 mM. The assay can successfully discriminate the four biflavonoids in proprietary products containing *G. kola*.

The three major biflavonones (GB1, GB2 and KF((kolaflavonone)) used as marker compounds/standards were isolated from a fraction of defatted methanolic extract *G. kola* seeds as described previously.[7] The structures these compounds are shown in Figure 2. The identities of GB1 and GB2 were further confirmed by MS.

IX. *G. kola*: the market, seeds of opportunity

In the late 1990s, the shelves of the typical dietary supplements department or store offered little variety. Practitioners, scientists, academicians, in the natural products field are familiar with a myriad of herbs and phytomedicines to maintain or restore

Table 1
US Botanical Consumer Sales ($million) 1998[8]

Ginkgo biloba	310
St. John's wort	290
Ginseng	250
Echinacea	230
Garlic	230
Saw palmetto	120
Goldenseal	70
Aloe	60
Valerian	50
Cranberry	50
Kava kava	50

health. However, when visiting the markets, one was faced with shelves jammed packed full of products based on mainstream, 'best selling' herbs consisting of a glut of Ginkgo, Echinacea and St. John's Wort and little else (Table 1).

Conspicuously absent or less pronounced were the less-known or new botanicals that are often overlooked.

G. kola falls into the latter category of new botanicals as an intriguing introduction to the dietary supplements market. While the seed has been used for an eternity throughout West Africa, its use as a dietary supplement in most other international markets has been limited. Yet, *Garcinia* has many of the earmarks of those blockbuster herbals. A herb of near ubiquitous use, it fits snuggly in that category of botanicals known as adaptogens. For retailers and manufactures, looking for something new to stimulate a sluggish dietary-supplements market, *G. kola* represents an opportunity.

These include:

- long history of use
- supportive and solid scientific research
- multiple use and indication
- ability to be manufactured into many forms
- safety.

X. *G. kola*'s history of use

G. kola enjoys a long history of use, as discussed earlier. This historical background supplies us with tips on use and safety, providing a firm foundation of information to build from. However, in today's dietary supplement market, a herbs history of use is only a beginning point, warranting further investigation and validation in order to create solid products for the US market. Owing to the escalating competitive nature of the market, increasingly, a long history of use is relying on significant scientific research to back up claims.

XI. Support by scientific research

One of the difficulties in utilizing herbs from traditional cultures in formulating dietary supplements is that they often lack or have limited scientific support of the use of the herbs. *G. kola* has a significant body of research to its credit with over 50 papers published about the herb. These works illustrate the chemistry and use, and in a limited way elucidate some of the mechanisms of actions.

XII. Multiple use and indications

Garcinia is categorized as an adaptogen similar to herbs such as ginseng, or eleutrococcus. It has a multiplicity of indications based on this and antimicrobial activity.[4]

XIII. Conclusion

Garcinia kola is a traditional food and medicinal plant with centuries of human use. Empirical evidence and emerging research suggest that it is safe and beneficial, especially when used in traditional dosages for traditional indications in traditional

Colds and flu
Antimicrobial
Antiviral
Antifungal
Antidiabetic
Internally and topically used
Gastritis
Dysentery
Hepatic conditions
Detoxification formulas
Alcoholic hangover remedy

Fig. 3. Indications and uses of *Garcinia kola*.

populations. It displays a wide range of traditional uses, and promises a similar scope for modern and clinical uses as new research progresses. The need for further research in chemical, pharmacological, in-vitro, animal and human research is clear, as is the need to capture apparently rare toxicity data. If developed as a concentrated or extracted product and used for needed but non-traditional indications, more meticulous study is needed; this unique 'bitter cola' promises to deliver significant benefits to some common human ailments, and merits world-wide attention and information dissemination.

References

1. Anslie JR. (1937) List of plants used in Native medicine in Nigeria. Oxford: Imperial Forestry Institute, p. 42.
2. Ayensu EA. (1978) Medicinal plants of W Africa. Algonac, MI: Reference Publications, p. 162.
3. Iwu MM. (1993) Handbook of African medicinal plants. Boca Raton, FL: CRC Press.
4. Taiwo O, Xu HX, Lee SF. (1999) Antibacterial activities of extracts from Nigerian chewing sticks. Phytother Res 13(8):675–679.
5. Orie NN, Ekon EU. (1993) The bronchodilator effect of *Garcinia kola*. East Afr Med J 70(3):143–145.
6. Braide VB. (1990) Histological alterations by a diet containing seeds of *Garcinia kola*: effect on liver, kidney, and intestine in the rat. Gegenbaurs Morpho Jahrb 136(1):95–101.
7. Iwu MM, Igboko OA. (1982) Flavoncids of Garcinia Kola Seed. J. of Natural Product. Lloydia 45(5):650–651.
8. Bailey GM. (1999) Nutritional outlook Nov/Dec.

Iwu and Wootton (eds.), Ethnomedicine and Drug Discovery

CHAPTER 18

Linking intellectual property rights with traditional medicine

MICHAEL A GOLLIN
magollin@venable.com

Abstract

Traditional knowledge regarding herbal medicines is in demand by the natural products industry and so has commercial value. However, indigenous peoples and local communities are mostly shut out from benefits that derive from use of their traditional knowledge and resources. Classic intellectual property strategies are of limited relevance to such communities, for economic, social and legal reasons. A major obstacle to traditional communities seeking to capture value from their knowledge is the ready availability of ethnobotanical knowledge in the public domain. Nonetheless, intellectual property can be used to help traditional communities maintain, build on, and benefit from their knowledge. Furthermore, despite criticism of 'biopiracy', those downstream in the chain of production may obtain intellectual property rights, which in turn may help capture benefits that can be shared with the traditional community through new linkages.

Keywords: *intellectual property, traditional knowledge, patents, natural products*

I. Introduction

High hopes have been placed on intellectual property as a way of protecting traditional medicinal knowledge in the selection, processing and development of pharmaceuticals, botanicals, and cosmetics. However, efforts to adapt existing intellectual property regimes to traditional medicine have run up against seemingly intractable obstacles. These obstacles arise from legal, social, ethical, economic, technical, and policy issues related to the value of ethnomedical knowledge and how to share benefits between traditional communities and industrial markets. This paper discusses the value of traditional medicinal knowledge, problems that have kept the holders of such knowledge from sharing its benefits, and some broader strategies for using intellectual property to help capture the value of traditional knowledge in ethnomedicines and other herbal medicines.

II. The value of traditional knowledge

Traditional knowledge regarding herbal medicines is in high demand by the natural products industry. This industry encompasses not only pharmaceuticals, where mass screening and combinatorial chemistry dominate, but also botanicals such as alternative medicines, nutraceuticals, dietary supplements, and cosmetics. In simple economic terms, the demand in these industries arises because traditional knowledge has a high commercial value. As analyzed by ten Kate and Laird, there are several sources of such value.[1]

The first source of value is the ability of traditional knowledge to provide leads for discovery of new products. A tradition of medicinal use for a plant increases the possibility that important bioactive substances are present, thus providing a lead for rigorous pharmaceutical research. In the botanicals market, the traditional use for a plant product like kava kava may be a sufficient lead on its own to launch sales of products based on the plant in industrial countries.

A second source of value is the evidence provided by a tradition of medicinal use for a plant that is non-toxic. This evidence may avoid $1–2 million in toxicity testing. At least, it indicates that the prospects are good for passing such rigorous testing. A third source of value is the potential that the traditional use may be a valuable or even central aspect of a marketing campaign for a herbal medicine or cosmetic.

However, the disturbing fact is that indigenous peoples and local communities are largely shut out from the benefits that derive from the use of their traditional knowledge. This paper briefly looks at causes for this disconnect (e.g. see Ref. 2) and explores ways to forge a linkage that could use benefits as an incentive to protect and enhance ethnomedicine.

III. Problems with sharing

The single largest obstacle for traditional communities seeking to capture value from their knowledge is the ready availability of ethnobotanical knowledge in the public domain.[1] Such public information can be found in scientific journals, databases, internet services, and university research communities.

More broadly, traditional communities do not share the value of ethnomedicinal knowledge because classic intellectual property strategies are of limited relevance to such communities, for economic, social and legal reasons. Much of traditional medicine is based on long-standing, widely held, unpublished information about plants found in many locations. Such diffuse information is extremely difficult to protect under conventional intellectual property regimes. Of course, many holders of traditional knowledge cannot afford to obtain formal intellectual property protection, even if it is available.

Some definitions and background discussion may help the reader understand the dimensions of the problem. To begin with, the term 'intellectual property' has different meanings to different people, and its meaning has evolved rapidly in the past decade. There is a culture clash between these various approaches. Some useful definitions from different sources are as follows.

- Dictionary: something intangible, created by the use of mental ability, to which legal rights attach.

- Law: a combination of doctrines from industrial property (patents, trade secrets, trademarks) and literary property (copyright).
- Business: a tool for converting human capital into value by defining and capturing new knowledge.
- Human rights: an ethical principal valuing all knowledge, including old and collective knowledge.

The last definition is the most recently developed. It has been expansively treated in many publications relating to the Convention on Biological Diversity, and has been most broadly defined in terms of 'Traditional Resource Rights' by the late Darrell Posey.[3] Traditional resource rights are seen as a link between cultural and biological diversity. They encompass a pot-pourri of human-rights principles including the following: basic human rights; the right to self-determination; collective rights; land and territorial rights; religious freedom; the right to development; the right to privacy and prior informed consent; environmental integrity; intellectual property rights; neighboring rights; the right to enter into legal agreements; rights to protection of cultural property, folklore and cultural heritage; the recognition of cultural landscapes; recognition of customary law and practise; and farmers' rights.

Traditional resources are said to include plants, animals, and other material objects that may have sacred, ceremonial, heritage, or aesthetic qualities, in addition to those economic values of the market economy noted above. The goals of traditional resource rights include establishing the identity of indigenous and local communities embodying traditional lifestyles, helping ensure their control over land, territory and resources. A mechanism by which that could be accomplished is equitable sharing of benefits from the wider use and application of the knowledge, innovations and practises of indigenous peoples and local communities, as well as the biological resources conserved on their lands and territories. Benefit-sharing requires the protection of both (1) conventional intellectual property rights and (2) traditional community rights (conservation of identity, culture and territory).

Thus, traditional resource rights analysis goes beyond existing legal doctrine to encompass values outside the legal and economic system of western market economies. As a practical matter, such concepts are important in developing new legal approaches, as with the World Intellectual Property Organization's Inter-governmental Committee on Intellectuctual Property and Genetic Resources, Traditional Knowledge and Folklore, but are of limited relevance in dealing with problems existing today.

IV. Practical linkages with intellectual property

Can existing doctrines of intellectual property law and practises be adapted to provide strategies for linking traditional ethnomedicinal knowledge to the value it provides? Although the possibilities are limited at present, there are some practical approaches.

First, traditional communities may, and some already do, treat their knowledge as trade secrets. This permits control over who shares value. For example, there may be trade secrets with respect to sources of supply of a plant, such as locations of wild groves. Other trade secrets may include methods and timing of harvesting to maximize activity, and varieties to cultivate. Those who are able to identify medicinal species and collect specimens at peak activity (e.g. traditional healers) hold valuable know-how. Given that 10% of the plant material in herbal medicines is incorrectly identified, and 20% adulterated,[1] such know-how should be sought after and compensated. Second, traditional communities may be able to use geographic indicators of origin to create value in the distinctive varieties or herbal formulations they produce, e.g. Fiji Kava. Third, traditional communities' and indigenous peoples' names, images, and photos are often used in marketing botanicals and cosmetics, another example of where intellectual property can be applied. Such publicity, if done without consent, may violate the rights of the individuals involved. Thus, the people can protect the publicity value of their names and images under theories of copyright, trademark, right of publicity, and use these IP tools to bring about benefit sharing. Fourth, plant varieties that are newly discovered in the course of ethnomedicinal activity may be protected via patents.

Other forceful intellectual property strategies apply to activities further downstream from ethnomedicine practitioners. Intellectual property opponents might say that such strategies are irrelevant to a traditional community. However, to share the benefits of traditional knowledge, there must be benefits to share. The more benefits there are, the more there is to share. The issue then becomes how to forge linkages between the downstream benefits and the upstream originators.

To understand how intellectual property helps form downstream benefits and how these benefits may be linked to the sources of traditional knowledge, it is useful to view the chain of production as flowing from A→B→C→D. Generally, the rights and physical property transferred from A to D may include:

- biological material
- information and texts
- data
- software
- equipment
- chemical reagents
- legal rights (patents, trade secret, trademarks, copyrights).

As a traditional medicine project is developed, it passes hands, typically adding more value as it goes, and intellectual property may attach to the added value. Thus, those downstream in the chain of production may obtain intellectual property rights, which in turn may help capture benefits that can be shared with the traditional community.

For example, classic intellectual property rights in downstream technologies, such as patents and trade secrets, can help capture benefits that can be shared with the traditional community. Such technologies protectable by intellectual property include:

- extraction methods
- fractionation and purification methods
- analytical methods (quality control, bio-assay)
- delivery systems (sustained release systems, foods).

In addition, downstream marketing activities are subject to trademark protection, as European botanical manufacturers and Shaman Botanicals have learned, and such trademark goodwill in the marketplace helps build value in the quality of the company's products and its environmental and social standards of practise. This in turn creates a value that can be passed back through the chain of production to share benefits with those who provide the material or information ultimately marketed as an ethnomedicine.

Finally, despite some difficulties, there is no doubt that herbal medicines can be protected by patents. Moreover, such patents can be enforced and may be the subject of vigorous challenges.

As explained in more depth elsewhere,[4] patents may be obtained in many countries on a variety of new discovered substances or methods, such as the following:

- purified compound of known structure (vs. heterogeneous cellular goo)
- purified micro-organism culture (vs. ecosystem)
- purified cell line (vs. in organism)
- chemical analog (vs. original structure)
- new fractionated preparation having defined reproducible physical, chemical, and/ or bio-activity
- new combination (vs. individual components)
- unpredicted new use for known compound (vs. known uses)
- new method of preparation
- recombinant protein vs. naturally derived
- new micro-organisms
- new plants (in the United States with analogous protection in other countries)
- new animals (in the United States, Europe, and some other countries).

How relevant are these types of natural product patents to ethnomedicine? A crude keyword search of the United States Patent and Trademark Office patent database (www.uspto.gov) revealed the following patents referring to the phrase 'ethno' (see Table 1).

Searches for patents referring to the names of top selling ethnomedicines revealed anywhere from a few to many hundred hits. The results of these searches are shown in Table 2.

V. The patents people love to hate

While the pace of patenting in this area has quickened, the extent of opposition has also increased. This leads to the topic of 'biopiracy and the patents people love to hate'. By this, I mean that various groups have decided to challenge certain patents,

Table 1
Recent patents referring to 'ethno'

US Patent #	Description
5,958,421	Cosmetic or pharmaceutical composition containing an andiroba extract – Laboratoires de Biologie Vegetale Yves Rocher (La Gacilly, FR)
5,900,240	Herbal compositions and their use as hypoglycemic agents – Cromak Research, Inc. (Bound Brook, NJ)
5,861,415	Bioprotectant composition, method of use and extraction process of curcuminoids – Sami Chemicals & Extracts, Ltd. (Bangalore, IN)
5,653,981	Use of *Nigella sativa* to increase immune function (black cumin, Egyptian folk medicine as a diuretic and carminative)
5,525,594	Treatment of hyperandrogenic disorders – Roussel Uclaf (FR) – process for the preparation of oenotheine B comprising extracting dried samples of *Epilobium parviflorum*
5,135,745	Extracts of *Nerium* species, methods of preparation, and use therefore
4,552,775	Process for the production of animal feed stuff from a liquid residue obtained by fermentation and distillation of grain raw material

not for economic or business reasons as is the norm, but instead out of hostility to the patent system in general or to how it is applied to natural products, and in particular because these patents are seen to be tools of pirating value from traditional knowledge. These challenges are typically meant to satisfy moral or political goals, not to achieve a business or market advantage as in most patent disputes (see Ref. 2 and www.venable.com/tools/ip/naturalproducts.html).

Some examples of challenged patents are as follows:

- **US 5,281,618** – Storage stable azadirachtin (neem) – re-examination was requested, but was denied, apparently because traditional use of water extracts did not teach the patented non-aqueous extract.

Table 2
Patents issued between 1976 and 1999 referring to key words relevant to top selling botanicals/ethnomedicine

Name	Number of patents
Echinacea	56
Garlic	1617
Ginkgo	186
Golden Seal	5
Saw Palmetto	7
Ginseng	459
Aloe	1376
Cat's Claw	2
Astragalus	75
St. John's Wort	1
Garcinia	41 (*Garcinia kola* 3)
Capsaicin	373

- **US 5,401,504** – Turmeric in wound healing – patent revoked based on Indian publications.
- **US 5,663,484** – Basmati rice lines (RiceTec Texmati rice) – patent claims rice plant, seed, rice grains, method of selecting rice plant; RAFI called for a boycot because the rice is based on a regional variety (see www.rafi.org.ca). Some claims were withdrawn.
- **US plant patent 5,751** *Banisteriopsis caapi* (Ayahuasca) – Challenged by indigenous groups due to sacred uses of plant – initially revoked because the patented variety was the same as the wild variety but reinstated by the patent office.

Several conclusions may be drawn from the intensity of challenges to patents on natural products and other biological materials. These may be summarized in the following 'political taxonomy'.

1. Intellectual property in non-replicating material is less controversial than in replicating material.
2. Intellectual property in sub-organism material (cells, DNA) is less controversial than IP in living organisms.
3. Plant intellectual property is less controversial than animal IP (e.g. TRIPS (Trade Related Aspects of Intellectual Property Rights) requires protection of plants but not necessarily for animals).
4. Animal intellectual property is less controversial than intellectual property in human tissue (e.g. Human Genome Diversity Project embryonic stem cells).

IV. The untapped power of ethnomedicine patents

Aside from these political battles, whose economic significance is uncertain, there are examples of how natural products patents do exert a strong influence on healthcare markets. This strength supports the position that patents on natural products can help create value and benefits that may then be shared with holders of traditional knowledge by contractual linkage.

Such an example is found in *Genderm Corp. v. Thompson Medical Company et al.* (US District Court District of Arizona). This case involved US patent 4,486,450, which was issued in 1984. The patent included claims for a method and topical formulations for treating psoriasis comprising capsaicin. Capsaicin is the active ingredient in hot peppers, long known for medicinal properties.

The patent owner's product, Zostrix, retailed for about $19 a tube, and was labeled for use as an analgesic, not for psoriasis. Many generic companies launched competing products that sold for much less, $4–10 a tube. During the life of the patent, there were two patent lawsuits, which resulted in two re-examination proceedings in the US Patent and Trademark Office. Ultimately, narrowed claims were confirmed and re-issued. Then, in 1998, the patentee commenced a third suit against the generic manufacturers of competing products labeled as analgesics.

As can be seen from the one patent claim invoked in the lawsuit, the invention was limited to psoriasis treatment:

> An antipsoriatic composition consisting essentially of a pharmaceutically acceptable cream carrier and capsaicin present in an amount greater than 0.01% and less than 0.1% by weight of the carrier wherein said amount provides effective therapy after topical application to the skin.

The patent seemed weak given prior art including a 1934 British Pharmaceutical Codex regarding capsicum oleoresin (the semipure extract) and a 1979 US Food and Drug Administration approval decision comparing the efficacy of 'natural' capsicum analgesics with those based on purified capsaicin.

Nonetheless, the District Court ignored prior art teaching the equivalency of products having comparable standardized amounts of the natural extract, capsicum, and upheld the validity of the patent. The District Court went further, and ignored the 'psoriasis' limitation in the claim to find that it was likely that the generic analgesic versions of Zostrix infringed, even though they were not labelled for psoriasis. The court therefore entered a preliminary injunction against all the generic companies marketing capsaicin cream products as analgesics.

This ruling required the generics to quit the market during a long trial, or to appeal against the decision. The result was that all but one generic company quit the market, given that the profit margin at the competitive generic price was too low. The median cost to defend a patent infringement suit with $1–10 million at stake is about $1 million. It would require sales of many tubes to justify that cost, even assuming that the generic company would win.

The one exception was a company that took a license from the patentee. Thus, due specifically to the patent, the owner of Zostrix was left with only one generic competitor, paying royalties, rather than a dozen competitors paying no royalties, and had two and a half years left of patent term to exploit the market.

Given the still limited scope of the patent, other parties remain free to sell some competing products such as lotions, ointments, cruder capsicum oleoresin products (which smell a little and are yellowish), and other types of analgesics (e.g. BenGay). Also, competitors can re-enter the market after the Zostrix patent expires in 2001.

There are several lessons for those interested in ethnomedicine and natural products. First, a natural product patent can be quite valuable. Second, even a weak patent can be enforceable in practise. For defendants, it may not be worth it to 'win' a lawsuit. Third, prior art regarding impure compounds does not necessarily invalidate claims in a later patent focused on a pure compound. Fourth, exclusivity due to a patent is worth a lot of money, because a lower court may favor the owner even of a weak patent. However, the ability of competitors to sell competing products makes the point that a patent does not actually provide a monopoly – it merely creates the right to exclude others from copying the claimed invention.

Finally, and sadly for the purposes of this paper, no benefits flowed to anybody but the patent owner. No linkage to conservation of hot pepper plantations, etc. was made. Furthermore, there was no involvement with, or benefit to, traditional users of hot pepper products. Considering how ubiquitous hot peppers are in every corner of the world, it is apparent how difficult such benefit sharing would be.

In conclusion, this paper has suggested how intellectual property can be used to help traditional communities maintain, build on, and benefit from their knowledge. Simply put, we need to link traditional knowledge with the value it creates in a much tighter and more effective fashion.

Existing legal strategies based on intellectual property can be used or adapted to help provide that link and to create benefits that can be shared. However, implementing these strategies can be complicated, expensive and offensive to some communities. New models would help.

Acknowledgments

The assistance of Zayd Alathari is gratefully acknowledged.

References

1. ten Kate K, Laird SA. (1999) The commercial use of biodiversity: access to genetic resources and benefit-sharing. London: Earthscan.
2. Gollin MA. (1999) Legal and practical consequences of biopiracy. Diversity 15:7–9.
3. Program for Traditional Resource Rights, Oxford Center for the Environment, Ethics and Society (OCEES), Mansfield College, University of Oxford, http://users.ox.ac.uk/~wgtrr/trr.htm (updated September 23, 1999, last acessed October 2000).
4. Gollin MA. (1994) Patenting recipes from nature's kitchen, Bio/Technology 12:406.

CHAPTER 19

The use of conservation trust funds for sharing financial benefits in bioprospecting projects

MARIANNE GUERIN-MCMANUS, KENT C NNADOZIE, SARAH A LAIRD

Abstract

Central among the thorniest aspects of implementing the Convention on Biological Diversity (CBD) is how to make progress on the Convention's objectives of guaranteeing appropriate access to genetic sources and arranging for the equitable sharing of benefits derived from these sources. Although a fundamental aspect of the Convention is the requirement for implementation through national legislation, strategies, plans and programs for both access and benefit-sharing, arriving at them has not been easy. In spite of several existing legislations with some bearing on the issues, including the trade and intellectual property aspects, the existing provisions are either too vague or too narrow to address the relevant issues. Sharing the benefits associated with the commercialization of genetic and other biological resources involves not only the determination of the nature and quantum of the benefits, but also the mode of distribution of those benefits to a range of stakeholders, and in a fashion that serves defined objectives over time. One possible model for distribution builds upon the delivery mechanisms employed in other areas of conservation and development such as conservation trust funds. The paper discusses the origins of the trust fund concept, its development in the context of conservation, and its application to biodiversity prospecting. It highlights the basic principles that should be taken into consideration and the issues to be addressed in designing the framework. It will describe concrete steps to developing and implementing a biodiversity prospecting trust fund and relate through specific case studies some of the lessons learned from past experiences.

Keywords: *benefit-sharing, access, intellectual property, stakeholders, conservation, trust fund*

I. Introduction

Sharing the benefits associated with the commercialization of genetic and other biological resources involves determination not only of the nature and size of the benefits, but also the mode of distribution of those benefits to a range of stakeholders, and in a fashion that serves defined objectives over time. Several different models have been advocated or are already in place to share financial benefits. Trust funds – the subject of this chapter – are one model that has been employed to date, building upon experience with delivery mechanisms employed in other areas of conservation and development. There is a short but well-documented history of the use of trust funds in the conservation and development community upon which biodiversity prospecting projects can draw.

The trust-fund model is well suited to the bioprospecting field because it can accommodate the long-term time frame of biodiversity prospecting projects, which typically encompass several years of collection, research and drug development and where some of the financial benefits might accrue as long as decades after field activities were completed. Trust funds provide a stable and enduring structure that can last over these long periods of time for the purpose of channeling benefits in a controled and consistent manner. Trusts can also be structured to operate under specific guiding principles, overseen by a diverse board of trustees, in such a way that the long-term interests of a group, or country, rather than the short-term gain of individuals, are served.

The process of developing a trust fund may also be an invaluable exercise in itself, in that it requires the definition of goals and objectives and the identification of potential beneficiaries. It also promotes a dialog on what constitutes equitable benefit sharing. A trust-fund charter might also provide a much-needed public record of the principles and objectives biodiversity prospecting activities are intended to serve, thereby offering a window into the complexity of projects and making the logic of commercial agreements available to a wider public.

This chapter will discuss the origins of the trust fund concept, its development in the context of conservation, and its application in biodiversity prospecting. It will highlight the basic principles that should be taken into consideration and the issues to be addressed in designing a trust fund framework. Finally, the chapter will describe concrete steps to developing and implementing a biodiversity prospecting trust fund and relate through specific case studies some of the lessons learned from past experiences.

II. Trust funds: general background

II.A. What is a trust fund?

A trust fund is a 'sum of money that is legally set aside and whose use is restricted to specific purposes for designated beneficiaries'.[1] The concept behind a trust is that assets are managed by a person or a group (the trustees) for the identified goals or benefits of a second group (the beneficiaries). Funds may be organized in several different ways including non-profit corporations, foundations, government-formed trusts and common law trusts. While this chapter focuses primarily on common law trusts, the concepts involved in their creation apply to the other forms as well (refer to Box 1).

The law of trusts grows from the Anglo-American legal tradition, including current or former members of the Commonwealth and the United States of America (common law countries). Trusts have been employed for centuries, primarily by families to ensure the financial health of future generations; in England, for example, their first reported use dates back to the 11th century. During the Crusades, English men going to war would 'give' their property to someone to hold in trust until their return from the Holy Land or to pass on to heirs if necessary.

Box 1. Key terms

- *Conservation trust fund* – a trust fund whose designated beneficiaries are conservation programs and activities identified, selected and developed by the fund trustees.
- *Non-profit corporation* – a corporation whose income is not distributed to its members, directors or officers but is used instead to fund ongoing activities of the corporation.
- *Common law trust* – a trust organized so that the trustees have responsibility for the fund's assets and manage its affairs, while benefits accrue to either private parties (individuals who are specifically identified) or the public (not specific individuals, but a community or segment of a community).
- *Trust established by an act of national government* – a trust made possible by a national government for the benefit of the people in that country. Consequently, the aims and objectives of the trust are considered to be in the national interest.
- *Foundation* – a trust-like arrangement used in civil law countries, where the resulting institution is a legal entity able to own assets (as opposed to the trust where assets are held by the trustees).
- *Debt-for-nature swap* – cancellation of foreign debt in exchange for a commitment to mobilize domestic resources for the environment.
- *Common law* – a body of law based primarily on judicial decisions employed, for example, in the USA, UK, and former Commonwealth countries.
- *Civil law* – a body of law based upon legislative enactments (laws created by statute) employed, for example, in France and Switzerland, and most former colonies of France, Spain, Belgium and Portugal.
- *By-laws* – a document that sets out the governing and operating rules to be followed by a board of trustees.
- *Charter* – a document issued by the government to a corporation or non-profit corporation assuring them certain rights, liberties, or powers in exchange for fulfilling certain requirements. In the context of trust funds, the charter is analogous to a deed, by-laws or a constitution.
- *Deed* – a document that records the goals of a trust, its structure, the identities of the beneficiaries, the trustees and the obligations of the trustees to the trust and beneficiaries.
- *Board of trustees* – the individual or group of individuals responsible for managing a trust's assets and affairs and distributing revenues to beneficiaries.

The most analogous counterpart to common law charitable trusts in civil law systems is the *foundation*, which exists in most modern legal systems in continental Europe and is widely used by environmental institutions in these countries. A well-known example of a foundation is the World Wild Life Fund for Nature (WWF), which was established in 1961 as a foundation in Switzerland, a civil law country.[1]

While foundations are essentially functional alternatives to common law trusts and adopt most or all of their principles, a major difference is that a foundation legally acquires a separate personality and has the capacity to own property, as opposed to the trust, where legal title is held by the trustees.[1]

Latin American countries employ the trust concept, using the term *fideicomiso* in national legislation. In the 1920s, Mexico introduced the trust concept in modified form, by way of national legislation patterned on Anglo-American experience. In other parts of Latin America, the trust concept can be traced back to Roman civil law, as transferred by the Spanish. The original intent of *fideicomisos* was to provide for future generations – much as in Europe – so early laws limited beneficiaries to private individuals and did not permit public charities. Today in Latin America, however, trusts can be employed as public charities.

The trusts described above for Latin America grow directly from European legal traditions. There are numerous non-Western trust-like concepts, as well. For example, parallels have been drawn to institutions such as the Islamic *waaf*, which historically served as a legal device for the establishment of perpetual public charities, and to the Asian *yayasan*, which are non-profit organizations treated as legal institutions capable of entering into legal contracts and civil actions. The scope of these trust-like concepts, however, can be more limited than the common law trust, especially with regard to beneficiaries.

II.B. The history of conservation trust funds

Trust funds have only recently been employed to achieve conservation objectives. They arose in response to various issues that surfaced in the late 1980s, in particular 'debt-for-nature swaps' or transactions in which a developing country's foreign debt is canceled in return for a commitment to domestic conservation investment.[2] Debt-for-nature transactions generated large amounts of local currency, which local beneficiaries could not adequately absorb. At the same time, there existed an increasing need for long-term financing of conservation projects, such as the recurrent costs of park management, but a visible, transparent and intermediary structure between various sources of financing and numerous biodiversity conservation projects was lacking.

In response to these concerns, the trust fund concept was carried from the estate-planning field to the conservation world. The benefits of the trust fund concept included: promotion of financial security unaffected by fluctuations in international donor or foundation money; building of local institutional capacity on the part of countries, projects, and others; increased community confidence in the longevity of conservation projects; and, by including representative stakeholders in the governance and management of funds, the promotion of consensus-building and a sense of ownership over a nation's natural resources (Box 2).

There is clearly a great deal of diversity in the manner and way in which trust funds can operate, and the breadth of the objectives they serve. By building upon expertise acquired through conservation trust funds, 'biodiversity prospecting funds' should be able to bypass some of the mistakes made earlier on, and draw from these

Box 2. Way in which conservation trust funds have proven valuable

- The broad private and public sector participation involved.
- They meet recurrent costs that might otherwise be very difficult to fund, although this requires a system of careful monitoring and evaluation.
- They improve absorptive capacity, i.e. the ability to hold and use large sums of money over an extended period of time.
- They provide a small grant-making capacity by 'retailing' large international grants to a wide range of smaller projects.
- They provide sustained funding, mitigating risks of unexpected stoppage of funds due to political changes, budget cuts, economic austerity programs, etc.
- They enjoy privileges like tax exemption that enable full application of available funds to designated beneficiaries.

models to facilitate quicker implementation and to establish better track records. Further, considering that trust funds, in practically all jurisdictions, can acquire a tax-exempt status and also enjoy certain other immunities and privileges, they present an added advantage because all the monies contributed to the fund can be applied entirely for the benefit of the intended beneficiaries. In the same vein, profits derived from the investment of part or all of the fund are fully utilized without any tax burden or deductions (refer to Box 2).

II.C. Size and scope of trust funds

Trust funds can range in size and scope. For example, Suriname's Forest People's Fund, which was established to facilitate benefit-sharing from biodiversity prospecting and foster biodiversity conservation in a small community, began with a start-up capital of $50,000 (Case Study 1). Nigeria's Fund for Integrated Rural Development and Traditional Medicine received original financing of $40,000 (Case Study 2). Although the start-up funding in these cases is relatively small, both funds are designed to receive additional financial benefits over time. At the other end of the spectrum, Colombia's ECOFONDO, designed to promote the nation's environmental conservation and sustainable development, received $41.6 million in local currency, paid as a counterpart to debt cancellation, over four years, by the Colombian government.[3]

Trust funds are also administered at different levels. The Suriname trust fund operates at the community level; Nigeria's is a national fund. Trust funds can be administered by governments, research institutions, non-profit organizations or community associations. For example, in Fiji, a community-based trust fund – the Verata Tikina Biodiversity Trust Fund – has been developed as part of a three-year project of The University of the South Pacific (USP), the Verata Tikina communities, and commercial partners. This project is intended to link pharmaceutical drug development with conservation and community development. Initiated in 1995, with commercial partner SmithKline Beecham, the fund subsequently changed its partnership to the Strathclyde Institute of Drug Research. (SIDR), which works with numerous commercial companies. The Verata community members receive per-sample fees, totaling as much as $100,000, as a short-term financial benefit, which they are managing through the community-based trust fund.[4]

III. Biodiversity prospecting trust fund

Biodiversity prospecting projects can generate financial benefits over many years. Usually, an up-front payment, or fees per sample, are made by a company to the collector or local community for the right to prospect. Then, over the course of what may be many years, payments are made in connection with each research milestone – these are known as 'milestone payments'. Eventually, royalties may be paid when products are commercialized. This course of financing sets up several potential problems if a framework for the distribution of the benefits, and an overarching plan for their use, is not in place. The establishment of a biodiversity prospecting trust fund must take these factors into consideration.

Establishment of an environmental trust fund generally moves through three phases of development:

- a feasibility study
- a design phase
- an implementation phase.

The following sections describe the considerations and actions that should take place during each phase.

III.A. Feasibility study

The feasibility phase involves consultations with a representative group of stakeholders, including government representatives, scientists, conservationists, business representatives and community leaders. The group will determine the priority needs of the community/region/country, the objectives of the fund, and the types of conservation and development projects it will finance. An international expert on trust funds, in conjunction with a local counterpart, might facilitate this exercise in order to help assess the feasibility of the trust fund model in the domestic context. The local counterpart can provide technical assistance regarding the national legal framework and determination of the appropriate form of trust (e.g. a foundation vs. a common law trust according to national law). Experts and stakeholder groups will also need to define the financial and banking structures most appropriate to the situation, and make preliminary determinations about potential board members, and the availability of management expertise for fund staff. The feasibility study will seek to answer the following questions.

III.B. Will the trust fund concept work in the target community?

Given the legalistic and seemingly bureaucratic nature of trust funds, some have suggested that they run counter to the ways in which local communities traditionally manage and distribute benefits. In fact, in traditional societies that communally own resources, the use of community development associations and cooperatives for rural development and renewal is common. Biodiversity prospecting is likely a culturally foreign activity. However, trust funds are generally not an unfamiliar concept.

III.C. Is there a legal framework in the host country to support the trust?

A trust fund must operate within the context of national and international law and policy. As an initial step, local and national laws must be analyzed to determine if provisions are made for trusts or trust-like devices. Depending upon national laws, other options might need to be considered, such as government and private partnerships, or a trust established by an act of the legislature. Another option to consider is the integration of a biodiversity prospecting fund within an existing institutional structure, e.g. creation of a biodiversity prospecting sub-account as part of a national conservation fund (Case Study 3).

The feasibility study should also include an analysis of other relevant law and policy, including intellectual property, trade, environment, natural resources, and access and benefit-sharing measures. The activities that generate revenue for the fund, and the objectives served by the fund, should fit within the wider legal and policy context of a country and region. A conservation trust fund should also endeavor to support a country's national environmental strategy, but should not be a substitute for governmental financial support for environmental management and enforcement of existing laws. Some trusts have explicitly addressed the issue in their by-laws, by including a pledge from the government to maintain its current level of support for conservation and development programs.

A feasibility phase should also pay attention to, and try to maximize, the wider objectives trust funds can serve – beyond the specific goals of the fund itself. A well-designed trust fund can generate a range of benefits that contribute in non-specific ways to the development of sustainable societies. For example, by making financing available in absorbable amounts, funds build the capacity of local implementing organizations. The transparent, participatory processes that characterize trust funds can strengthen civil society.[5] Additional benefits of current conservation trust funds include the creation of new parks, and strengthened capacity among NGOs or governments for generating and managing financial resources, a feeling of 'ownership' among stakeholders resulting from direct participation, increased conservation awareness, and increased community involvement (Box 3).[6]

Box 3. Key factors for success

- A well-balanced board, reflecting a range of society – government, scientists, local communities, NGOs, etc.
- Dialog and regular communication between involved parties.
- Transparency, which allows independent monitoring, avoidance of mistakes, and discourages misuse of funds.
- High visibility of the structure to the public, as well as domestic and foreign donors.
- Government commitment to development and conservation.
- Adequate funding.
- Dynamic and skilled organizers and managers.
- Diverse representation and involvement of stakeholders.
- Well-defined objectives.
- Ongoing monitoring and evaluation.
- Regular reporting and feedback on activities to the public.

IV. Designing the trust fund: 'nuts and bolts'

IV.A. Goals and objectives

Defining a fund's objective(s) is crucial to a trust fund's success, since the structure of a fund depends on, and responds to, its goals and objectives. This should be the first step in the process of fund establishment. One of the key lessons learned from the establishment of conservation funds is that it is critical to have the basic vision of the fund in place before making decisions on design. In deciding on the scope of the fund, it will be necessary to first define the issues to be addressed, and following on this, the types of activities or beneficiaries that a fund could support in order to respond to these issues.

A common objective of most conservation funds is to provide a stable source of financing to meet the recurrent costs of operating and maintaining protected areas and/or to ensure the sustainable use of natural resources through community support.[1] However, these objectives have varied in breadth and depth. They may be narrowly focused, such as the maintenance of a park or park system. For example, the Jamaica National Parks Trust Fund was established to fund two pilot national parks and the establishment of a National Parks and Protected Areas System in Jamaica.[7] One of the principal goals of Suriname's ICBG project and conservation trust fund is to record and secure the value of tribal knowledge (Case Study 1). Advocates of specific, clear-cut goals such as these indicate that when the objectives of a trust fund are narrow, they are easy to understand and communicate, and room for disagreement among governing board members is limited.[8]

At the other end of the spectrum are funds that incorporate broad goals. For example, the main objective of Peru's FONANPE is to finance protected area projects in Peru.[3] Such all-encompassing goals can tie into national environmental agendas and allow experimentation with new forms of partnerships between the public and private sectors.[8]

Generally, funds that focus goals and objectives on a specific range of activities selected for strategic impact, feasibility, and which can be carried out quickly to build a track record, do better than those that start out with an 'open door' policy that reacts to whatever is proposed. The scope of a fund can always be broadened later, if appropriate. Alternatively, if a fund starts out with a fairly broad mission and objectives, a 'pilot phase' can be declared in which the fund concentrates on a focused area before accepting proposals from other areas. There are several practical reasons for taking this approach:

1. A fund can only process so many proposals, and finance even fewer. It is better for a fund to narrow its focus, receive fewer proposals, and select as many high-quality proposals as can be funded, in order to establish a track record.
2. A narrow focus will allow selection of fund staff and advisory committees with specific technical skills (taxonomy, pharmacology, sustainable development) in mind, avoiding the necessity of staffing for multiple disciplines.
3. Fund-raising for a fund with a narrow focus will be more directed, and will therefore be able to achieve quicker results.
4. A narrower focus will enable the fund's management or trustees to acquire expertise and competencies, which, over time, will translate into greater efficiency in the handling of its operations. Lessons and skills acquired could be subsequently applied to other areas.

Reaching agreement on the goals and objectives of a trust fund will normally entail stakeholder meetings to discuss the fund's focus.

However, if consensus cannot be reached in a collegial setting, an outside facilitator may be hired to help build consensus and arrive at a set of goals and objectives.

IV.B. Goals and objectives specific to funds financed through bioprospecting activities

Biodiversity prospecting funds are distinct from conservation funds in that financial benefits are directly linked to a set of commercial activities. They seek to facilitate and ensure the equitable distribution of the benefits derived from biodiversity prospecting activities and the sustainable use of biological resources associated with prospecting. Their goals might include:

1. To serve as the channel through which the benefits and economic rewards are distributed to the areas from which source species for drug or other product development are found.

2. To improve the standard of living in target areas through community development initiatives, information, healthcare, education and communication. This would include support and assistance to rural families, particularly women and children and to other activities that will help alleviate poverty.
3. To apply revenues to projects or ventures that will promote conservation of biological diversity. For example, in the case of the Nigerian trust fund, biodiversity prospecting activities are intended to promote conservation, but through sustainable management of species in ways that promote biodiversity conservation, rather than support for protected areas (Case Study 2).
4. To promote improved ways of seeking the prior informed consent of stakeholders. This would include collaboration and consultation with government; research institutions; and town associations, village heads and professional guilds of healers in order to determine the nature of compensation and priority projects in their localities.
5. To improve the domestic capacity to research local biodiversity, including the capacity to conduct research on, and standardize, traditional medical systems, participate in drug development efforts at a higher level, and research tropical diseases. This would include technology transfer and capacity building, in forms such as the provision of equipment, know-how and training.

IV.C. Origination document (constitution, charter, deed, articles of incorporation)

The four kinds of origination documents are:

*   constitution
*   charter
*   deed
*   articles of incorporation, and/or by-laws.

Despite the differences in name, the substance of the documents is the same. Origination documents are the legal documents that set up the trust, establishing the fund's goals and objectives, and institute the mechanisms by which grants will be awarded and other benefits distributed. Fundamentally, the origination document is the 'law' by which the new fund will be administered and by which the activities of the board and the fund will be directed and ultimately judged. While the document establishes the legal right in the board to initiate suits on behalf of the trust to affect its objectives, it also forms the basis for removing the board, or perhaps dissolving the trust when the goals and objectives are not carried out or there is wrong doing.

The *Constitution of the Verata Tikina Biodiversity Trust Fund and the Trust Fund Committee*, created by the Verata Tikina Council in Fiji, specifies that the objective of the fund is to promote the sustainable development of Verata Tikina and its people. Specifically, the Trust Fund and its Committee shall: invest the funds to ensure long-term sources of funds for sustainable development projects; select projects in priority areas of conservation, education and training, health, and microenterprise, providing financial and technical resources for their implementation; and provide advice to the Tikina Council on sustainable development issues.

The Healing Forest Conservancy (Case Study 4) has developed a model constitution based upon its work with Shaman Pharmaceuticals' biodiversity prospecting activities in several countries. The HFC constitution includes general information about biodiversity prospecting fund structures as well as specific guidelines for the distribution of benefits and suggestions for the creation of technical committees, which can be helpful in meeting the goals and objectives of the fund. Such committees are particularly important because boards typically only meet two or three times a year and need to rely on other bodies meeting more frequently to perform the groundwork, or to make on-the-spot decisions.

V. Governing structure

The governing board is made up of stakeholder representatives who make important decisions about the fund, such as defining the guiding principles for proposal selection and grant-making. The governing board's decisions should be open and transparent, and an internal checks-and-balances system should be in place (Box 4).

The composition of the board is of the utmost importance because it can make or break the fund. A primary requirement of governing boards is that they represent the interests of all stakeholders, including government, community, industry, and NGO members. As mentioned earlier, the board members must be dynamic, enthusiastic, well connected to the constituency they represent on the board (e.g. government, communities, scientists, private sector), and committed to conservation and sustainable development. Some consideration should also be given to how well the board members will work together. In some cases, successful boards have one 'super-board-member' (usually the chairman) who leads and shepherds the board and the fund to success during its infancy.

Box 4. Possible checks and balances for fund management

- Advisory committees should include outside participants who will provide a fresh perspective.
- Give certain board members veto power or require super-majorities (75%, 80% or 100%) on certain issues.
- Stagger board membership terms, with members serving terms that expire at different times.
- Institute international arbitration and dispute resolution provisions.
- Include detailed provisions on auditing, accounting and reporting requirements.

Source: Spergel.[9]

The board of the Nigerian 'Fund for Integrated Development and Traditional Medicine' (FIRD-TM) includes one traditional healer, four traditional medical practitioners, one representative from the Federal Ministry of Health, a representative from the National Agency for Science and Engineering Infrastructure in the Ministry of Science and Technology, a professor of medicine with extensive experience in traditional medicine and a professor of pharmacology associated with the pharmaceutical industry. The professional and sectoral, as well as ethnic diversity of the membership is part of a deliberate policy to ensure that as many relevant national constituents are represented as possible. The board's composition is intentionally diverse to encourage the meaningful exchange of ideas and to capture the range of experiences represented by individual members (Case Study 2).

In comparison, the board of the Forest People's Fund in Suriname is relatively small, with only five members, including two representatives from the Saramaka Maroons, two representatives from Conservation International (CI), and one representative from the Surinamese pharmaceutical partner. The smaller size of this board reflects the regional and community focus of the fund (Case Study 1).

Equally important when deciding upon the composition of the board is the fact that the structure of the board affects outside perceptions of, and attitudes towards, the fund. For example, if a board does not have government representation, the government may distrust the organization and believe that it is trying to usurp its right to determine the disposition of natural resources. However, if there are too many government representatives, NGOs, communities, researchers, and other stakeholders, may feel that the fund serves only the national government's agenda. The United States Agency for International Development (USAID) and the Global Environment Fund (GEF), for example – both donors to conservation funds – will not contribute to the capital of a fund whose board has more than 50% government representatives. These structural dynamics reflect some aid agencies' desire to promote and build civil societies.

Securing the explicit support and goodwill of the national government is very desirable and could be achieved by affording it limited or nominal participation without sacrificing the independence or objectivity of the board. Operating procedures for boards should also be clarified to ensure transparency, checks and balances, and maintenance of standards over time. While an emphasis may naturally be placed on the financial activities of the fund, provisions for the monitoring and evaluation of the grant making process should also be instituted. Mechanisms, such as audits and annual financial reporting, must be designed and implemented to measure both financial management and grant-making aspects of the fund's work. In many cases, provisions to that effect are imposed by the donors in the grant agreement.

V.A. Financial structure

The financial structure of a trust fund can vary depending on the time period and goals of the fund. There are four main options for the structure of a trust fund:

• endowment

- revolving fund
- sinking fund; or
- a combination of two or more of these structures.

An *endowment* is a fund that maintains a bulk sum of money as principal and only disburses the income earned on that amount.[8] Only under specific circumstances can the capital (*corpus*) of an endowment be invaded. The Mexican Fund for Nature Conservation. (FMCN) is an endowment whose one main objective is to support and strengthen the capacity of Mexican NGOs through mid- and long-term financing of initiatives for conservation and sustainable natural resource use. The initial capital of the fund, was $36 million in 1994, including $16 million granted by the GEF for protected area management and $20 million from USAID for sustainable development.[3]

A *revolving fund* is a fund to which new assets are added periodically (annually, for example) through fees, taxes or levies collected.[8] The Belize Protected Areas Conservation Trust is a revolving fund whose capital comes partially from a US$3.75 visitor fee as well as 20% of all site entry fees, recreation licenses and permits, and cruise ship fees. Five per cent of the collected revenues are managed as a permanent endowment for emergency purposes.[3]

A *sinking fund* is designed to disburse its entire capital plus its income over a designated period of time.[8] This type of structure can be well adapted to the funding of projects with development or income-generating potential that are expected to become self-sufficient after an initial seed money or startup phase. The Dominican Republic's PRONATURA exemplifies a sinking fund: donations are converted to national currency as they are received and immediately deposited into separate accounts for each project.[1] Sinking funds are rare, however, partially because of the perception that the time and effort necessary for their creation merit a more permanent structure. Furthermore, most conservation-oriented projects require long-term funding that sinking funds cannot guarantee. Therefore, most sinking components end up being one composite feature of more complex structures.

As expertise builds in the field of conservation funds, the people in charge of their design have come to realize that the most useful structures may involve a combination of two or three funding mechanisms. For example, it is good to bear in mind that, at the onset of most structures, the fund will be under pressure to demonstrate results and success quite rapidly, but it must not lose sight of, or sacrifice, its long-term sustainability. It might therefore be advisable to sink a percentage of the fund and finance some priority exercises or projects so that an immediate impact can be felt by the different stakeholders, while the remainder of the funds remain as an endowment. In the case of biodiversity prospecting funds, staggered revenues, such as milestone payments or royalties can contribute to the fund in a revolving manner, but can also be expended to increase the endowment.

V.B. Sources of funding

Conservation trust funds traditionally receive funding from three categories of donors: multilateral donors, bilateral donors, and private and NGO donors. In some cases, a fourth category is represented by host government donations. Examples of multilateral donor organizations are the World Bank/Global Environment Facility and the United Nations Development Program. The United States and Canada are examples of countries that contribute bilaterally to environmental funds through agencies such as the USAID and the Canadian Agency for International Development. (CIDA). The MacArthur Foundation has supported training and design work for the creation of conservation funds, as well as contributing to capitalization. Some national governments have also committed specific amounts to funds in their own countries. For example, the Royal Thai Government has earmarked specific budgetary items to be disbursed directly to the Thailand Environmental Fund and indirectly through support programs. Most of these institutions are potential sources of initial, start-up funding, or seed money, for funds.

Although start-up funds are crucial to any fund's development, ongoing capitalization is equally important. Sustained funding ensures that activities and programs continue and that there are funds available for new projects and increased needs of existing projects. Biodiversity prospecting funds derive income across time from sample fees and up-front payments, milestone payments, and royalties (Chapter 9). Milestone payments are attached to various stages of drug discovery (e.g. screening, identification of active compounds) and development (e.g. pharmacology, safety studies, Phase I, II and III clinical trials, or other steps linked to government regulatory requirements). As a promising sample moves through discovery and development, payments can be made automatically to a fund. Long-term fund revenues might come from licensing fees and royalties on net sales of a commercial product (Chapter 9).

Funds created to be capitalized primarily from biodiversity prospecting projects might also receive additional financing from donors, as long as the fund's goals match the donor's priorities. In some situations, it may be advisable to widen the funds' goals to cover general conservation objectives, or create a biodiversity prospecting 'sub-account' in an existing conservation trust fund. By linking biodiversity prospecting trust funds to conservation funds, a track record of successful programming and fiscal responsibility might be established more easily, which would increase chances of funding from the donor community.

V.C. Location of trust and assets

Trusts must be physically located in a selected country. Two main components of trusts that must have a physical base are the trust's board of trustees and the trust's assets. The components may be located in different countries, depending on various factors, and result in either of the following two possible scenarios (Box 5):

- a domestic trust with a domestic and/or off-shore asset management account;
- an off-shore trust with off-shore asset management.

Box 5. Pros and cons of trust location

- *Domestic funds* are local institutions, whose boards hold title to their assets. Their capital, however, can be invested domestically or off-shore. Domestic management of funds can increase local management capacity as well as the perception of national ownership, and can even contribute to domestic awareness and community participation in environmental issues.[1] However, domestically managed funds can suffer from political instability, thin capital markets, currency devaluation, or legal status conflicts with other countries. Bolivia's domestically managed National Environmental Fund, for example, lost much of its autonomy when a new government took control in 1993.[3] Domestic management of the fund, along with the fund's close ties to the national government, led to significant political influence on the fund, which impaired the fund's activities and undermined its principles.
- *Off-shore funds* can be advantageous because they provide a secure, hard currency market and access to professional asset managers, both of which foster donor confidence. Offshore management, however, does not foster domestic capacity-building or a sense of national ownership of assets, as can be the case with domestically managed funds. In addition, this type of management may not respond as promptly and effectively to the needs of the designated beneficiaries. It may also result in a lost 'connection' with the intended objectives and targets if the line of communication is not properly established.
- *Domestic fund with off-shore asset management* or *a 'two-tier' structure* allows a domestic fund to be paired with an offshore trust. The offshore trust holds title to the assets invested offshore, insuring that hard currency investments are located in an account in a secure market. The local fund is designated as the sole beneficiary of the trust and ensures that local stakeholders are fully represented. The local fund holds title to local assets (e.g. proceeds from a debt swap) and can choose to invest some of the benefits domestically (e.g. government bonds or interest-bearing accounts in local bank).

Box 6. Some lessons learned in the design of trust funds

Case-specific lessons learned

- Belize, Protected Areas Conservation Trust. (PACT): Recurrent costs – salaries for existing and additional park guards, forest rangers and other field staff – were not accounted for in the original list of funded activities.[3] The lesson learned was that planning must include consideration for existing short-term, ongoing, and long-term costs, in addition to those of new programs.
- Peru, National Fund for the Natural Areas Protected by the State (PROFONANPE): Without a tradition of charitable giving to environmental or other non-profit organizations, organizers had to look beyond in-country giving and rely on outside sources of funds, primarily debt-swaps.[3] The lesson

learned was that debt swaps are excellent opportunities for countries where there is great economic hardship. Debt swap proceeds, however, were restricted to certain activities, and did not include the fund's own operating costs. It is important to secure ongoing funding for day-to-day operations of the fund until income begins to accrue. Peru's PROFONANPE has also demonstrated that small administrative bodies can run successful funds. In 1996, the permanent staff included only four personnel, a Coordinator, Accounting Coordinator, Secretary-Assistant, and Driver-Messenger.

- Nigeria, Fund for Integrated Rural Development and Traditional Medicine (FIRD-TM): The principal lesson learned in development of the Fund – still in its early stages – is that a fully participatory process, involving all stakeholders, is the key to the successful design of a fund. Further, a fund that considers the local/political peculiarities on the ground is essential to fund stability and helps reduce potential areas of conflict or friction (Case Study 2).
- Suriname, the Forest People's Fund: The biodiversity prospecting project as a whole, including the agreement process among the NGOs and private sector, must be based upon clear communication, reliability, honesty, and trust. Experience has also shown that some communities may need significant help from fund staff to develop and write proposals, and to participate in the application process itself (Case Study 1).
- Panama, National Environment Fund: This project yielded lessons on the value of collaborating with existing funds. The bioprospecting fund is capitalized by a biodiversity prospecting project, currently organized as a sub-account within a larger national fund, NATURA. The fund contributes overhead to the larger institution, in exchange for the use of their administrative structure. This arrangement allows the bioprospecting fund to focus on fundraising and implementation of other activities (Case Study 3).

A general lesson learned: the importance of champions
At the early stages of organization, there needs to be one person who shoulders the responsibility of driving the process forward. This person may or may not be from the host country, but should be chosen by the organizing group and have skills that help catalyze the process. This person will:

- contact local legal counsel and assess the requirements for setting up a trust and engage his or her services to advise the forming trust;
- scout for additional organizers, qualified personnel and facilities, and begin to pre-select a possible board, keeping in mind the guidelines mentioned below;
- coordinate efforts by the group to enlist donors, including making contacts with international agencies and NGOs;
- hire a consultant skilled in the development of conservation trust funds.

Once the initial stages are complete, this person would likely become the executive director of the fund.

In determining where to locate a trust, the following should be considered:

1. Are there good reasons not to locate the trust in the country? For example, is the government stable? Does the local economy offer sufficient investment opportunities?
2. Even if the country's government is stable, is there a legal framework to support a trust, foundation, etc.?
3. What types of investment laws exist in the country? Will the country prohibit off-shore investment? Is the local economy stable? Is there enough technical expertise to manage the assets domestically?
4. Are the intended beneficiaries located in only one country?

If the answers to these questions are affirmative, a local trust would be advisable. If negative, an off-shore fund would be best. Cases will not always be clear, however, and it may be necessary to come up with a creative solution. Other options may include: establishing a trust by national act; obtaining a government exemption to invest abroad; establishing a trust under the auspices of the United Nations or other international agency; and/or establishing a two-tier trust. This last mechanism may work particularly well because it allows an off-shore trust to be combined with local beneficiaries.

V.D. Implementation

There are several important steps that must be completed to implement a biodiversity prospecting fund. They include:

1. Draft the foundation's deed and its bylaws. These legal documents establish the legal identity of the fund, state its purpose and goals, and create its governing structure.
2. Select the members of the board. The members should represent different backgrounds – culturally, professionally, sectorally (private, government, non-profit) – and be willing to use their expertise as advocates for the fund and not to act simply as managers.
3. Develop operation manuals. Of particular importance is the drafting of guidelines and rules for the grant process, including required materials, criteria for successful applicants, and defined areas of funding. The fund may need to engage a consultant to help draft these manuals and to implement the guidelines through training of staff. With more than 40 new or emerging conservation funds throughout the world, there is now a good selection of existing manuals in several languages that can be used as models.
4. Select administrative staff and management. These professionals will prepare budgets and work plans; develop and administer the grant process; develop and implement financial management systems; arrange for external auditing; and develop fund-raising strategies.
5. Establish technical committees to advise the board. For example, a finance committee would advise and inform the board about the fund's economic health

and potential investments, and can aid the management team. Scientific committees might advise on research priorities.
6. Train the board, managers, and administrative staff. These groups will need training in fund management and identification of priorities for the fund's constituents.
7. Inform potential fund beneficiaries about the funds' activities and grant application process. Provisions in the operations manual may need to include outreach to local communities to inform them, help them develop proposals, and help them with the grant making process as a whole.
8. Draft a monitoring and evaluation plan. These plans are essential to ensure that the fund meets its goals and continues to be responsive to changing needs.

V.E. Criteria for fund disbursement and compensation

Once the feasibility and design phases are settled, criteria for disbursement of income must be agreed upon. In the case of biodiversity prospecting funds, the relative contribution of different stakeholders must be assessed and difficult issues addressed such as: sharing of benefits with individuals vs. communities/institutions; distribution of benefits across communities and society, including to those not directly involved in research; and the most effective ways to promote conservation and sustainable development objectives (Box 7).

Box 7. Fund distribution

Suriname Fund distribution mechanisms
A multifaceted approach was developed in the Suriname project to distribute benefits, including:

- a long-term Research Agreement that controls the ownership, licensing, and royalty fee structure for any potential drug development;
- a Statement of Understanding that further defines the parties' intentions regarding the distribution of royalties among Surinamese institutions;
- a trust document establishing the Forest People's Fund (which is capitalized in part with up-front payments);
- provision for the transfer of technology and other forms of non-monetary compensation to Suriname.

FIRD-TM formula and criteria for fund disbursement
Apart from deciding the sectoral allocation of funds to targets and beneficiaries, the Board of Management has some general guidelines and criteria stipulated to guide it in the allocation and disbursement of funds. These include:
Project funding:

- the project must have clear, definite and identifiable results;
- the project must be sustainable and should have a reasonably attainable endpoint;

- the Board should, as much as possible, maintain a fairly balanced geographical spread in approving projects for funding.

Funding of individuals:

- will be based upon relative contribution to discovery of useful materials or drug development.

Criteria such as the following can act as a starting point for the development of more detailed criteria used in the evaluation of grant proposals:

- Is the project in conformity with the underlying principles of the fund?
- Will it help to promote the conservation of biodiversity and sustainable development?
- Will it meet the priority needs of target communities/institutions/stakeholders, as defined by these groups?
- Does it recognize and reward the contributions of stakeholders (e.g. communities, scientists, government, research institutions, conservation projects) to the biodiversity prospecting project?
- Will it promote the development of domestic and local capacity to study biodiversity, to conduct research on tropical or other locally important diseases, to standardize traditional medical systems, and to improve capacity to participate at a higher level in drug discovery and development?

A clearly defined set of criteria, a reasonably simple application and transparent evaluation process, are all necessary in order to facilitate prompt response to potential grantees and the release of funds to approved beneficiaries or projects (refer to Box 7).

V.F. Flexibility and efficiency

The creation of a trust fund will not be the answer to all organizational needs. Sometimes, it may be more efficient to channel money through a local NGO or to integrate a biodiversity prospecting trust fund into an existing conservation fund as a 'sub-account', thereby reducing the costs associated with building a fund from the ground up. An integrated fund might be particularly attractive in the context of biodiversity prospecting because financial benefits often arrive after a number of years (sometimes decades) or in smaller sums spread across many years. Without a steady source of income, a biodiversity prospecting fund risks running out of operating finances and jeopardizing its new programs. By associating with an already-established fund, a biodiversity prospecting fund could focus its resources on substantive activities and is thereby allowed a greater flexibility in the formation process.

VI. Conclusion

One of the principal lessons learned from existing funds is that a system predicated on a wide consultative and participatory process will increase the chances of acceptance by the principal stakeholders and, therefore, of the fund's success. Furthermore, since many countries have yet to enact relevant access and benefit-sharing measures, parties need not wait for legislation in order to commence a process of compensating relevant stakeholders, especially where the prospects of such enactment are not immediate. Even in cases where national legislation has been passed, the sharing of financial benefits continues to present numerous challenges, and many governments are calling for the establishment of trust funds. Benefit-sharing systems are sought that are based on fairness, are transparent, and support the objectives of the Convention on Biological Diversity. These systems should also be flexible and should allow incorporation of the multi-sectoral, multi-stakeholder agendas that characterize biodiversity prospecting relationships. The trust fund model presents an opportunity to develop this type of system for the sharing of financial benefits resulting from biodiversity prospecting partnerships.

VII. Case Study 1

VII.A. The Forest People Fund of Suriname

Marianne Guérin-McManus, Lisa M. Famolare, Ian A. Bowles, Stanley A.J. Malone, Russell A. Mittermeier, and Amy B. Rosenfeld – Conservation International (USA and Surinam)
The creation of the Forest People Fund of Suriname arose out of the International Cooperative Biodiversity Group* (ICBG) project, which was started in 1993 in Suriname, and is expected to provide a long-term compensation-sharing mechanism for revenues arising out of genetic resources and ethnobotanical knowledge.

The Suriname ICBG involves five different institutions, including Virginia Polytechnic Institute and State University (VPISU), Conservation International (CI), an international non-governmental conservation organization, Bedrijf Geneesmiddelen Voorziening Suriname. (BGVS), a pharmaceutical company owned by the Surinamese government, the Missouri Botanical Gardens (MBG), an American botanical research institution, two American pharmaceutical companies, Bristol-Myers Squibb Pharmaceutical Research Institute (B-MS) and, since 1998, Dow, each institution carrying out a specific role, including botanical and ethnobotanical collections and inventory, extraction, screening, chemistry, and drug development. The main focus of the Suriname ICBG project is to promote the discovery of biologically active plants for drug development and biodiversity conservation, as

---

*The International Cooperative Biodiversity Group (ICBG) is a US government-funded program sponsored by the National Institutes of Health (NIH), the National Science Foundation (NSF), and the United States Agency for International Development (USAID).

well as ensuring that communities and the source country derive maximum benefits for their biological resources and their intellectual contribution.

The Suriname ICBG group works with local tribal people to conduct some of the bioprospecting activities. The majority of the local participants are Bushnegros, or Maroons, who are descendants of runaway African slaves that escaped Dutch plantations on the coast over 300 years ago and settled along the river in central Suriname. Six distinct Maroon tribes live in the interior and depend on their extensive knowledge of forest resources for their survival. When the Maroons first fled into the forest, they experimented with medicinal uses for the plants and, through a process of trial and error, identified plants that were effective for various illnesses. They based their experiments in part on their memories of the healing traditions and plants in their native Africa and on information learned from Amerindians in Suriname's interior. This knowledge has developed into a rich and expansive understanding of the medicinal qualities of Suriname's forest plants.

While the ICBG contract and a Statement of Understanding govern the means by which future financial gains from bioprospecting are to be distributed, a separate trust fund was established to ensure that the tribal communities would benefit immediately from the access granted to their forest resources as well as provide an instrument meant to capture additional longer-term revenues. The fund compensates these communities for their ethnobotanical contributions to the ICBG project, creates conservation incentives, finances sustainable management projects, provides research and training exchanges, and supports other socially and environmentally sound projects.

The Forest People's Fund was established in 1994 with a $50,000 contribution from BMS, followed by another $10,000 donation in 1996. Additional contributions will be made of $20,000 each year as part of the renewed ICBG project until 2003. The Forest People's Fund Foundation is headquartered in Paramaribo, Suriname, and administers the Forest People's Fund according to the Foundation's by-laws. These by-laws were written by the Surinamese participants and are governed by the laws of Suriname.

The by-laws require the Board to meet at least four times a year and, whenever deemed necessary, to manage the fund's day-to-day operations, finances and handle legal arrangements. The Board of Directors comprises five members, including two representatives at large, two representatives from CI, and one who is nominated by BGVS. One Amerindian and one Maroon must fill the position of the members at large. CI's representatives are the President of CI, based in Washington, and the director of CI-Suriname. Richene Libretto, a District Representative of the people to Suriname's central government for the interior and part-Maroon, is the current representative of BGVS. Paul Abena is the representative of the Maroon communities, and Armand Karwafodi is the representative of the Amerindian communities. Each member is limited to a five-year term and may cast one vote in the Board's decisions.

The main activity of the Board of Directors is to review project proposals. Any tribal person in Suriname, community or foundation that has an idea for a project can submit a proposal. CI-Suriname staff members are available to assist interested parties in their project design and proposal. The Board then determines whether to grant funding according to whether the project advances the purpose of the fund, which is to 'stimulate residents of the interior and related living persons who contribute to and participate in the preservation and long-term protection of biodiversity and to provide them with social, educational, and economic assistance'.

The Forest People's Fund supports local communities in the interior of Suriname in projects involving community development, biodiversity conservation, and healthcare. To date, projects funded by the FPF include:

1. A project designed to transport people and goods bound for Paramaribo by boat to Ajonia, the furthest village accessible by road from Paramaribo. This project facilitated travel for people living in the interior while avoiding the creation of new roads, which cause environmental damage in the forest.
2. A sewing project, organized by Afinga, acquired sewing machines and material to make clothes.

VIII. Case Study 2

VIII.A. The Fund for Integrates Rural Development and Traditional Medicine (FIRD-TM) in Nigeria

Kent C. Nnadozie; Maurice M. Iwu; Elijah N. Sokomba and Cosmas Obialor 'The Fund for Integrated Rural Development and Traditional Medicine' (FIRD-TM or the Fund) was established in Nigeria at the initiative of Bioresources Development and Conservation Program. (BDCP) as an autonomous body to address the issues relating to bioprospecting and equitable benefit sharing aspects of the CBD within the framework of existing laws. It was a response to the major institutional gap, and lack of appropriate and effective vehicle to receive and channel benefits in an equitable and consistent manner, to source communities from which commercially useful genetic resources and specialized knowledge are derived. In establishing the Fund as an independent body with constituents from across all sectors, including the government and grassroots, it is anticipated that the principal problem of getting benefits to the localities that are the sources and custodians of the relevant biotic materials as well as knowledge will be overcome.

VIII.B. Background

BDCP is primarily focused on the establishment of integrated programs for the discovery of biologically active plants for drug development and the promotion of the conservation and sustainable use of biodiversity while ensuring that local source communities derive maximum benefits for their resources and their intellectual contribution. The basis for the establishment of the Fund is also linked to the International Co-operative Biodiversity Groups. (ICBG), which are networks of bioprospecting projects involving several countries, national and international institutions. The African ICBG, one of the networks, is a collaboration of the BDCP, Walter Reed Army Institute of Research, Washington, DC, the Smithsonian Tropical Research Institute, the University of Dschang (Cameroon), the International center for Ethnomedicine and Drug Development (Nigeria) and 13 other institutions in Africa and the United States.

VIII.C. Goals and objectives

The Fund was registered under the relevant laws in order to give it a legal personality capable of owning property, maintaining and defending actions. This position also bestowed it with a tax-exempt status under the laws as a not-for-profit organization. The Fund was established as a private, non-governmental and non-profit body primarily to facilitate and ensure the equitable distribution of benefits derived from biological diversity and knowledge of rural communities. It will provide short- and medium-term benefits in the form of immediate cash payments to individuals or groups and the sponsorship of development projects in communities, while long-term benefits in the form of royalties will depend on the final outcome of the bioprospecting activities. It will also apply revenues available to it to such projects or ventures that will promote conservation and sustainable use of biodiversity. It is expected to create the interface that ensures the establishment of mutual respect and bridge the gap of misunderstanding through the inclusive participation of the relevant community and traditional medical practitioners along with conventional scientists in the benefit distribution process and sharing of information.

VIII.D. Structure and governance

FIRD-TM is governed by a Constitution, which stipulates its aims and objectives, structure the nature of its principal organs, financial matters and conditions for dissolution.
The Fund has three principal organs:

- *The Board of Trustees*, in whom the property of the Fund legally resides but has no executive capacity with respect to the day-to-day running of the body.
- *The Advisory Board*, whose capacity is purely advisory and consists of distinguished experts in fields that are related to the objects of the Fund as well as eminent leaders and individuals who can contribute positively to the fulfillment of those objects.
- *The Board of Management*, which is the executive/administrative organ of the Fund. The members serve for a period of five years each and may be re-appointed for a further five-year term. Members of the Board of management are currently 10 in number drawn from a wide spectrum of interests and constituencies, including traditional leaders, traditional healers, representation from the government, and independent experts. All the members are currently serving on voluntary bases and receive neither remuneration nor allowance for their input

except for costs incurred directly in the performance of their duties with respect to the Fund.

There is also a provision in the Constitution for a full-time *Administrative Secretary*, who will administer the secretariat and the day-to-day businesses of the Fund including maintaining record of the Fund's activities and overseeing all other staff of the Fund.

VIII.E. Funding

The principal source of the Fund's funding at the initial stage came from BDCP and its collaborators, especially through the ICBG program. At the inauguration of the Board, an initial donation of $40,000 was received from the Healing Forest Conservancy as part of the Benefit Sharing Program of Shaman Pharmaceuticals Inc., and this is yet to be disbursed. Further substantial donations were received from Orange Drugs Limited, an indigenous pharmaceutical company, as well as from the Indigenous Pharmaceutical Manufacturers Association of Nigeria. There have also been pledges of further support and assistance from various sources, in both the public and private sectors. Although the law requires the auditing of its accounts at regular intervals, the Constitution further stipulates for the annual audit of the accounts of the Fund by external auditors, and the preparation of the annual report, copies of which will be made available to the necessary and interested parties. There is a statutory prohibition against the distribution of profits or dividends to the members or Trustees even upon dissolution. Upon dissolution or winding up, the assets can only be transferred to some other organization having similar objects or applied to some other charitable objects as may be determined by the Board of Trustees.

VIII.F. Criteria for Fund Disbursement and Compensation

The Board of Management (the Board) manages all the affairs of the Fund, decides on issues of policy and budget, reviews and approves proposals and work plans but does not undertake direct participation in the implementation of approved projects. It does, however, exercise direct supervisory roles with regard to approved projects. The Board has adopted a fund allocation formula for disbursement to the various targets within the mandate of the Fund. These include biodiversity conservation activities/national interest; education; traditional healer's associations for group projects or as micro-credit fund; community development associations/village projects; women (especially widows), and children's welfare. The Board has ongoing consultations with village heads and a professional guild of healers in determining the nature of compensation to apply or projects to embark upon in any given locality. In executing its mandate, priority is to be given to such projects and activities that promote or encourage biodiversity conservation and sustainable development.

VIII.G. Prospects and anticipated problems

One of the initial problems experienced at its inception, was the misconception in some circles about the actual roles of the Fund especially with respect to the target communities. But efforts are being made to educate the relevant parties of the true roles and position of the Fund, especially its community development and conservation-oriented focus.

Other key problems that have arisen or anticipated and which might impede progress include:

1. Lack of adequate resources – because of the nature of expected benefits to be shared (often uncertain and long term), there is bound to be pressure on the Fund, especially in the light of pressing needs and pervasive poverty. Apart from external funding, the Fund is ultimately expected to become self-financing, and part of the approach to address this is the development of a reasonable secure and sustainable investment portfolio from the part of the funds available to it. The Board is also engaged in local fund-raising drives to supplement external funding and to broaden its resource base.
2. Competing demands. With respect to this, the Board intends to follow clearly outlined modalities and criteria to assist it in choosing projects to sponsor and communities to benefit. It had, however, resolved, especially at this initial stage, to embark on small manageable projects that touch the people closely rather than undertaking any major or large project that might prove wasteful or get stuck along the way. It is currently evaluating several projects already commenced or planned by some local community development associations and herbalist unions for the purpose of advancing additional support or sponsorships, where appropriate.

VIII.H. Key lessons

- A participatory process, exemplified by a cross-cultural membership of the Board of Trustees/Management is key to acceptance and co-operation by stakeholders, as well as necessary for balanced and informed decision-making.
- Recognition of and adaptation to local socio-political realities is essential to stability and sustainability.
- Continuing communication and consultation with stakeholders enhances the profile and enduring relevance of the trust fund.

IX. Case Study 3

IX.A. The Panama ICBG Trust Fund: The National Environment Fund of the Fundación NATURA, Todd Capson, Smithsonian Tropical Institute, Panama

To couple biodiversity conservation with bioprospecting, we chose to establish a fund that works through a Panama-based foundation that promotes the study, conservation, and sustainable use of biological diversity. The foundation we work with, Fundación NATURA, has supported projects throughout Panama. The fund we established, the National Environment Fund, will provide financing that will be available for conservation and development projects, including biodiversity prospecting, accessible through Fundación NATURA's existing competitive grants program. The National Environment Fund will receive a portion of all access fees, milestones and royalties that are generated by ICBG bioprospecting activities in Panama.

IX.B. Fundación NATURA

To appreciate the context in which the National Environment Fund will operate, background information on Fundación NATURA is helpful. Fundación NATURA was legally incorporated in 1991 with endowments from the US Agency for International Development (as principal donor), the Government of Panama and The Nature Conservancy, as donors to a permanent fund (personal communication, IUCN). The Nature Conservancy serves as fiduciary. The objective of Fundación NATURA is to manage this permanent fund, known as FIDECO. The fund is used to finance projects that promote knowledge, management and conservation of the environment in Panama, with a special emphasis in the Panama Canal watershed.

An additional goal of Fundación NATURA is the enhancement of infrastructure within Panama for conservation and sustainable development, primarily through strengthening of the institutions, such as non-governmental organizations, that implement those projects. The endowment for Fundación NATURA is currently $33 million, the interest from which provides Fundación NATURA's operating budget and revenue for the programs it supports. Fundación NATURA has 10 full-time staff that work on project-funding programs, accounting and administration.

All proposals submitted to Fundación NATURA for funding are reviewed by a Technical Committee, a group of highly capable volunteer professionals with different areas of expertise. The Technical Committee reviews applications and makes recommendations with regard to the proposals received by Fundación NATURA, in addition to assisting Fundación NATURA's technical staff in the supervision of ongoing projects. The Technical Committee makes recommendations for funding to Fundación NATURA's Board of Directors, the foundation's highest authority.

IX.C. How the National Environment Fund will work with Fundación NATURA

As the majority of the revenue received by Fundación NATURA from FIDECO must be spent in the Panama Canal watershed, Fundación NATURA is actively seeking new donors, in particular, donors that will allow the foundation to support projects aimed at conservation and sustainable development in areas outside of the Panama Canal watershed. There are ecosystems in Panama of global importance that reside outside the watershed, such as those found in the provinces of Bocas del Toro, Chiriquí, San Blas and Darién. Donors to the National Environment Fund can specify that projects be supported in regions of Panama that would otherwise receive a small fraction of Fundación NATURA's support.

In the contract that has been developed between the Smithsonian Tropical Research Institute and the Government of Panama, 30% of the revenues flowing to Panama such as access fees, royalties and milestones has been committed to the National Environment Fund. We envision that when the National Environment Fund has enough capital to generate significant amounts of interest, the principal will be invested, while the interest will be used to support projects. Until that point is reached, meritorious projects can be directly supported by simply providing funds directly to Fundación NATURA, to be spent according to well-defined criteria to be provided by both Fundación NATURA and the coordinators of the National Environment Fund. Among Fundación NATURA's most successful projects are those that were funded through their Small Grants Program in which no award is greater than $5000. Thus, even if our initial contributions to Fundación NATURA are through access fees (US$75,000 per year), we will be in a position to make substantive contributions to biodiversity conservation in Panama.

In summary, Fundación NATURA plays an innovative and important role in the ICBG Panama program. The infrastructure of Fundación NATURA, in particular the rigorous peer-review process by qualified professionals that is the heart of its competitive grants program, makes it an extremely attractive beneficiary for funds that may result from our biological prospecting work in Panama. The National Environment Fund was established explicitly to receive funds from biological prospecting initiatives in Panama. As donors to Fundación NATURA, we can explicitly include sustainable biological prospecting projects as among those that are supported by our fund. Thus, the National Environment Fund is an important step along the road to a long-term sustainable biological prospecting program. As a well-known, highly regarded foundation, both within Panama and internationally, our association with Fundación NATURA also provides legitimacy, transparency and credibility to our biological prospecting work in Panama.

X. Case Study 4

X.A. The Healing Forest Conservancy: Trust Fund Constitution: Katy Moran and Tom Mays

At the time of its incorporation in 1990 as a for-profit corporation, Shaman Pharmaceuticals, Inc. founded the Healing Forest Conservancy (the Conservancy), an independent non-profit foundation. The Conservancy was established specifically to develop and implement a process to return long-term benefits to Shaman's collaborating countries and cultures after a product is commercialized.

The company provides immediate and medium-term benefits to collaborating cultures and countries during the drug-discovery process. Through the Conservancy, Shaman will donate a percentage of profits from commercial products to all long as Shaman has a profit. The Conservancy will distribute these long-term benefits equally to all the countries and cultures that are Shaman collaborators, regardless of where the plant sample or traditional knowledge that was commercialized originated. In such a financially unpredictable industry, spreading the benefits and risks among all Shaman collaborators increases opportunities for benefits.

Trust funds are proposed as financial mechanisms to receive and disburse long-term revenues generated from the commercial use of bioresources to a variety of stakeholders whose representatives serve as board members. Critical to the success of a fund is a constitution that serves as the general operative document establishing the goals, objectives, rights and duties of the fund. A constitution also supplies a legally enforceable mechanism, under domestic law, for a trust fund. Trustees of the fund, as a collective entity, may institute legal proceedings in their capacity to achieve trust fund objectives.

The Conservancy Constitution is a template to use for the benefit-sharing actions of the Conservancy in many different countries. As stated under the mission of the Conservancy, the model Constitution is a legal document that is flexible enough to respond to unique conditions in the numerous countries where Shaman collaborates. It supplies a legal mechanism to widely disburse financial resources, over a long time frame, and within varied sectors of society. The highlights are as follows:

X.B. Sponsoring entity

The use of a sponsoring entity, such as an NGO, in the Constitution is intended to facilitate the establishment of the Fund. Such an entity may not exist in every country, however, and it may be necessary in certain cases to rely on other sponsors, such as governmental entities or universities, to establish the Fund. It may also be possible to have the Fund established independently by various groups joining together as founding members.

X.C. Healers' associations and culture groups

It is also possible to structure the Fund without a Membership and have the Fund managed directly by a Board of Directors. The template Constitution incorporates a membership component in order to promote greater community participation in the activities of the Fund. The Fund must be an open forum, and it should be easy to become a Member. The provision requiring a two-thirds majority vote of all members ensures general agreement for the admission of a new Member. It is possible that a sponsoring entity of the Fund may insist on having ultimate approval for the admission of new members. While the sponsoring entity should be advised on who is made a new member of the Fund, the sponsor should not be able to veto the admission of new Members.

Unlike the admission of a Member, removal of a Member is intentionally made difficult and subject to a unanimous vote requirement to ensure that a single Member is not removed by a simple majority of the Members due to some disagreement over one or more policies of the Fund. Permitting non-Members to serve on the Board is designed to have the same effect as an outside Director that can provide objective advice to the Board. Inclusion of the reference to the selection of Board members without regard to their ethnic, political or other background is particularly important for countries where the absence of such a provision may result in disparate treatment among various ethnic or other groups.

X.D. Governments

It is important to acknowledge the consent and support of each host country government for the activities of the Fund in order to minimize the risk that the Fund is perceived as a threat to the sovereign right of the host country government to exploit its own natural resources. However, this acknowledgment should not be viewed as undermining the independence of the Fund and its autonomous operating authority as a non-governmental organization. The precise legal status of the Fund will depend on laws of the host country wherein the Fund is established. It is important that the Fund be able to obtain the benefits normally associated with non-profit and charitable entities, such as exemption from taxation. The international character of the Fund is intended to permit the Fund to collaborate with similar entities established in other countries. The NGO status of the Fund is equally important, as it reinforces the independence of the Fund from the host government and permits the Fund to participate in other NGO fora.

X.E. Distribution of benefits

Article VIII of the model Constitution provides for the distribution of benefits to all stakeholders in accordance with the following guidelines:

1. At least 50%, but not more than 70%, of available funds shall be distributed to traditional healers' organizations and community development funds.
2. At least 10%, but not more than 15%, of available funds shall be distributed to national universities and other national institutions that share a commitment to the aims and objectives of the Fund.
3. At least 10%, but not more than 15%, of available funds shall be distributed to the sponsoring entity for its furtherance of conservation and development activities.

X.F. Committees

The model Constitution designates Committees that are intended to assist the Board of Directors in allocating the benefits to be distributed by the Fund. It is not intended that the committee membership be limited to Members of the Fund.

X.G. Benefits Allocation Committee

This Committee shall ensure that the benefits provided by the Fund are allocated consistent with Article VIII of this Constitution.

X.H. Training Committee

This Committee shall serve as the liaison between the Fund and scientists, traditional healers and any other individuals or organizations as the Committee sees appropriate. It shall actively promote the implementation of programs or other mechanisms designed to train individuals in the areas of biodiversity conservation and traditional medicinal knowledge.

X.I. Educational Committee

This Committee shall serve as the liaison between the Fund and universities and other educational institutions and support university departments and other individuals or groups that are committed to the education of individuals in the areas of biodiversity conservation and traditional medicinal knowledge.

X.J. Credit Union Committee

This Committee shall supervise the activities of the Fund with respect to its lending programs and make recommendations to the Board of Directors as to appropriate credit activities of the Fund.

X.K. Other Committees

The Board of Directors may, by resolution, approve the establishment of such other committees as may be required to achieve the objectives of the Fund and as permitted by law.

References

1. Mikitin K. (1995) Issues and options in the design of GEF supported trust funds for biodiversity conservation. Environment Department Papers, Biodiversity Series, Paper No. 011. Global Environment Division, The World Bank.
2. Conservation International. (1991) The debt-for-nature exchange: a tool for intertnational conservation.
3. ECOFONDO. (1996) Regional consultation on National Environmental funds (NEFs) in Latin America and the Caribbean. Case studies on the in-country resource Mobilization, June 11–14, 1996. Cartegena, Columbia.
4. Putzel LR, Zerner C. (1998) Community-based conservation and biodiversity prospecting in Verata, Fiji: a history and review. Rainforest Alliance Natural Resources and Rights program, New York.
5. Curtis R et al. (1998) Designing a Fond Haitien Pour L'Environement et Le Developpement; issues and options. The Nature Conservancy.
6. Global Environment Facility (GEF). (1998) GEF evaluation of experience with conservation funds. GEF Council, GEF/C.12/Inf.6 (Sept. 10, 1998).
7. United States Agency for International Development (USAID). (1996) Endowments as a tool for sustainable development. USAID Working Paper No. 221. Center for Development Information and Evaluation.
8. International Planning Group on Environmental Funds (IPG). (1995) Environmental funds: the first five years; issues to address in designing and supporting green funds. A preliminary analysis for the OECF/DAC Working Party on Development Assistance and Environment.
9. Spergel G. (1993) Trust funds for conservation. World Wildlife Fund.

Iwu and Wootton (eds.), Ethnomedicine and Drug Discovery
© 2002 Elsevier Science B.V. All rights reserved.

CHAPTER 20

The regulation of botanicals as drugs and dietary supplements in Europe

JOERG GRUENWALD

Abstract

There are considerable differences in the classification of botanical products in the overall world market. In the US, the category of dietary supplements holds the largest share of the whole botanical market with over 80%. The over-the-counter (OTC) status plays a minor role with only about 3%. In Europe, however, the dietary supplement segment only accounts for 15–20% of the European herbal market. The rest are registered drugs. In some countries like Germany and France, they are still often prescribed and reimbursed. Under the restrictions in the reimbursement situation, this segment of dietary supplements, as well as the self-medicated OTC drugs, has recently started to grow, while the group of prescribed botanicals is decreasing. In several European countries, botanicals have always been part of mainstream medicine and therefore have already found their way into the regulations of each country from the beginning. Continuous scientific research with many clinical studies gave botanicals the necessary scientific background to enter the market on the drug level. The European Union (EU) legislation has established a common basis for botanicals in Europe for approximately 30 years, but is still not active in all member states. In the US, however, botanicals have only started to broaden the market, recently after the legislation of dietary supplements was modified to make better claims. In addition, new drug applications can be expected soon since the first INDs have been passed. Research based on a lot of science 'borrowed' from European sources has just started in the US.

Keywords: *regulation, classification, drugs, food, dietary supplements, cosmetics, herbal medicinal products*

I. Introduction

I.A. Classification of botanicals in Europe

In Europe, botanicals are marketed under several different categories from full drug status to foodstuffs. The categories are:

- herbal medicinal products
- prescription drugs
- OTC drugs (self-medication or prescription and reimbursement)
- traditional medicines
- dietary supplements, dietetic foods, foodstuffs

- herbs (as food/nutraceuticals or food additives)
- cosmetics.

The first classification of a product is always the determination of whether it is considered a medicinal product or not.

I.B. Herbal medicinal products

The European Guideline 65/65/EEC regulates the registration and marketing of medicinal products in Europe, because the European harmonization has still not been fully implemented, and national products dominate the market, regulated according to national laws. Only a few multinational pharmaceutical companies, like Boehringer Ingelheim/Pharmaton, Bayer or Novartis, and some phytomedicine companies, like Lichtwer, Schaper & Brümmer, etc., have Europe-wide strategies. It can be expected that this trend will be followed by the majority of the multinational pharmaceutical companies, which will develop Europe-wide registered products. Some countries refer mainly to the European Council directive 65/65 for medicinal plants; others still refer to their own national legislation, which varies from country to country.

I.C. Dietary supplements, foodstuffs and cosmetics

Other products like dietary supplements, foodstuffs or cosmetics fall under the regulation of the national food legislation. So far, there is no common EU legislation of these products other than as spices and flavorings. A report to find a common regulation has been filed, and is under discussion. The classification of botanicals into the different categories is not easy, because the definition of herbal products varies from country to country. Some countries have special dietary supplement categories, while other countries treat all herbals as drugs. The main criterion for a differentiation is the application of specific medical or therapeutic claims to the product. This brings another problem into focus: there is a wide variety in the indications of botanicals, ranging from the use of garlic for coughs and colds in the United Kingdom to the use of garlic for arteriosclerosis in Germany.

If the product is considered a medicinal product, full registration with a full dossier on quality, safety and efficacy is necessary. Bibliographic documentation can be based on clinical studies and the monographs of the ESCOP,[2] the WHO[3] and the German Commission E.[4] A possibility exists for a simplified proof of efficacy for traditional medicinal products in some countries like Germany, Austria, Belgium and France. Spain is considering this regulation.[5,6] If full medical claims are added to these products, they require a full application. An exemption of drug registration of botanicals can also be made for products, which a practitioner makes for or supplies to the patient directly. This is specifically the case in the United Kingdom. Powdered herbs with no claims and brand name can also be marketed without a licence in the UK.

II. Definition and classification of herbal medicinal products in the different countries

II.A. Germany

In Germany, botanicals can be marketed as drugs, dietary supplements or food. In accordance with the WHO definition, the term 'herbal remedies' applies to medicinal products whose active ingredients consist exclusively of plant material or herbal medicinal product preparations (e.g. powdered herbal medicinal products, herbal medicinal product secretions, essential oils or extracts). Homeopathic preparations are excluded as well as chemically defined substances (e.g. thymol, menthol).

Herbal medicines marketed as drugs must submit a full registration according to the German Drug Law and the council directive 65/65/EEC. References to the Commission E are useful support but not sufficient.

Traditionally used botanicals can be marketed with mild claims if the products have a long history on the market. This application is not accepted by EU criteria. They are still considered medicinal products.

Dietary supplements are regulated by the 'Lebensmittel-Bedarfsgegenständegesetz'. The products do not need a registration process if they do not state medicinal claims and if the plants are not monographed or generally regarded as medicines.

The main criterion of classification is the objective purpose of the product as it is seen by the trade. The criteria for this classification are composition, posology and claims. Sometimes, dosage form, labeling and retail outlets or price are considered if necessary.[7-10] The legal status is required on the label.[11]

II.B. France

Plant-based medicinal products, with the exception of isolated chemical constituents, are considered herbal medicinal products by the French Medicines Agency 'Agence du Médicament'. There is a distinction in prescription status between poisonous plants and others.[12]

Herbal products that state therapeutic claims or have pharmaceutical activity are considered medicinal products. They have to be sold by pharmacists or authorized herbalists. There is a list of 34 plants and seven combinations, which make exceptions, and can be sold directly to the consumer as a therapeutic medical product or refreshing beverage. For some historically used herbal preparations, there is an abbreviated registration procedure according to the 'Avis aux Fabricants'[13] of 1987, a list of 170 plants and indications.

Other than these exceptions, there is no category of dietary supplement. Outside pharmacies, there are a number of herbal products sold as 'para-pharmaceuticals' despite their questionable legal status. The French pharmaceutical industry has also denounced the sale of 'pseudo-medicinal products', which could be dangerous to public health.[14]

II.C. Italy

Herbal products are regulated by the Council directive 65/65/EEC. They can be marketed as medicinal products, dietary supplements, dietetic foods or food. A full registration is needed for medicinal products.

Some single herbs without indications can be sold freely in pharmacies without registration. Products classified as health foods can state health claims unless they are referring to therapeutic qualities. These products have to be traditionally used as food or flavoring. They can be marketed as dietetic products, as replacement for meals with dietetic indications, or as food supplements, which need to be notified with the Ministry of Health and show their status on the label. A list of narcotic substances, which can be used in prescription-only products, exists. These plants cannot be part of dietary supplements or food products.[15,16]

II.D. United Kingdom

In the United Kingdom, The Medicines Control Agency is responsible for the classification of herbal medicines. In section 132 of the Medicines Act 1968, a herbal remedy is defined as solely consisting of plant material and an inert ingredient or water. Botanicals can be marketed under the categories of registered products in three different lists determining the OTC or prescription status, as dietary supplement or as non-licensed products (either given to the patient on an individual basis or sold without claims).

The classification of herbal products is made with consideration to claims, properties, labeling, advertisement, product form and comparison to similar products on the market. There is no list of plants that distinguishes between medicinal plants and foodstuffs. They have to be considered by their function at the presented dosage if they do not have exclusive pharmaceutical usage. There are no lists of official monographs or indications.[17,18]

II.E. Ireland

Herbal drugs are generally considered medicinal products. Products that state medicinal claims or contain herbs that are recognized as having medicinal properties, are medicinal products. A list of these plants exists. Products containing herbal extracts, essential oils, herbal teas, etc. have to be authorized by the Irish Medicines Board.

Herbal teas can be marketed as foodstuffs, if they are considered 'Generally Recognized As Safe (GRAS)' and do not state any medicinal claims.

Essential oils with a general use as flavoring or natural source of flavouring and which have the GRAS status may be marketed without a licence. Any other essential oil has to be registered.

There is a not yet fully implemented report on Food Supplements and Health Foods by the Food Safety Advisory Committee of 1993. This includes a list of substances that are toxic under normal circumstances of use.

II.F. Austria

In Austria, the Austrian Drug Act of 1983 and 1988 regulates herbal medicinal products. Herbal products that make therapeutic claims or have particular pharmaceutical activity are regarded as medicinal products and require a full drug registration. For some non-prescription medicines, an abridged registration procedure is possible.

In Austria, herbal products that are supposed to be sold as food supplements can be defined in two different categories:

They are either 'Verzehrprodukte' or dietetic products. Both have to be announced at the Federal Ministry of Labor, Health and Social Affairs. They are subject to the judgment of the Ministry. The marketing can be prohibited within three months after notification. Both can state health claims. With a later addition of a medical claim, the Ministry may decide that the product has to be switched from a food supplement or dietetic food to a medicinal product.

A 'Verzehrprodukt' is a product intended for oral intake and is not intended to be food or medicinal product. These are 'not preponderantly serving nutritional or delight reasons'. Dietetic products are suppliers of nutrients and vitamins and not regular foodstuffs. The legal status of a product has to be shown on the label.[11]

II.G. The Netherlands

Herbal medicinal products are regulated according to the EC guideline 65/65/EEC. A full registration is necessary as long as claims are made.

They can also be marketed as food supplements or foodstuffs that are not registered.

Herbal products are defined as products containing plants, plant parts, extracts, gums, juices, etc. as active ingredient.

II.H. Luxemburg

In Luxemburg, there is no legal definition of herbal medicinal products. The general definition of the European guideline applies.

However, there is a list of medicinal plants that can be sold single or in combination as teas outside of pharmacies, if they do not state therapeutic claims. For this, they need to be pre-packaged.[19]

II.I. Belgium

The Belgian Notice to Applicants is the basis for classification and regulation of botanicals in Belgium. Herbal medicinal products are products consisting only of plant material.

The presentation or destination of the product defines it as a medicinal product or foodstuff. If claims are made, the product is considered a medicinal product.

There are three lists of plants that help define herbal products and their applicability under an abbreviated registration procedure.

- List 1 contains dangerous plants that cannot be used as food or be contained in food.
- List 2 contains non-toxic mushrooms.
- List 3 contains plants that have to be notified if they are sold in any standard dosage form like capsules, tablets, powders, sachets, ampoules, etc.

The plants from list 3 are not considered as medicinal products if no claims are made. It is not permissible to produce pre-dosed forms of plants that are not on the list.[20,21]

II.J. Denmark

In Denmark, there is a new guideline concerning 'Natural remedies'. The active component of the preparation has to be exclusively from naturally occurring substances in concentrations that are not substantially greater than those in which they are found in nature. Chemical substances can be added as inert galenic aid substances. The natural material can be derived from plants animals or minerals, and only minor illnesses can be addressed. Exceptions are natural remedies containing prescription-only or homeopathic medicinal products. If any of these factors do not apply, the product is treated as a regular medicinal product.[22]

II.K. Finland

Here, we also have the definition of 'Natural remedies'. Similar to the definition in Denmark, the active ingredient can be derived from plant, animal, mineral or bacteria. The natural remedy has to contain substances that are traditionally used for medical purposes. Medical purposes are basically therapeutic or preventive functions, which also classifies the product as a medicinal product. There is a list of traditionally used herbs for medicinal purposes. If the herb is not included, it can be marketed as foodstuff, as long as no medical claims are made.

Food supplements are sold under the legislation of the Food Act and cannot state any medicinal claims. They can be named as Food supplements, health products, herbal products or special preparations.

II.L. Sweden

Medicinal products have to be authorized by the Medical Products Agency prior to marketing. The have to be registered with full documentation.

Similarly to the other Scandinavian countries, herbal products can also be defined as a 'natural remedy'. The Medical Products Agency (MPA) defines natural remedies as medicinal products (with claims of therapeutic use or preventative use) in which the active ingredient is derived from natural sources of plant, mineral, bacteria or animal. Constituents may not be chemically modified or isolated chemically. They can only be intended for self-medication and conform to traditional use. Exceptions are homeopathic preparations or products for injection.

The main criterion for the classification of herbal products is the intended use and intention indicated in the marketing presentation as well as the dosage form. Some products fall under the general food legislation like camomile or peppermint teas.[23,24]

II.M. Spain

Botanicals can be registered with a full documentation of quality, safety and efficacy. Medicinal plants in tea form, powder in capsules, or tablets can be registered on the basis of the abridged procedure according to directive 65/65 EEC. The Ministerial Order of October 3, 1973 established a special registration for medicinal plants. Products such as extracts and tinctures require complete documentation and can be registered in the OTC category.

The Ministerial order of 1973 mentions a second classification of medicinal plants called phytotraditional medicinal products. Phytotraditional products cannot state claims of therapeutic use and do not have to file for marketing authorization as long as they contain only one species. Furthermore, the Ministry of Health has to be notified, and the legal status and reimbursement conditions are required on the label.[11]

II.N. Portugal

In Portugal, no legal definition exists for herbal medicines. They are controled by the Ministry of Health and are regulated in the same way as chemical drugs. No specific registration rules exist for herbal medicines, and products that state no claims are not considered medicinal products.

Herbal products can also be regulated as health products. These products can state health claims or maintenance claims. Health products include cosmetics, medicinal plants, dietetic products with therapeutic use, and homeopathic preparations. The controlling Agency for this category is the National Institute of Pharmacy and Medicines (Inframed).[25] The reimbursement conditions and the legal status are required on the label.[11]

II.O. Greece

In Greece, herbal drugs are regarded as medicinal products. Herbal medicines have not been monographed, and the marketing authorization is granted on the basis of the evaluation of the specific finished product. Bibliographic data can be used. Herbal medicinal products contain only plants or plant preparations as active ingredients.

Products that state no claims and have no pharmaceutical activity can be sold as dietary supplements. Food supplements have to be authorized by the Greek Health

Authority (EOF). They have no claims for indications,[26] but the legal status is required on the label.[11]

References

1. AESGP Study. (1999) Herbal medicinal products in the European Union; Internet: http://www.aegsp.be, email info@aegsp.be.
2. ESCOP (1996–1997) ESCOP monographs on the medicinal use of plant drugs; European Scientific Cooperative on Phytotherapy.
3. WHO. (1999) List of monographs on selected medical plants, Vol. 1. Geneva: World Health Organization.
4. Blumenthal M, Goldberg A, Gruenwald J, et al. (1998) American Botanical Council, German Commission E Monographs. Boston, MA: Integrative Medicine Communications.
5. Commission of the European Community. (1996) Guidelines on the quality, safety and efficacy of medicinal products for human use. In: The rules governing medicinal products in the European Community, Vol. III, Parts I and II.
6. Commission of the European Community. (1996) Guidelines on the quality, safety and efficacy of medicinal products for human use. In: The rules governing medicinal products in the European Community: quality of herbal remedies, Vol. III, Part I.
7. Lebensmittel- and Bedarfsgegenständegesetz. (LMBG). (1994) BGBl. I. p. 1169; 8 June 1993; revised by BGBl. I p. 3538.
8. Verordnung über diätetische Lebensmittel (Diätverordnung). (1988) Bekanntmachung zur Neufassung des BGBl. I. p. 1713.
9. Arzneimittel der biologischen Medizin nach der 5. AMG-Novelle. (1994) Gesundheitspolitische Umschau 45; pp. 265–268.
10. Verlängerung der Zulassung nach Artikel 3 § 7 AMNG von außerhalb der Apotheke freigegebenen Human-Arzneimitttel nach § 44 Abs. 1 (1.7.1993). AMG, Bundesanzeiger Nr. 119.
11. Commission of the European Community. (1997) Guidelines on the packaging information of medicinal products for human use authorized by the Community.
12. Ministry of Social Affairs and Solidarity. (1990) Notice to Manufacturers Concerning Marketing Authorization Applications for Plant-Based Medicinal Products. Issue 90/22b.
13. Médicaments à base de plantes: Les Cahiers de l'Ágence No.3. Agence du Médicament (replacement for the 'Avis aux Fabricants', Bulletin officiel No. 90/22 1990).
14. Organization Charts of The Agence Du Medicament: French National Medicines Evaluation Agency and Direction of Drug Evaluation. Regul Affairs J 166–168; February 1994 and 901; October 1996.
15. Organization Chart of The Italian Ministry of Health. (1996) The Regul Affairs J 712.
16. AESGP. (1989). Economic and legal framework for non-prescription medicines; an overview on 11 European countries; AESGP; 1st Edition.
17. Organization Chart of The UK Medicines Control Agency (MCA). (1998) Regul Affairs J 640.
18. Medicines Control Agency. (1995) The Medicines for Human Use (Marketing Authorisations etc.) Regulations 1994: Medicines Act Leaflet 81.
19. Organization Chart of The Division of Pharmacy and Medicines, Luxembourg. (1994) Regul Affairs J 166.
20. Ministere Des Affaires Sociales, de la Sante Publique et de l'énvironment: Belgisch Staatsblad vom 21.11.1997: Arzneipflanzenzubereitungen in Belgien als Nahrungsergänzungsmittel verkehrsfähig; Anlage zum BAH-Rundschreiben 5/1998.
21. Organization Chart of The Belgian Ministry of Health. (1996) Regul Affairs J 342.
22. Organization Chart of The Danish National Board of Health Medicines Division. (1994) Regul Affairs J 73.
23. Organization Chart of The Swedish Medical Products Agency (Läkemedelsverket). (1994) Regul Affairs J 988.
24. Sweden: Regulatory Information Page. http://www.medmarket.com/tenants/rainfo/sweden.htm
25. Organization Chart of The Portuguese Institute of Pharmacy and Medicines (INFARMED). (1994) Regul Affairs J 256.
26. Organization Chart of The Greek National Drug Organization (EOF). (1995) Regul Affairs J 903.

Iwu and Wootton (eds.), Ethnomedicine and Drug Discovery

Regulation of herbal medicines in Nigeria: the role of the National Agency for Food and Drug Administration and Control (NAFDAC)

GABRIEL E OSUIDE

Abstract

In most developing countries, including Nigeria, the majority of the populace lives in the rural areas, where the use of herbal medicines is common. The use of herbal medicines in the urban areas is on the increase, arising from the global inflationary trend, which hampers the sustainable supply of orthodox medicines and reduces the purchasing power of the populace. The Nigerian Government has recognized the need and shown political will by approving and adopting guidelines for the practice of traditional medicine. The regulatory authority, the National Agency for Food and Drug Administration and Control (NAFDAC), has also taken steps to protect the health of consumers by drafting the 'Guidelines for the Registration and Control of Herbal Medicinal Products and Related Substances in Nigeria'. Three broad classes are defined in the Guidelines, and preparations will be considered under four categories, each of which has its protocol. Extemporaneous preparations are only to be listed and not registered or advertised. Post-listing evaluation/monitoring is, however, mandatory. Herbal medicinal products manufactured on a large scale, whether imported or locally manufactured, must be registered and their advertisement messages and scripts approved by NAFDAC prior to their marketing. Homeopathic medicinal products must be registered and their advertisement messages approved prior to marketing. Post-registration evaluation/ monitoring is also mandatory for both large-scale herbal medicinal products and homeopathic products.

Keywords: *sustainable, orthodox medicine, traditional medicine, National Food and Drug Administration and Control, homeopathic*

I. Introduction

Medicinal plants and their preparations have been used since the earliest history of mankind, and they have formed one of the foundations for healthcare in virtually all cultures throughout the world. The use of herbal remedies is an integral part of traditional medicine, which is practiced in different forms in Nigeria as well as in other regions of the world. In spite of their long tradition of use as medicinal products, the public's interest in herbal remedies is still growing with particular emphasis on their clinical, pharmaceutical and economic value. A significant shift has occurred from the use of orthodox medicines to herbal medicinal products globally, and this trend is similarly observed in Nigeria. In many rural communities in Nigeria and other developing countries, the use of herbal remedies forms an important component of primary healthcare, and in some cases, it is the only accessible healthcare.

In recognition of this fact, over the years, the World Health Organization (WHO) has been playing a major role in the promotion of traditional medicine. Accordingly, the WHO resolution WHA42.43 of 1989 urged member states to introduce measures for the regulation and control of medicinal plant products and for the establishment and maintenance of suitable standards. The International Conference on Primary Healthcare, held in Alma-Ata, USSR, in 1978, recommended inter alia the accommodation of proven traditional remedies in national drug policies and regulatory measures. In pursuance of its various resolutions, the WHO has prepared guidelines for the assessment of Herbal Medicines by member states to assist national regulatory authorities, scientific organizations, and manufacturers to undertake an assessment of the documentation and dossiers in respect to such products. These guidelines provide basic criteria for the assessment of quality, safety and efficacy and the requirements for the labeling of herbal medicinal products.

I.A. Regulation of herbal medicines

In spite of the importance of herbal medicines to healthcare delivery, not much attention has been given to their regulation and control. This poses public health problems regarding the quality, safety and efficacy of the products. The public is also exposed to toxic substances resulting from either some inherent property of the product or possible adulteration of the herbal preparations with potential toxic substances. For these and other reasons, it is necessary to regulate the production and use of herbal medicines in all countries.

I.B. Responsibilities of regulatory authorities

The basic responsibilities of a regulatory authority are to ensure that all products subject to its control conform to acceptable standards of quality, safety and efficacy, and that all premises and practices employed to manufacture, store and distribute these products comply with requirements to ensure the continued compliance of the products with these standards until they are delivered to the end-user. These responsibilities are being carried out in respect to pharmaceutical products in most countries but the same cannot be said of herbal medicinal products, as their control has not been given adequate attention. The situation is, however, changing, and many countries have taken necessary steps to address the issue in the interest of public health.

In Europe, significant progress has been made with regard to the licensing of herbal medicinal products, with proof of quality, safety and efficacy as basic requirements for registration. An attempt is being made to establish a centralized system of marketing authorization, which may be extended to phytomedicines.

The European scientific cooperative on phytotherapy was formed in 1989 to establish harmonized criteria for the assessment of phytomedicines, to support scientific research and to contribute to the acceptance of phytotherapy at a European level. A number of herbal medicinal products have been approved for use in some of the countries.

India and China are well-known examples of countries where traditional medicine has been integrated into the official healthcare delivery system, and the use of herbal medicines is governed by appropriate regulatory provisions.

In the United States, most herbal health products are referred to as dietary supplements with structural/functional or health claims and therefore not subject to FDA approval for marketing authorization. However, herbal products with therapeutic claims are classified as drugs and require FDA approval.

In Africa, the Organization of the Unity Scientific, Technical and Research Commission has played a leading role in promoting medicinal plant research in the region. The efforts of the commission resulted in the publication of the first edition of the African pharmacopoeia in 1985. The need to regulate the use of herbal medicinal products has been recognized by many countries in the region, and appropriate regulatory measures are being put in place.

I.C. The role of NAFDAC

The National Agency for Food and Drug Administration and Control (NAFDAC) is the regulatory authority in Nigeria. The Agency was established in 1993 with the mandate to regulate and control the importation, exportation, manufacture, advertisement, distribution, sale and use of food, drugs, cosmetics, medical devices, bottled water and chemicals. These functions are carried out through the enactment of relevant laws and regulations. The Agency has made a significant impact through its operations too.

As earlier mentioned, herbal medicinal products are widely used in Nigeria as part of traditional medicine practices, and this trend is growing probably as a result of the current unfavorable economic situation in the country, which has made orthodox medicines unaffordable and inaccessible to a large proportion of the population. The government has also shown a serious commitment towards the regulation of traditional medicine practices in pursuance of its National Health Policy. The recent approval of the National Policy on Traditional Medicine Code of Ethics and draft legislation by the National Council on Health is a significant step towards the establishment of National and State Traditional Medicine Boards, which will enhance the regulation of traditional medicine practices and promote cooperation and research in traditional medicine. In response to these developments, NAFDAC has taken necessary steps to regulate and control the use of these products with the hopes of ensuring their quality, safety and efficacy. Accordingly, the Agency in 1997, organized an International Workshop on Standardization and Regulation of Herbal Medicines in collaboration with the Bioresources Development and Conservation Programme and the West African Pharmaceutical Federation. The Workshop made far-reaching recommendations to facilitate the regulation and control of herbal medicines by NAFDAC.

The Agency subsequently organized a consultative meeting with experts in 1998 to develop draft guidelines for the registration of herbal medicinal products in Nigeria. This meeting was followed by another consultative meeting comprising participants from various interest groups such as researchers from universities and research institutes who specialize in different areas of medicinal plant research, traditional healers, and orthodox medical practitioners. This meeting reviewed the draft guidelines prepared by the previous group and produced the blue print for regulatory control of herbal medicinal products in Nigeria. The National Council on Health recently approved this document. The Agency has received applications for registration of herbal medicinal products and related substances that are currently being processed for listing based on the following criteria:

- evidence of proven efficacy based on evaluation of documentation;
- laboratory evaluation;
- submission of certificate of manufacture and free sale from the country of origin;
- expert opinion on the product.

The Agency will commence full registration of herbal medicinal products in line with the recently developed protocols, as the necessary facilities are already in place.

I.D. Definitions of terms

As expressed by the WHO in its definitions, different terminologies and overlapping definitions for 'herbal products' exist in different countries. Such terminologies include alternative medicine, phytopharmaceuticals, traditional remedies, dietary supplements, nutraceuticals, herbal remedies, etc. Bearing in mind the need for simplicity and suitability and taking cognizance of the Nigerian situation, three classes of traditional medicines have therefore been defined as follows:

Herbal medicinal products are defined as finished and labeled medicinal products containing plant material and/or they make therapeutic or prophylactic claims. This definition includes homeopathic preparations derived wholly or partly from plant sources. All preparations containing a plant in part or wholly are also included. All herbal preparations that are presented with no therapeutic or prophylactic claims are considered to be cosmetics or food items verses medicinal products. However, in Nigeria, some traditional medicine preparations contain only animal or mineral material with or without plant parts. In such cases, the two definitions below apply:

Animal medicinal products are defined as finished and labeled medicinal products containing only animal material and/or their preparations are presented with therapeutic or prophylactic claims.

Mineral medicinal products are defined as finished and labeled medicinal products containing inorganic minerals and/or their preparations are presented with therapeutic or prophylactic claims.

I.E. Classification of herbal medicinal products

The following classification of herbal medicinal products has been adopted by NAFDAC.

I.F. Extemporaneous preparations

These are herbal medicinal products that should be used in a locality by the patient within a short period of a few days only. They are compounded and dispensed by the traditional medical practitioner for patients on a one-to-one basis in the clinic. Any regulation for such products is difficult to implement, as the products are not usually available for general distribution and sale. However, their documentation should be encouraged.

I.G. Large-scale manufacture of herbal medicinal products

A large-scale manufacturer produces units of a preparation in quantities beyond the usual one-to-one quantities made by a traditional medical practitioner for their own patient's use within a short period of time. Large-scale manufacture is intended for distribution to a very wide area through various outlets and for storage for long periods of time. The issues of preservation and shelf life are significant in the consideration of approval for registration of such products.

II. Approved registration procedure for herbal medicinal products

The approved registration procedure for herbal medicinal products is summarized as follows:

1. Application by applicant in writing.
2. Payment of prescribed application fees.
3. On satisfactory assessment of the duly completed application form and relevant documents, a pre-registration inspection of the production facilities of the applicant will be required for locally manufactured products on payment of prescribed fees. Documents to be submitted with completed form are as follows:
   i For locally manufactured products:
     a Authorization as manufacturer of products to be registered
     b Evidence of registration of applicant by relevant local competent professional body
   ii For imported herbal products:
     It is necessary to apply to have an import permit to bring samples for registration:
     • Current Registration License of superintendent Pharmacist and Pharmaceutical premises issued by the Pharmacist Council of Nigeria.

- Power of Attorney issued by the manufacturer and certified by a public notary.
- Certificate of manufacture and free sale in the country of origin.
- Certificate of analysis of the product from country of manufacture.
- Evidence of satisfactory clinical trials conducted in Nigeria.

4. Dossier and samples will be submitted for all products for vetting.
5. Samples required for laboratory analysis will be submitted, and prescribed processing fees will be paid.
6. On satisfactory inspection (for local products) and laboratory reports, a brief for registration of the product will be presented for consideration by the Product Registration Committee of the Agency.
7. On approval by the committee, the applicant will be required to pay the prescribed fee for the Registration Certificate, which is renewable yearly for products listed and every five years for the other categories of herbal medicinal products.
8. Mandatory post-marketing surveillance and reporting on the use (including adverse reactions) of the herbal medicinal product shall be required. Such a report should be submitted to relevant health authority (Local Government, State Ministry of Health, Federal Ministry of Health) for forwarding to NAFDAC.
9. All advertisements (promotion) of registered herbal medicinal products must be approved by NAFDAC. The detailed protocol for registration of herbal medicinal products is outlined in Appendix A.

II.A. Other considerations

Highlighted briefly are some areas that are relevant to the development of herbal medicinal products and their regulatory control.

II.B. Research and development (R&D)

In order to improve on the general acceptability of herbal medicines and provide them with some scientific validity, concerted R&D work is needed to standardize the nomenclature, collection, extraction process, formulation procedures, quality, safety, dosage, indications, contra-indication, etc. There is a call for the establishment of the actual need and appropriateness of the medicinal substances used. We must also ensure that they are rationally used and that the requirements for their use are assessed as accurately as possible. Basic criteria for the evaluation of the quality, safety and efficacy of traditional medicines need to be defined in order to assist national and state authorities with control and regulation. The aim of herbal medicine R&D should be directed at providing rationale for the continued use of efficacious and safe products, and excluding inactive or toxic effects. Despite the drawback of low funding of the R&D programs in both the public and private sectors in Africa, field studies and evaluation should be focused on.

II.C. Conservation

As demands for herbal medicine increase, pressure on the medicinal plant resources also becomes greater. Collection of wild plants for traditional medical use is extremely detrimental to the existence of certain species. Cultivation of plants in gardens or private farms must be encouraged. Alternative technology must be applied. Traditional medicine practitioners lack the means or know-how for large-scale cultivation of medicinal plants, but they can manage small herbal gardens in their localities. It is also absolutely necessary to design alternative cultivation methods that will help local traditional medicine practitioners harvest medicinal plants without endangering them. The large demands made by major importers for high volumes of plant material also contribute to the decline of medicinal plants species in Africa.

II.D. Education and training

Well-trained practitioners of traditional medicine will greatly improve the quality of herbal products and stimulate research into many of our medicinal plants and products. It is absolutely essential to determine an appropriate development caliber for herbal practice at all levels. A formal training program should be designed to complement the hereditary mode of knowledge transfer. Such trainings should be organized by the relevant government institutions or by Non-Governmental Organizations. (NGOs). It may also be necessary to organize a forum for the exchange of ideas and information on a routine basis. This may be in the form of workshops, seminars or courses to educate the traditional healers on some important aspects of their practice and update them on latest developments.

III. Conclusion

In closing, this has highlighted the measures that NAFDAC has put in place to facilitate the regulatory control of herbal medicinal products in Nigeria. NAFDAC will no doubt check the current indiscriminate use and advertisement of these products and build confidence in their beneficial effects. Furthermore, the Agency will establish a database on herbal medicinal products with the help of compiled information on medicinal plants that have been studied by various research institutions in Nigeria. A database will also be established on poisonous plants, which will require the continued support and assistance of the Bioresources Development and Conservation Programme, and other relevant organizations within and outside the country.

Appendix A

Extemporaneous preparations

Information about the applicant

- name and location address of the applicant (not PO Box)
- name and location address of the herbal clinic, if different from above (not PO Box)
- registration number with the State Board of Traditional medicine
- site description.

Technical information on product

- name of medicinal product
- type
- single or multi-component
- herbal medicinal product, animal medicinal product or mineral medicinal product.

Details of the plants (give the details below with respect to each plant used in the preparation)

- Common name (if any)
- Botanical name (including family and the authority)
- Local name (give all the possible ethnic names)
- Photograph of the whole plant (preferably in color and at the flowering stage).

General description of the plant

- macroscopical description of the plant parts (gross morphology, organoleptic including sensory characters)
- microscopical description of the plant part including its powdered form (while it is desirable to have micromorphological data, this should not prevent initial listing of the herbal medicinal product)
- evidence of Herbarius identification and authentication from a recognized herbarium/taxonomist (e.g. voucher specimen number or herbarium number).

Plant part used in the preparation (give details in respect to each plant)

- morphological parts used (leaves, stem, etc.)
- season of collection (dry or wet)
- geographical source (indicate place where collection was made)
- was it collected from the wild or cultivated? (if cultivated, state agricultural means used, e.g. fertilizer/manure type)
- date and time of collection
- method of collection (manual labor or mechanized).

Production process

- Form in which it is used (whole or power)
- Condition for plant collection (fresh or dry)
- Drying method (e.g. sun-drying, oven-drying, or by fire, if applicable)
- Duration of drying, if applicable
- Reduction of size (grinding or chopping, if applicable)
- Other post harvest treatments, if any.

Special, unique and necessary production requirements (if any)
Production environment (give a brief description)
Type of final presentation (tea bags, pills, etc.)

- in making any dosage form indicate excipients or adjutants used, if any
- packaging or container (plastic, glass bottle, calabash, etc.)
- solvent used (water, oil, local gin, etc.)
- temperature of extraction (cold, warm, or boiling, etc.)
- quantity of each plant material and solvent
- method of mixing of the materials (shaking, stirring etc., indicate sieving for powers, if need be)
- utensils for extraction and mixing (earthenware pot, metal pot, mortar and pestle, rod, etc.)
- duration of mixing with solvent
- total time of plant material staying in contact with solvent
- actual extraction process (sifting, sieving, decanting, etc.)
- other extraction process (maceration, percolation, etc.)
- dilution or concentration by evaporating
- volume reduction or making up to volume, if any.

Uses of the herbal medicinal product

- major indications for which it is used (diseases treated)
- contraindications (not to be used by pregnant women, etc.).

Methods of use

- dosage (specify for adult and children)
- storage conditions (state clearly especially if it affects the efficacy of the product)
- route of administration (oral, anal and topical)
- duration recommended for treatment
- antidote (in case of an overdose).

Standardization

- Since extemporaneous preparations are to be used within seven days, phytochemical and pharmaceutical data on the raw materials may suffice.

Determination of contaminants

- residual organic solvent
- extraneous material (insect, animal parts, stones, etc. should not be present if not part of the product composition listed above).

Pharmacognostic standards

- ethnomedical information to support efficacy and safety (describe the popularity, frequency and extent of use). Include any incantation available or relevant literature.

Acknowledgments

I would like to express my profound gratitude to the Bioresources Development and Conservation Programme for inviting me to participate in this conference. I consider the conference very timely and relevant, especially with the renewed global interest in the search for remedies from plant sources for emerging diseases that currently have no known cure.

References

1. WHO. (1989) Drug Inform 3:43–50.
2. WHO. (1991) Guidelines for the assessment of herbal medicines.
3. WHO. (1998) Regulatory situation of herbal medicines: a worldwide review.

Iwu and Wootton (eds.), Ethnomedicine and Drug Discovery
© 2002 Elsevier Science B.V. All rights reserved.

CHAPTER 22

Considerations in the development of public standards for botanicals and their dosage forms

V Srini Srinivasan

Abstract

With the widespread use of herbal supplements by the American public, there is an increasing need for public standards, which manufacturers can adopt to assure proper identity of the article contained in the container and thus safety of the public. An important element involved in the development of public standards for botanicals and their dosage forms is the intended use. Articles recognized as having fitness for use could be defined adequately only by a full range of standards – namely identity, strength, quality, purity packaging, labeling and, where necessary, performance standard. This paper focuses on the approaches to the development of pharmacopeial standards for botanicals and their preparations treated as dietary supplements. Botanicals like biologicals are complex in nature, and standardization of the marketed products offers unique challenges to the standard setting organization.

Keywords: *herbal supplements, pharmacopeial standards, botanicals, standardization, dietary supplements*

I. Introduction

With the widespread and increasing use of herbal supplements by American consumers, United States Pharmacopeia is currently focusing on the development of public standards for botanicals treated as dietary supplements. Following a survey of the list of botanicals that are currently in the market as dietary supplements, a short list of 20 botanicals was identified and prioritized for development of information and standards (see Table 1). Among the criteria used in arriving at the above, were concerns of safety as demonstrated through documented traditional use and some reasonable evidence supportive of the reported pharmacological action and efficacy based on clinical trials reported in the literature. Our review of the literature pertaining to reported clinical trials on botanical dosage forms of interest to us, and the attendant statistical evaluation of the trials results, are primarily to ensure that the evidence of efficacy supports our standards development work. The above selection criteria have a direct impact on the policy that governs the admission of botanicals in the United States Pharmacopeia and National Formulary.

Table 1

List of botanicals prioritized by the USP Subcommittee on Natural Products for Monograph Development

1. *Zingiber officinale* Roscoe (Ginger)
2. *Valeriana officinalis* L. (Valerian)
3. *Allium sativum* L. (Garlic)
4. *Panax ginseng* C.A. Meyer (Asign Ginseng)
5. *Ginkgo biloba* L. (Ginkgo)
6. *Tanacetum parthenium* L. (Feverfew)
7. *Hypericum perforatum* L. (St. John's Wort)
8. *Panax quinquefolius* L. (American ginseng)
9. *Echinaceae angustifolia* DC (Echinacea)
10. *Chamomilla recutica* (L.) Rauschert (Matricaria Flower)
11. *Serenoa repens* (Bartr.) Small (Saw Palmetto berries)
12. *Crataegus laevigata* (Poir.) DC (Hawthorn)
13. *Hydrastis canadenis* L. (Golden Seal root)
14. *Silybum marianum* (L.) Gaertn. (Milk Thistel)
15. *Vaccinium macrocarpon* Ait. (Cranberry)
16. *Urtica dioica* L. (Nettle root)
17. *Piper methysticum* L. (Kava Kava)
18. *Eleutherococcus senticosus* (Rupr. Et Maxim) Harms (Siberian ginseng)
19. *Glycyrrhiza glabra* L. (Licorice)
20. *Angelica archangelica* L. (Angelica)

I.A. Placement of botanicals in the United States Pharmacopeia-National Formulary (USP-NF)

The criteria that govern the admission of botanicals in the USP and NF are outlined below:

Evidence supporting use of standard:	Publication
FDA-approved use	USP
USP-accepted use	
no FDA-approved use	NF
no USP-accepted use	
descriptive evidence for use for a material time and material extent	
absence of significant safety risk associated with use	
no FDA-approved use	
no USP-accepted use standard	No
report of significant safety risk associated with use	

I.B. Information issues

The guiding principle is the well-known definition of evidence-based medicine (EBM), which 'involves integrating current best evidence with clinical expertise, pathophysiological knowledge and patient preferences in making decisions about the care of individual patients'. Unlike synthetic organic medicinal compounds, which are often very pure, homogenous and well characterized, botanicals have some unique characteristics that differ from the former. Botanical dosage forms are usually prepared from standardized extracts of the plant material that often contain chemical constituents of varying ratios. This heterogeneity imparts batch-to-batch variability of the quality of both the extract and the dosage form derived from it. The term standardized that is used here is commonly believed to mean that every batch of the extract prepared contains a marker (or more than one marker substances) substance at some constant level or range when analyzed by a specified analytical method.

I.C. Devising a methodology for evaluation

In our approach, each botanical is evaluated on an individual basis, and the available research will govern the framework of the evaluation. The most common databases that are routinely searched for information include Medline, Embase, Agricola, Napralert, and Chemical Abstracts, etc. Our review of the literature is focused on the identification of the level of evidence of the safe and effective use.

The information generated includes the following:

- nomenclature
- plant description
- commercial preparations
- chemistry
- reported uses (folk medicines, global contemporary practise, medical literature reports)
- dosage range (traditional and clinical trials preparations)

Products available

- pharmacology/pharmacokinetics
- potential risks/precautions
- side-effects/adverse effects/interactions
- regulatory status
- evidence tables.

With regard to the level, type of evidence and the grading of recommendations, we are guided by the definitions of the above as enumerated by the US Agency for Health Care Policy and Research (AHCPR) and World Health Organization (WHO) (see Tables 2 and 3). The development of public quality standards for botanicals and their dosage forms raises further issues.

I.D. Standards issues

In view of enormous variations in the quality of plant materials such as root, rhizome, leaves, flowers, etc., arising out of factors such as variations in seasons, soil

Table 2
Levels of evidence (AHCPR and WHO guidelines)

Level	Type of evidence[2]
Ia	Evidence obtained from meta-analysis of randomized controled trials
Ib	Evidence obtained from at least one randomized controled trial
IIa	Evidence obtained from at least one well-designed controled study without randomization
IIb	Evidence obtained from at least one other type of well-designed quasi-experimental study
III	Evidence obtained from well-designed non-experimental descriptive studies, such as comparative studies, correlation studies, and case control studies
IV	Evidence obtained from expert committee reports or opinions and/or clinical experience of respected authorities

Table 3
Grading of recommendations (AHCPR and WHO guidelines)

Grade	Recommendation[3]
A (evidence levels quality Ia, Ib)	Requires at least one randomized controled trial as part of the body of the literature of overall good and consistency addressing the specific recommendation.
B (evidence levels IIa, IIb, III)	Requires availability of well-conducted clinical studies but no randomized clinical trials on the topic of recommendation.
C (evidence level IV)	Requires evidence from expert committee reports or opinions and/or clinical experience of respected authorities. Indicates an absence of directly applicable studies of good quality.

conditions, etc., the USP Subcommittee considers it appropriate to develop standards for the plant material first, to be followed by development of standards for the extracts derived from such standardized plant raw materials and finally the botanical dosage forms that are made from standardized plant extracts. The logic of this approach lies in the fact that controls instituted at each stage of manufacture of dosage forms assure the safety and quality of the end product consumed by the consumers.

II. Plant raw materials

II.A. Definition

The definition includes the name of the part(s) of the botanical used, common name as specified in the Herbs of Commerce (published by the American Herbal Products Association), and Latin binomial name with family name.

II.B. Labeling

Labeling requires the Latin binomial name and the part of the plant contained in the article along with the compendial name.

II.C. Packaging and storage

Since compendial specifications are intrinsically shelf-life specifications, appropriate container closure systems as well as temperature conditions of storage are specified to assure the article retains its stated strength, quality and purity until the declared expiration date.

II.D. Identification

Botanical identification and chemical identification tests are both critical. Botanical identification – macroscopic and microscopic – assures proper identity of the genus and species of the family specified in the definition. Chemical identification of the characteristic components of the article is usually achieved by employing separation techniques such as thin-layer chromatography, gas–liquid chromatography, and liquid chromatography in comparison with appropriate reference standards of the known chemical constituent.

II.E. Microbial limits

Botanicals are products of nature and therefore require rigorous control for pathogenic organisms. The tests provided ensure absence of pathogenic organisms such as *Salmonella* species, and *Escherichia coli*. Limits are specified on the total combined molds and yeasts count. When needed, tests for aflatoxins will be specified.

II.F. Limit tests

Since botanicals are products of nature, soil contamination is almost a predictable feature. Tests for total ash content, water-soluble ash, acid-insoluble ash, etc. would address this contaminant.

II.G. Heavy metals

This test ensures that contamination of the plant material with harmful heavy metals such as lead is detected and limited to low levels. The primary sources of this contamination, besides ground water, are milling operations.

II.H. Pesticide residues

Botanicals as dietary supplements are subject to food law. The Environmental Food Agency (EFA) determines food pesticide limits. When no limit is set, the limit is zero.

USP standards do not modify statutory requirements. With the intent to harmonize the requirements for pesticides testing with those of the European Pharmacopoeia, the limits and test procedures adopted by the USP in the established monographs are so far the same as those in European Pharmacopoeia. Pesticide limits in USP botanicals monographs are not applicable in the United States when the articles are labeled for food purposes. The limits may be applicable in other countries where the presence of pesticides residues is permitted or where botanicals are marketed as drugs.

II.I. Tests for marker substances

It is generally recognized that the reported pharmacological action is due to more than one chemical component acting synergistically with other chemical components present in the plant material. Therefore, no *assay* procedure for any single chemical constituent is specified in USP monographs on botanicals and their preparations. However, quantitative test procedures for more than one chemical constituent, commonly termed in the botanical world a marker compound, are specified. Generally, the chosen marker compounds are representatives of the different groups to which they belong. For example, the monograph on Ginkgo specifies procedures for flavone glycosides (flavonoids), and ginkgolides (terpene lactones).

II.J. USP reference standards

In support of every monograph established by the USP Committee of Revision, reference standards for unambiguous quantitative determination of the marker compounds as well as positive identification by qualitative tests will be developed and distributed for public use. Botanical reference standards offer totally different types of challenges than chemical standards. The reference standards could be of different types such as pure chemical compounds isolated from the plant, dried powdered plant material, or an extract. The main issues to be addressed here are those of availability, replaceability, stability and shipping conditions. In view of the expensive nature of pure chemical compounds, reference standards vials supplied may contain as little as 20 mg or so. To facilitate easy handling, single-use ampules are also under active consideration.

II.K. Botanical extracts

The definition of botanical extracts monographs specifies the solvent(s) used, the ratio of the crude plant material to extract, the content or range of marker substances and the plant part from which the extract is derived. The labeling requirements in the extract monographs call for specifying the above in labels.

II.L. Botanical dosage forms

Botanical dosage forms used as dietary supplements or as drugs are formulated into tablets or capsules using the same manufacturing technology as that used for vitamin-mineral combination products or drugs. Thus, one would expect the botanical dosage forms used as dietary supplements to exhibit the same performance characteristics as those of multivitamin-mineral preparations or drugs. Accordingly a *dissolution* requirement is a sine qua non for all oral solid dosage forms whether they are intended for drug or dietary supplement use. Unless a tablet or capsule releases the nutrients or active ingredients, further dissolution of the same in the biological fluid may be impaired. Since a typical botanical preparation formulated using an extract contains several ingredients, it is manifestly impossible and probably unnecessary to measure the dissolution of every ingredient. USP's approach is to require dissolution of only one marker compound for which a quantitative test procedure is specified in the dosage form monograph. The USP Subcommittee is encouraging manufacturers and other interested parties to submit suggested specifications and dissolution data for botanical dosage forms that are of interest to USP.

Iwu and Wootton (eds.), Ethnomedicine and Drug Discovery

The Belize Ethnobotany Project: safeguarding medicinal plants and traditional knowledge in Belize

MICHAEL J BALICK, ROSITA ARVIGO, GREGORY SHROPSHIRE, JAY WALKER, DAVID CAMPBELL, LEOPOLDO ROMERO

Abstract

The Belize Ethnobotany Project was initiated in 1988, through a collaborative effort of a number of individuals and institutions. This paper discusses some of the components of the project, and its accomplishments and challenges. A checklist of the flora has been produced and includes 3408 native and cultivated species found in Belize. The multiple-use curve is introduced as a way of determining the most appropriate sample size for ethnobotanical interviews/collections. Valuation studies of medicinal plants found in two areas of local forest are described and compared with values of traditional uses for farming, using a net present value analysis. Studies to determine sustainable levels of harvest of medicinal plants were also initiated in Belize and are ongoing. The link between conservation, drug development and local utilization of medicinal plants is discussed, and the various impacts on conservation considered. Our experience with the production of a traditional healer's manual is detailed, and the benefit-sharing program it resulted in is described. Various local efforts at developing forest-based traditional medicine products are discussed, as is the natural products research program based on Belizean plants. Other results of this project include the development of a medicinal plant forest reserve and a video documentation and teaching program. Ethnobotanical and related studies in Belize are continuing.

Keywords: ethnobotany, medicinal plants, traditional knowledge, benefit-sharing, valuation

I. Introduction

The Belize Ethnobotany Project (BEP) was initiated in 1988 as a collaborative endeavor between the Ix Chel Tropical Research Foundation and Belize Center for Environmental Studies, both Belizean non-governmental organizations, and the Institute of Economic Botany of The New York Botanical Garden. The main goal of the project has been to conduct an inventory of the ethnobotanical diversity of Belize, a country with significant tracts of intact forest. The project has carried out over 100 collection trips to various locales, and has gathered over 8000 plant specimens as of the end of 2000. The specimens have been deposited at the Belize College of Agriculture, the Belize Forestry Department Herbarium, as well The New

York Botanical Garden. (NYBG) and US National Herbarium. The BEP involves the cataloging of traditional knowledge provided by dozens of traditional healers and bushmasters of Mopan, Yucatec, and Kekchi Maya, Ladino, Garifuna, Creole, East Indian, and Mennonite descent. This paper provides an outline of some of the components and results of this project.

II. Checklist of the flora

In order to properly understand the ethnobotany of a region, it is essential to have a listing of its flora, with accurate botanical names. In view of this, a project was initiated in 1989 to produce a checklist of the flora of Belize. This was published in December 2000. The checklist is an aggregate of what is known about the plants of Belize (listing 3408 species) and their importance to people. The checklist, entitled *Checklist of the Vascular Plants of Belize, with common names and uses,*[1] encompasses all native and naturalized vascular plants, including ferns, gymnosperms, and cultivated plants, and is the foundation for much future work on the Belizean flora. This work recognizes 1219 genera and 209 families in Belize. The largest family is the Fabaceae *sensu lato* with 295 species plus eight subspecific taxa. The Orchidaceae is next, with 279 species, followed by Poaceae (248 spp.), Asteraceae (153 spp.), Cyperaceae (146 spp.), Rubiaceae (142 spp.), Euphorbiaceae (104 spp.), Melastomataceae (96 spp.) and Aspleniaceae (58 spp.). A total of 41 species in 24 families are endemic to Belize, comprising 1.2% of the flora. The taxonomic treatments are based on a thorough review of the literature, and the study and citation of approximately 17,000 specimens collected in Belize and examined at various herbaria, and includes the plants collected by the BEP.

Each species entry (see below) includes the currently accepted Latin binomial with author. Synonyms are included when they have been applied in the literature of Belize or the region. Following the synonym are references used to support the nomenclature and occurrence in Belize. Common names (Nv) used in Belize are reported next. The common names are in English, Spanish, and Maya, when known. Local and regional uses are reported from the literature and our fieldwork, and divided into 19 broadly defined categories, including medicinal uses, food, ornamental uses, poison, and product (a catch-all term for any use from a child's toy to a household implement). The growth form of each plant is also reported (e.g. erect shrub, small tree, climbing vine, creeping herb). The last entry, vouchers, are specimens collected in Belize and deposited in herbaria. A sample species entry is as follows:

Solanum nudum Dunal – Syn: *Solanum antillarum* O.E. Schultz – **Ref**: FG 10:131. 1974. **Local use**: PRD, MED, POIS; **regional use**: MED. Nv: diaper wash, lava paêal, lava pañal, maya washing soap, nightshade, sak-kol, yierba de barrer. – **Habit**: Shrub. – **Voucher**: *Arvigo 46, 503, 799; Atha 1021, 1334; Balick 1737, 2530, 2720, 3102; Bartlett 12962; Brokaw 31,368a; Dwyer 10802a; Gentle 2531, 4767, 6627, 6646, 8752; Lentz 2381; Lundell 435, 469; McDaniel 14339; Peck 808; Ramamoorthy 3022* MEXU!; *Ratter 4579, 4593; Schipp 312, 429, 959; Warrior 1862.*

III. The concept of the multiple-use curve

Essential to understanding the ethnobotany of a country is obtaining the confidence that the entirety of knowledge surrounding each plant has been collected. Establishing this confidence is dependent not only on interviewing as many individuals as possible, but on interviewing people from as many ethnic groups and regions within the country as possible. We have placed priority into obtaining as many 'collections' of ethnobotanical data from Belizean peoples as possible. For one group of plants, we obtained 143 interviews from individuals from four regions and five ethnic groups in Belize about their knowledge of 14 widespread medicinal plants. Using this, we could attempt to answer the question that ethnobotanists must constantly face – how many collections/interviews are sufficient to give the researcher an idea of the totality of a plant's uses in a particular area. In order to avoid problems often encountered by outside interviewers in foreign countries, Belizean nationals not only conducted all the interviews, but also aided in their conception and design.

To interpret and categorize the information, we have used the standards of ethnobotanical collections as established by The Royal Botanic Gardens, Kew. Kew's standards separate medicinal uses and treatments into 24 categories such as 'Circulatory System Disorders', 'Nutritional Disorders', etc. (Table 1). In using these

Table 1
Categories of medicinal uses of plants[a]

1. Unspecified medicinal disorders
2. Abnormalities
3. Blood-system disorders
4. Circulatory-system disorders
5. Digestive-system disorders
6. Endocrine-system disorders
7. Genitourinary-system disorders
8. Ill-defined symptoms
9. Immune-system disorders
10. Infections/infestations
11. Inflammation
12. Injuries
13. Mental disorders
14. Metabolic-system disorders
15. Muscular-skeletal-system disorders
16. Neoplasms
17. Nervous-system disorders
18. Nutritional disorders
19. Pain
20. Poisonings
21. Pregnancy/birth/puerperium disorders
22. Respiratory-system disorders
23. Sensory-system disorders
24. Skin/subcutaneous cellular tissue disorders

[a] Categories adopted from FEM Cook (1995) Economic botany: data collection standard. Royal Botanic Gardens, Kew.

categories, we have been able to establish, in this particular case, not only the totality of how medicinal plants are used, but how many interviews are necessary to capture that knowledge.

In attempting to describe the number of interviews necessary, we have applied the concept of a species-area curve to ethnobotanical collections. The species-area curve is used in ecological surveys as a method of estimating what area of forest must be inventoried before all species in that forest type have been located (Figure 1). We use the same concept to describe the rate at which ethnobotanical data are collected, going on the assumption that once asymptote of the graph has been reached, little information remains undiscovered (Figure 2). As is demonstrated by Figure 2, it is not uncommon to find new information about a plant's uses after as many as 100 interviews. This graph compares the rates at which the knowledge surrounding three species of medicinally useful plants was discovered. The graph shows that we can be confident that we have effectively captured all the manners in which *Vitex gaumeri* is used within the area. *Ruta graveolens*, as a new use is described at the 140th interview, demonstrates the importance of a large sample size. In addition, the slope of the *Bursera simaruba* line leads us to believe more interviews are necessary for a complete description of its uses.

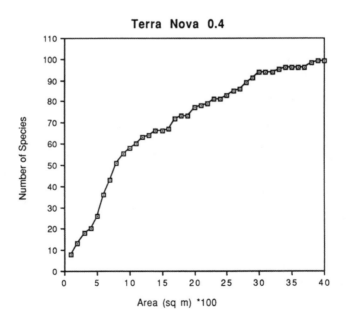

Fig. 1. Species area curve for 0.4 ha of forest at the Terra Nova Site, Belize, showing ca. 100 species per hectare.

Number of Medicinal Uses Described vs. Number of Interviews

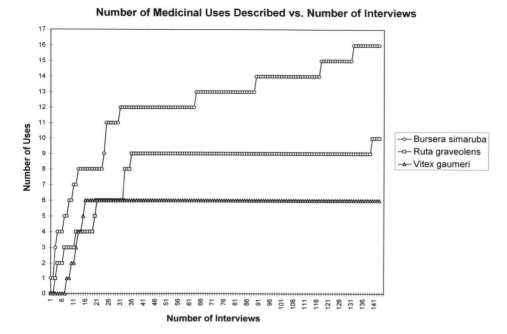

Fig. 2. Multiple use curve for three species of plants used in traditional medicine in Belize, with 141 interviews/collections.

IV. Valuation studies

A great deal of attention has been given recently to the value of non-timber products in the tropical forest. One method of ascertaining this value is to inventory a clearly defined area and estimate the economic value of the species found there. Peters et al.[2] were the first to elucidate the commercial value of non-timber forest products found within a hectare of forest in the Peruvian Amazon. This study did not include medicinal plants in their inventory, and, at the suggestion of the authors, this aspect was evaluated in Belize. From two separate plots, a 30- and 50-year-old forest, respectively, a total biomass of 308.6 and 1433.6 kg (dry weight) of medicines whose value could be judged by local market forces was collected. Local herbal pharmacists and healers purchase and process medicinal plants from herb gatherers and small farmers at an average price of US$ 2.80/kg. Multiplying the quantity of medicine found per hectare above by this price suggests that harvesting the medicinal plants from a hectare would yield the collector between US$ 864 and US$ 4014 of gross revenue. Subtracting the costs required to harvest, process and ship the plants, the net revenue from clearing a hectare was calculated to be US$ 564 and US$ 3054 on each of the two plots. Details of the study can be found in the original article.[3] The lists of plants and their uses are presented in Table 2 and Table 3. Not enough

Table 2
Medicinal plants harvested from a 30-year old valley forest plot (No. 1) in Cayo, Belize

Common name	Scientific name	Use[a]
Bejuco Verde	*Agonandra racemosa* (DC.) Standl.	Sedative, laxative, 'gastritis', analgesic
Calawalla	*Phlebodium decumanum* (Wild.) J. Smith	Ulcers, pain, 'gastritis', chronic indigestion, high blood pressure, 'cancer'
China Root	*Smilax lanceolata* L.	Blood tonic, fatigue, 'anemia', acid stomach ache, rheumatism, skin conditions
Cocomecca	*Dioscorea* sp.	Urinary-tract ailments, bladder infection, stoppage of urine, kidney sluggishness and malfunction, to loosen mucus in coughs and colds, febrifuge, blood tonic
Contribo	*Aristolochia trilobata* L.	Flu, colds, constipation, fevers, stomach ache, indigestion, 'gastritis', parasites

[a] Uses listed are based on disease concepts recognized in Belize, primarily of Maya origin, that may or may not have equivalent states in Western medicine. For example, kidney sluggishness is not a condition commonly recognized by Western-trained physicians but is a common complaint among people in this region.

information is available to understand the life cycles and regeneration time needed for each species, so we cannot comment on the frequency and extent of collection involved in sustainable harvest. However, assuming the current age of the forest in each plot as a rotation length, we calculated an estimate of the present value of harvesting plants sustainably into the future using the standard Faustman formula $V = R/(1 - e^{-rt})$, where R is the net revenue from a single harvest, and r is the real interest rate; t is the length of the rotation in years. Given a 30-year rotation in plot 1, this suggests that the present value of medicine is US$ 726 per hectare. Making a similar calculation for plot 2, with a 50-year rotation, yielded a present value of US$ 3327 per hectare. These calculations assume a 5% interest rate.

Table 3
Medicinal plants harvested from a 50-year-old Ridge Forest Plot (No. 2) in Cayo, Belize

Common Name	Scientific Name	Use
Negrito	*Simarouba glauca* DC.	Dysentery and diarrhea, dysmenorrhea, skin conditions, stomach and bowel tonic
Gumbolimbo	*Bursera simaruba* (L.) Sarg.	Antipruritic, stomach cramps, kidney infections, diuretic
China Root	*Smilax lanceolata* L.	Blood tonic, fatigue, stomach ache, rheumatism, skin conditions
Cocomecca	*Dioscorea* sp.	Urinary tract ailments, bladder infection, stoppage of urine, kidney sluggishness and malfunction, to loosen mucus in coughs and colds, febrifuge, blood tonic

These estimates of the value of using tropical forests for the harvest of medicinal plants compared favorably with alternative land uses in the region such as milpa (corn, bean and squash cultivation) in Guatemalan rain forest, which yielded US$ 288 per hectare. We also identified commercial products such as allspice, copal, chicle, and construction materials in the plots that could be harvested and added to their total value. Thus, this study suggested that protection of at least some areas of rain forest as extractive reserves from medicinal plants appears to be economically justified. It seems that a periodic harvest strategy is a realistic and sustainable method of utilizing the forest. Based on our evaluation of the forest similar to the second plot analyzed, it would appear that one could harvest and clear one hectare per year indefinitely, assuming that all of the species found in each plot would regenerate at similar rates. More than likely, however, some species such as *Bursera simaruba* would become more dominant in the ecosystem, while others such as *Dioscorea* could become rare.

The analysis used in this study is based on current market data. The estimates of the worth of the forest could change, based on local market forces. For example, if knowledge about tropical herbal medicines becomes even more widespread and their collection increases, prices for specific medicines would fall. Similarly, if more consumers became aware of the potential of some of these medicines or if the cost of commercially produced pharmaceuticals becomes too great, demand for herbal medicines could increase, substantially driving up prices. Finally, destruction of the tropical forest habitats of many of these important plants would increase their scarcity, driving up local prices. This scenario has already been observed in Belize with some species. It seems that the value of tropical forests for the harvest of non-timber forest products will increase relative to other land uses over time, as these forests become scarcer.

IV.A. Establishing sustainable harvest rates for locally used medicinal plants

In attempting to establish sustainable harvest rates of trees that produce bark of value in local traditional medicine, we performed a series of bark cuttings on approximately 40 individuals representing six species. With these individuals, we varied the width of bark cut, the length of the cut, and the alignment of the cut (north, east, south or west facing) hoping to establish the manner of cutting which removes the most bark with the least damage to the tree (see Figure 3). These trees were then marked and returned to every 6 months to monitor regeneration. As can be seen in the photos, the bark mends itself from the sides of the cut or in asymmetrical patterns, eventually entirely covering the cut area and becoming harvestable once again.

These preliminary cuts led us to the conclusion that the width of the cut is most influential in the rate of regeneration, the length of the cut being nearly inconsequential. As much as half of the circumference of the bark can be removed without permanently harming the tree, provided the removal is distributed in several cuts rather than one large cut. For example, the most sustainable and productive manner to harvest a tree with a circumference of 60 cm would be three 10-cm-wide

Fig. 3. Upper left and upper right: *Bursera simaruba*, bark regeneration over a 24-month period; lower left: *Simarouba glauca*, fresh cut for bark harvest; lower right: *Vitex gaumeri*, showing an irregular regeneration pattern after 6 months.

cuts, as long as desired, all 10 cm from the next cut. Having established the most efficient 'design' of the cut, we are now, through removing samples from several individuals of similar size of one medicinally valuable species, attempting to establish the exact rate at which bark regenerates. These studies are continuing at various sites in Belize, under the direction of Daniel Atha, in collaboration with Charles Peters and the primary author.

V. The link between medicinal plants, drug development and conservation

There is an often-stated assumption that the discovery of a new plant drug will help in conservation efforts, especially in tropical forest regions. This notion is based on the profit potential and economic impact, as well as the feeling that governments and people will somehow impose a greater value on a resource if it can produce a product with a multinational market. Table 4 is a summary of the distribution of value and potential of medicinal plants to support conservation efforts, viewed from three levels or perspectives: regional traditional medicine, the international herbal industry and the international pharmaceutical industry. Within each level, the distribution of economic benefits varies greatly. In traditional medical systems, the economic benefits accrue to professional collectors who sell the plants to traditional healers, or to the healers themselves. The local and international herbal industries produce value for a broad range of people and institutions, including collectors, wholesalers, brokers as well as companies that produce and sell herbal formulations. Proportionally, the bulk of the economic value in the international pharmaceutical industry is to be found in the upper end of the economic stratum, at the corporate level, as well as to those involved in wholesale and retail sales.

A comparison of the market value of these products reveals an interesting point – that the value of traditional medical products, which are used by billions of people around the world, comprises billions of dollars each year. Whether or not it is comparable to the US$ 80–90 million of global retail sales of pharmaceutical products has not been calculated, to the best of our knowledge. However, it can be argued that commerce in traditional plant medicines, consisting primarily of local activity such as that previously described, comprises a significant economic force. If it is assumed that three billion people use traditional plants for their primary healthcare, and each person utilizes US$ 2.50–5.00 worth annually (whether harvested, bartered or purchased), then the annual value of these plants could range between US$ 7.5 and US$ 15 billion, a sum that is significant and comparable to the two other sectors of the global pharmacopoeia. It is roughly estimated that the international herbal industry is about 10 times the size of the US herbal industry, which is about US$ 1.3 billion annually (M Blumenthal, personal communication, American Botanical Council).

Those who promote the linkage between conservation and the search for new pharmaceutical products often fail to point out that the time frame from collection of a plant in the forest to its sale on the pharmacist's shelves is 8–12 years, and that programs initiated today must be viewed as having long-term benefits, at best. Exceptions to this are agreements such as that between Merck, Sharp and Dohme and INBIO, the National Biodiversity Institute of Costa Rica. This agreement provides a substantial 'up front' payment from Merck for infrastructure development at INBIO and for the national parks system in Costa Rica and, hopefully, will be a model on which to base such North/South collaborations in the future. In traditional medicine and the herbal industry, the yields are immediate, and the economic impact to the individual, community and region can be quite significant.

The potential for strengthening conservation efforts ranges from low to high, depending on whether or not the extraction of the resource can be sustainably managed over the long-term, or is simply exploited for short-term benefits by collectors and an industry that has little interest in ensuring a reliable supply into the future. The conservation potential is minimal if the end-products are derived from synthetic processes, or from plantations developed outside of the original area of

Table 4
The economic value and conservation potential of plant medicines

Sector	Distribution	Market value	Pitfalls	Conservation potential
International Pharmaceutical Industry	Upper end of economic system	High – in the billions	• Over harvest • Synthesis (if no provision for benefits included) • Plantations established outside of area discovered	Low to high
National and International Herbal Industry	Full spectrum of economic system	High – in the billions	• Over harvest • Plantations established outside of area discovered	Low to high
Regional Traditional Medicine	Lower end of economic system	High – in the billions	• Over harvest-(sustainability)	Low to high

collection. To address this issue, the National Cancer Institute's Developmental Therapeutics Program seeks to ensure that the primary country of origin of the plant will have the first opportunity to produce the plant, should commercially valuable products arise as a result of their program (G Cragg, personal communication).

Finally, Table 4 summarizes the pitfalls inherent to each level, including overharvest, synthesis with no provision for benefits, land-tenure issues, and, as previously mentioned, plantations established outside the range of the species. In any attempt to plan for the maximum conservation potential of a discovery, these pitfalls must be kept in mind.

Further, harvest itself is not without pitfalls. One of the primary concerns about extraction is sustainability. A case in point is the extraction of a drug used in the treatment of glaucoma, pilocarpine. The source of pilocarpine is several species of trees in the genus *Pilocarpus* that occur naturally in the Northeast of Brazil: *P. pinnatifolius*, *P. microphylla* and *P. jaborandi*. Leaves have been harvested from the trees for many decades, usually under subcontract from chemical companies. Limited attempts at sustainable management were undertaken in the 1980s but, for the most part, harvest continued in a destructive fashion. Extinction – at the population level in many areas – has been the fate of these plants. Finally, over the last few years, cultivated plantations of *Pilocarpus* species have been developed, which will reduce the value of the remaining wild stands, as well as eliminate any incentive there was for conserving them.

VI. Rainforest remedies: a traditional healers' manual

One of the early requests received from the traditional healers we worked with was that the project prepare a semi-technical book on the uses of Belizean plants in traditional medicine, that could be used by local people, for their healthcare and in teaching their children. The result was *Rainforest Remedies: 100 Healing Herbs of Belize*, co-authored by Arvigo and Balick,[4] with line art by Laura Evans. The book contains sections on the common and scientific names, plant family, a simple botanical description, and information on habitat, traditional uses, and research results. Included in the latter category is information on clinical trials that might have been undertaken as well as any contraindications (cautions) known for the use of this particular plant medicine. The book was published by Lotus Press, Twin Lakes, Wisconsin, and distributed in the United States as well as Belize. A portion of the sales price is donated to a traditional healer's fund established by the Ix Chel Tropical Research Foundation and The New York Botanical Garden, and has benefited the healers who collaborated with the authors on this book. Proceeds are distributed twice yearly, in July and December, through the 'Traditional Healers Foundation'. The total value distributed as of 2000 was over US$ 20,000. The manual has gained widespread acceptance amongst people in Belize interested in traditional healing, as well as tourists looking for information on the use of local plants in medicine. A second edition of *Rainforest Remedies* was published in 1997. A summary of the program, and how the individual healers used their royalty payments, is presented in Johnston.[5]

VII. Development of a forest-based traditional medicine industry

One of the primary dilemmas in a development of a program of extraction of non-timber forest products (NTFPs) has been the long history of over-collecting of the resources, with their resultant decline, as well as the export of raw materials to centers and countries far from their origin. Rattan is a classic example of this over-exploitation, with people in producing countries who are closest to the resource receiving the smallest percentage of the profits involved in its production into high quality furniture. At least three locally developed brands of commercialized traditional medicine are now being marketed in Belize. These brands include 'Agapi', 'Rainforest Remedies', and 'Triple Moon' and are all entrepreneurial ventures. A key difference in these types of endeavors is that the 'value-added' component of the product is added in the country and region of origin of the raw material. As these particular product brands develop, and as new brands and products appear, based on the success of the original endeavors, a greater demand for ingredients from rainforest species will result. This could potentially contribute to preservation of tropical forest ecosystems, if people carefully manage the production or extraction of the plant species that are primary ingredients in these unrelated products. In addition, it is expected that small farmers will cultivate some of the native species, for sale to both local herbalists and for commerce. To address this latter possibility, the Belize Ethnobotany Project worked with the Belize College of Agriculture, Central Farms, in learning how to propagate and grow over two dozen different plants currently utilized in traditional medicine in Belize. Individual students took on particular species and carried out various agronomic experiments on propagation and growth. Hugh O'Brien, Professor of Horticulture at BCA has coordinated this effort, which has included the following genera: *Achras*, *Aristolochia*, *Brosimum*, *Bursera*, *Cedrela*, *Croton*, *Jatropha*, *Myroxylon*, *Neurolaena*, *Piscidia*, *Psidium*, *Senna*, *Simarouba*, *Smilax*, *Stachytarpheta*, and *Swietenia*.

VII.A. An ethnobiomedical forest reserve

In June 1993, the Government of Belize designated a 6000 acre (2428 ha) parcel of tropical forest as a forest reserve, for the purpose of providing a source of native plants used locally in traditional medicine. This forest is rich in medicinally important plant species, as well as serving as a wildlife corridor joining nearby conservation reserves. The initial philosophy behind the development of this forest reserve was that programs in traditional medicine, scientific research, and ecological tourism should create a synergistic effect to translate into economic return for the surrounding community, as well as provide an interface where scientists and traditional healers can work together to develop state-of-the-art management strategies for the sustainable extraction of important plant products.

The reserve was designated specifically for the extraction of medicinal plants used locally as part of the primary healthcare network. Accordingly, we refer to this type of extractive reserve as an 'ethnobiomedical forest reserve', a term intended to convey a sense of the interaction between people, plants and animals, and the healthcare system in the region.

The reserve was initially championed by a local group of traditional healers known as the 'Belize Association of Traditional Healers'. Conflict arose during the early years of the reserve, with a second group of individuals forming a healer's association, curiously enough, having the same name as the initial group, and demanding control over the management of the reserve and, at one point, the utilization of the assets in the initial group's bank accounts that were raised to implement the reserve. The new group, having political support from a newly elected government at that time, took over management of the reserve (but not the bank accounts), and for several years, there was no activity in the reserve. In addition, loggers encroached upon the reserve during that time and, with no guards in the area, were able to log a portion of the mahogany in the protected area. It has been said locally that one of the members of the second group was selling phony deeds for 'retirement home' subdivisions in the reserve to Belizeans living in the United States, a scam that quickly fell apart when these people returned to Belize and wanted to inspect their 'property'. Finally, the management of the reserve was given to a third group that is currently looking for funds for the reserve's preservation and operation. In this case, conflict over the ethnobiomedical forest reserve was initially quite destructive, but in the end, members of the second group requested that another medicinal plant reserve be set up near the village of San Antonio and named as a memorial to Don Elijio Panti, one of the elders involved in this project. Demarcation and surveying of the Panti reserve are now just beginning. Despite the painful and often comical drama associated with the establishment of the first reserve, the final result has been that at least double the land area is now set aside for use by the traditional healers in these two reserves.

VIII. Natural products studies

Our initial journeys to Belize were sponsored by the collection contracts received from The National Cancer Institute (NCI). During the 10-year span of two contracts with the NCI, thousands of bulk samples of plants were collected under the supervision of various local government agencies, for study by the NCI. Data on initial screening results were returned to these agencies, given to the individuals who, collaborated in the collections, and discussed during various traditional healers' meetings and seminars offered in Belize. While a number of samples had an interesting initial activity, no samples screened to date have been selected for further study by the NCI research team (Gordon Cragg, personal communication).

Other collaborations with natural product chemists have taken place during this period, based on plant materials collected in Belize. One interesting example is found in Glinski et al.[6] After discussing the interest in identifying bioactive compounds with healer Don Elijio Panti, he suggested a group of plants for testing in various screens by the Glinski group. One of these, *Psychotria acuminata*, was identified as a source of phenophorbide a, a green pigment that inactivates cell surface receptors. According to the paper,[6] 'our investigations suggest that the inactivation of cell surface receptors contributes not only to the antitumor effect of PDT [photodynamic therapy], but also to the systematic immunosuppression, a serious side effect of PDT'. It was found that an extract of this plant inhibited cytokinine and monoclonal antibody binding to cell surfaces, and this was attributed to the presence of phenophorbide a and pryophenophorbide a. This discovery was a contribution to the corpus of scientific literature about plant natural products chemistry and bioactivity – it was not focused on the development of a new drug. What is interesting and important to note, however, is that Don Elijio Panti was a co-author of this paper, published in *Photochemistry and Photobiology*, recognizing, in the judgment of the research team, that his discovery and utilization of the plant for many decades constituted a crucial and significant intellectual contribution to this paper. This is the standard that we and increasingly more of our scientific colleagues have attempted to adhere to in our ethnobiological studies.

IX. Video interviews, documentation and teaching programs

Videography was an important tool in documenting the work in this project. This aspect of the work was initially directed by Francoise Pierrot, who, working with the first author, interviewed a number of healers about their backgrounds, training, healing practices, philosophy, ambitions and goals. Around a dozen hours of interviews with six healers were edited into a 29-minute tape, *Messages from the Gods: conversations with traditional healers of Belize*. This was aired on local television, and the footage from which it was drawn deposited at various places in Belize and provided to the healers' families. We were quite pleased with the wide circulation of these tapes amongst family members of the healers. It was clear that this technology is a powerful tool in helping to develop respect for traditional practices and values, as well as for the individual healers themselves. Following this experience, we decided to produce a second video, aimed at the source of future generations of healers children. A program was developed that included a video tape, *Diary of a Belizean Girl: Learning Herbal Wisdom From Our Elders*, for use in the middle schools of Belize. A teacher's guide, of the same name was written and published by a team led by Elysa Hammond, including Michael Balick, Charles Peters, Mee Young Choi, Don Lisowy, Glenn Phillips, Joy Runyon and Jan Stevenson. 'In this 23-minute video, Bertha Waight, a teenage girl from western Belize, talks about her desire to become a traditional healer like her mother Beatrice. She travels to meet several of her country's well-known healers, including Don Elijio Panti, Mr. Percival Reynolds, Miss Hortense Robinson, Mr. Polo Romero and Dona Juana Xix – in order to learn about the medicinal properties of many forest and field plants. The healers explain the use of different herbs in their medical practice to treat illnesses such as anemia, diabetes, diarrhea and migraine'.[7] She then takes her sister to a healer for the treatment of a headache, keeping a diary of her thoughts. Finally, she discusses her dream of becoming a healer, and a Western-trained health professional as well, with her friends.

This program and guide were distributed without charge to all of the middle schools in Belize, and for several years, a contest was held to determine the class that could make the best healer's manual, based on interviews with their elders. Competition was judged by the healers, and the prize that was offered was a television and video-tape player donated to the winning class.

X. Conclusion

In this paper, we have described only a portion of the project that has been ongoing in Belize for over a decade. What began as a simple ethnobotanical inventory in the late 1980s has evolved into a complex, multidisciplinary and inter-institutional program aimed at understanding better the relationship between plants and people in Belize. Some of the initial results beyond ethnobotanial inventory include: refinement of the valuation methodology for the study of traditional medicines; development of nursery protocol for valuable native plant species; progress towards an encyclopedia of the useful plants in the region as well as several major publications on the ethnobotany and floristics of the country; development of a teaching curriculum based on the appreciation and utilization of native plant species; the establishment of a program of pharmacological investigation linking a US governmental agency with a network of traditional healers; and, the establishment of a protected forest reserve. The BEP has also shown that ethnopharmacological investigation and ethnobotanical surveys can lead directly to the conservation of valuable ecosystems and, hopefully, contribute to their maintenance over the long term. One of the great priorities in ecosystem conservation today is developing economically sustainable strategies for maintaining reserves that involve human activity over the long term (measured in hundreds of years) long after initial enthusiasm and philanthropic support have subsided.

Acknowledgments

Gratitude is expressed to the multitude of individuals who have collaborated in the Belize Ethnobotany Project. The following organizations have provided support to The Belize Ethnobotany Project: The US National Institute of Health/National Cancer Institute; The US Agency for International Development; The Metropolitan Life Foundation; The Overbrook Foundation; Grinnell College; The Edward John Noble Foundation; The Rex Foundation; The Rockfeller Foundation; The Healing Forest Conservancy; The John and Catherine T MacArthur Foundation; The Gildea Foundation, The Nathan Cummings Foundation; as well as, the Philecology Trust, through the establishment of Philecology Curatorship of Economic Botany at the New York Botanical Garden. This paper includes information that has appeared in papers by the first author in *Conservation Biology* with R Mendelsohn and in M Balick, E Elisabetsky, S Laird, editors (1996) *Medicinal Resources of the Tropical Forest: Biodiversity and Human Health*. Columbia University Press.

References

1. Balick MJ, Nee MH, Atha DE. (2000) Checklist of the vascular plants of Belize: with common names and uses. Bronx, NY: New York Botanical Garden Press, 246 pp.
2. Peters CP, Gentry AH, Mendelsohn RO. (1989) Valuation of an Amazonian rain forest. Nature 339:666–656.
3. Balick MJ, Mendelsohn RO. (1992) Assessing the economic value of traditional medicines from tropical rain forests. Conserv Biol 6(1):128–130.
4. Arvigo R, Balick M. (1993) Rainforest remedies: one hundred healing herbs of Belize. Twin Lakes, WI: Lotus Press, 221 pp.
5. Johnston B. (1998) The new ethnobotany: sharing with those who shared. Herbalgram 42:60–63.
6. Glinski JA, David E, Warren TC, Hansen G, Leonard SF, Pitner P, Pav S, Arvigo R, Balick MJ, Panti E, Grob PM. (1995) Inactivation of cell surface receptors by phenophorbide a, a green pigment isolated from *Psychotria acuminata*. Photochem Photobiol 62(1):144–150.
7. Anonymous, no date. Diary of a Belizean girl: learning herbal wisdom from our elders. The New York Botanical Garden, The Ix Chel Tropical Research Foundation, 8 pp.

Iwu and Wootton (eds.), Ethnomedicine and Drug Discovery

CHAPTER 24

Ethnobotanical research into the 21st century

SUSAN TARKA NELSON-HARRISON, STEVEN R KING, CHARLES LIMBACH, CAREY JACKSON, ANDREW GALIWANGO, SIRIMANI KISINGI KATO, BR KANYEREZI

Abstract

In 1992, the Convention on Biological Diversity (CBD) was created at the Earth Summit in Brazil. Since that time, numerous individuals and organizations working on the sustainable development of biological and cultural diversity have faced both significant support and criticism for conducting field research with local healers throughout the world. This paper describes the specific approach and activities undertaken by a collaboration of two primary groups, Shaman Pharmaceuticals and the Buganda Traditional Healers Association. The political setting and languages spoken are presented along with a description of the incidence of diabetes. An ethnomedical approach to identifying plants to treat diabetes is also described. Details on the process of project development and prior informed consent are presented within the context of the CBD. Data on the multiple projects funded as part of immediate, medium and long-term benefit sharing are presented on two distinct research expeditions. A discussion on the return of data from plant screening and follow-up fieldwork is also presented.

Keywords: *medicinal plants, diabetes, ethnobotanical and ethnomedical field research, prior informed consent, benefit sharing*

I. Introduction

According to The World Health Organization, up to 80% of Africans – or more than a half billion people – visit traditional healers for some or all of their medical care.[9]

In Africa and in many developing nations, medical services are limited or unobtainable for the majority of the population. It is the traditional healers and birth attendants in rural and urban areas that have historically provided and continue to provide primary healthcare. They are the vital link to supplying the needed services in their communities, and yet their efforts must continue to expand as populations grow, and health concerns continue to increase in complexity and case numbers. One possibility to support and retain the bio-cultural and biological diversity in Africa, and elsewhere, is to initiate mutually beneficial collaborations to help preserve and develop traditional medicine. While a controversial subject in its own right, this paper provides an example of such a partnership between Shaman Pharmaceuticals and the healers of the Buganda Traditional Healers Association in Uganda.

A copious but ecologically threatened level of biological diversity supports this intricate web of African traditional medicine by providing valuable natural resources as fuel for herbal preparations. During the period 1990–1995, Africa lost 3.7 million hectares of forest every year or an annual deforesting rate of 0.7% (more than double the global average, which averages 0.3%). African forests cover 520 million hectares or close to 18% of the area of the continent, which hosts the worlds' second largest tropical forest cover after Latin America.[19] As traditional knowledge is the basis of primary healthcare in Africa, so is conservation of natural resources crucial to a healthy ecosystem to support health services.

This paper will focus on the East African nation of Uganda, where environmental factors, such as drought, along with political and social changes, are key issues that are continuously addressed in relation to the health of the population and environment. The idea of researching and developing new and much needed medicines from natural resources, also known as the science of ethnobotany, is not a new idea to Western medicine. In the USA, 57% of all prescriptions filled in the USA are natural product-derived pharmaceuticals. Plant medicine is not novel to traditional healers where herbal medicines have long been derived from species mostly found within the tropical regions of the world. Still, only 5–15% of the approximately 250,000–500,000 higher plant species have been investigated for the presence of bio-active compounds.[5]

What has changed, and will continue to change into the next century, is how ethnobotanical driven drug discovery is conducted within Uganda, and elsewhere in regions of the world where the extent of biological and cultural diversity is the most abundant on earth. Essential to this process is how collaborative agreements and strengthened regulations will embody a benefit sharing structure for indigenous communities where traditional medicine is integrated into rural, and often urban life. The Convention on Biological Diversity (CBD) is the most comprehensive global agreement to date, which 'can be seen as an instrument to promote the equitable exchange, on mutually agreed terms, of access to genetic resources and associated knowledge in return for finance, technology and the opportunity to participate.[22] In the 21st century, the CBD will pose a greater opportunity for mutually beneficial collaborations to be structured in cooperation with the people who directly use, and are affected by, the sharing of knowledge of how their natural resources are utilized for medicine. Hopefully, these collaborations can become aligned with those industries, governments, organizations, and individuals involved in the preservation of biological and cultural diversity. The word 'development' must include preservation from the onset. How this is accomplished is no simple task and ultimately will be criticized by some, and applauded by others.

In December, 1999, the World Health Organization (WHO) urged its African Member States to develop strategies for the integration of traditional medical practice into their national healthcare systems. The call was made by experts attending a three-day consultative meeting on the Strategy for Traditional Medicine for the African Region (2000–2001), which ended in Harare, Zimbabwe. It was projected that by 2020, traditional medicine will be an integrated component of the minimum healthcare package in countries in the region.

The meeting further recommended that action plans be developed for the implementation of the strategy at the regional and national levels. Participants requested WHO to provide technical and financial support for the development of traditional medicinal products in the region for the treatment of priority diseases such as malaria, hypertension, diabetes and HIV/AIDS.[18]

From a research standpoint, the question is how to conduct ethnobotanical research in a manner that can not only benefit collaborating communities, but also act as examples for future researchers, in academia, and industry alike. It is a fact that many indigenous communities, governments and organizations scrutinize ethnobotanical research. Part of the 'answer' clearly lies in the fact that indigenous communities can, and should, have an equal voice in creating the frameworks for adequate benefit sharing strategies and practices in ethnobotanical drug discovery. All agreements are dependent on many factors, including the location, the natural resources being utilized, international laws such as the CBD, as well as national and local laws, and many other local community issues and statutes.

Shaman Pharmaceuticals has worked for 10 years, with nearly 30 nations to develop natural products from the tropical regions of the world while at the same time providing benefit sharing (in three stages) to indigenous communities throughout all phases of research, discovery, and product development. In addition, the sustainable use and ecological monitoring of these resources have been integrated into Shaman's partnerships should any raw plant material be collected and taken into development stages. To date, only one species, *Croton lechleri*, from Latin America, has been taken through all stages of development and is the first product to market for Shaman Pharmaceuticals division Shaman Botanicals.com as a herbal preparation for the treatment of diarrhea.

This paper relays an example of the collaboration between Shaman Pharmaceuticals and the Buganda Traditional Healers of Uganda, East Africa, which began in 1997, and continued through to the year 2000, in order to research diabetes mellitus in Uganda. While there are numerous such published examples of Shaman Pharmaceuticals' collaborations in practice in countries such as Guinea, Nigeria, Belize, and elsewhere,[2–4,10,13] this discussion includes the changes occurring in policy in Uganda that ultimately will affect current and future research in Uganda. This partnership is ultimately the result of phases of meetings between Shaman field teams, Bugandan Healers, Buganda and Ugandan Government officials, Makerere University and community liaisons.

The unique nature of the government of Uganda and the Kingdom of Buganda, within Uganda, will be addressed initially. The many cultural groups and geographical setting will then be discussed. For proprietary and protective identity reasons, the majority of traditional healers' names will not be included in the text. The formation of the collaboration from the beginning stages to now will also be reviewed. Data from each expedition were returned to the healers and voucher specimens left to the herbaria where specimens were identified. Benefit sharing in the immediate and medium term will be presented. Lastly, a discussion regarding the Shaman – BTHA collaboration will conclude with input from Ugandan traditional healers, researchers, physicians, and policy-makers with input into a framework for future collaborations and long-term benefit sharing strategies.

I.A. Uganda: 'The Pearl Of Africa'

Traditional healers are under the auspices of the Buganda Government as the Buganda have a strong say in the Ugandan Government regarding cultural practices in Buganda, and this includes traditional medicine. Likewise, the Bugandan Ministry of Health plays a strong role in the healthcare system of the local people where traditional healers hold a significant role in primary healthcare, particularly outside the capital city of Kampala. Who ultimately is responsible for permitting who conducts research, and where, in Uganda, is the Ugandan Ministry of the Environment? This, and other related government policies affecting ethnobotanical research will be discussed.

The Ugandan economy is recovering after the Amin years, with the aid of the World Bank, International Monetary Fund (IMF), and many foreign non-governmental organizations. The stability of the Ugandan economy is almost completely dependent on coffee exports. However, the barter system does exist with foreign trading partners (usually trading coffee for much needed imports like fuel). The foreign-exchange 'black' market has been practically eliminated by Forex bureaus. Foreign aid is used for imported fuel and to get factories back to competitive production levels.[6]

I.B. Plant habitats/flora of Uganda

Uganda is divided into several biogeographical zones, for instance:

- Sudae-Congolean (North)
- Somali-Maasai (North-east)
- Guinea-Congolean (West, South-west)
- Afro-montane (Mountains)
- Transition (North-western)
- Lake Victoria basin (regional mosaic)

Although there are not many species that are strictly endemic to the country, the flora is still of great importance because of its major contribution to regional endemic species. Many species occur in the western valley and areas around Lake Edward and Lake Victoria, which do not exist anywhere else in the world. Uganda is a country of considerable climatic variation, with important consequences for the distribution of vegetation and land use. The country receives between 1015 mm and 1525 mm of precipitation. Climactic and physical conditions vary a great deal within short distances in Uganda. Areas at higher altitudes have a reliable rainfall that can support montane rainforests, and most areas of the country have sufficient rainfall to support agriculture.

The Ssese Islands in Lake Victoria, with over 2030 mm of rain, are also very wet, and there is a narrow zone with a relatively high precipitation on the mainland close to the Lake shore. The islands vary considerably from low prairieland and marshland, to dense jungle. The main island of Kalangala is the most forested island, with approximately 45% covered in secondary forest. An abundant population of endemic and migrating birds exists here as well.[6]

II. Traditional medicine in Uganda

II.A. Primary source of healthcare

It is the opinion of many Ugandan physicians and herbalists alike that traditional medicine in Uganda is the primary source of healthcare for the Ugandan people. The working relationship between traditional healers, birth attendants and Western physicians has become increasingly vital to the functioning of the healthcare system in Uganda and elsewhere in Africa. Many people not only do not have access to Western clinical healthcare, but they avoid it because of cost, spiritual reasons, and distance traveled to a clinic. In Uganda, the national AIDS control policy calls for active involvement of all sectors of society, and traditional cultural practices have played a critical role in the fight against the epidemic. Traditional health practitioners – including healers, herbalists, and birth attendants – have collaborated with Western-trained professionals in offering safe healthcare, counseling, education for HIV prevention, and research.[18]

However, it must be said that the coordination and overlap of medical efforts between healers and the local or city medical clinics have not been an easy or always prevalent road in many areas in Uganda. In general, the integration of two systems in Africa, as Dr. Maurice Iwu, an ethnopharmacologist from Nigeria points out, is vital:

> What is needed is the study and analyzes of the traditional healing methods by the orthodox or modern doctors, and the re-orientation of the native medicine man, so that he realizes his limitations and benefits from the knowledge of other cultures and civilizations. This should lead to an integrated and improved health care delivery system.[7]

An example of where traditional healers play a vital role in healthcare in Uganda is found in the Ssese Islands, which is a group of approximately 84 islands, with a population of more than 15,000 people. Here, there is one Western medical clinic, with two buildings, 12 beds and one full-time physician, two nurses (sometimes one) and one midwife. This clinic is located on the main island, in Kalangala. Therefore, the herbalists who live in the outer islands, and even on the main island Kalangala where many elderly Ugandans live, conduct primary healthcare activities. When a patient becomes very ill, if a boat is available, they can be transported to the main island and then the mainland to Entebbe or Kampala. However, the mainland is a four- to eight-hour boat ride, and weather conditions often do not permit a crossing. A few NGOs offer air evacuation for seriously ill patients from these islands, and there are even a few NGOs who have offered some medical supplies and sent nurses or doctors to administer patient care.

Ssese has a unique population in contrast to mainland Uganda. Many of the elders in Ssese are healers, and a large number of them have not left the islands in years, since the Amin dictatorship. Those who escaped the bloodshed in Uganda had set to live in Ssese for the remainder of their lives. Therefore, there is also a need for special care for the elderly, who are also limited in travelling to the capital island and main land. This was particularly important when visiting Ssese to work with the healers, as they could then decide on transportation for the seriously ill as a form of benefit sharing that Shaman could provide. In the main city of Kampala, there are some medical centers and private practitioners who are beginning to work more with traditional healers, mostly on HIV/AIDS-related treatments.

The Buganda Kingdom's Minister of Health, Mr Robert Ssebunya, works closely with the Minister of Health of Uganda on the many issues urban and non-urban people face. Mr Seebunya has established forums for practitioners and patients to discuss health issues, such as malaria, which varies in severity by the season. The King of Buganda has set aside some land and a building site for a traditional clinic where urban healers could come to practice and develop their herbal preparations. Still, these discussions are in the early stages as the discussions between Western practitioners, government, and healers continue as to when and how to build this healers' clinic.

II.B. Diabetes and the ethnomedical approach

Diabetes affects a large percentage of the population in Uganda. In 1998, a newspaper headline in a Ugandan daily paper claimed Diabetes to be the fifth killer in the country.[23] It appears that the increase coincides with traditional food and lifestyle being replaced with a more sedentary and Western way of life. However, much of Western fast-food and processed foods are still not found in Uganda, and many Ugandans prefer traditional Ugandan food to Western meals. So, the reason for Diabetes prevalence clearly envelops a more complex pathophysiology and is worthy of more research.

The ethnomedical approach to studying Diabetes in the developing world 'provides an interface between modern clinical medicine and the empirical facets of traditional medicine'.[17] According to Dr Charles Limbach, who was a Shaman team physician on the first excursion to Uganda in 1997:

> my overall impression is that diabetes is both common and well known to the healers. Most of the data I gathered showed a solid understanding of the disease. In a certain way, this makes data gathering more easy, because the word 'sukaali' or sugar can be used in direct translation. In fact, I think we were introduced in most places as researchers interested in diabetes. Of course, such situations can also prejudice the data, since signs and symptoms come after the diagnostic term is revealed. I modified my approach because of this. At each interview, I started by having the healer describe 'sukaali' in as much detail as possible. With gestational diabetes, the case presentation worked in more the normal fashion, giving the history of 'big baby' and how to prevent a second one.

This was also the opinion of Dr Carey Jackson, who was part of the physician–ethnobotanist team on the second expedition to Uganda. The Bugandan healers are astute in detection of diabetes and know many subtleties of diabetes:

> There are clearly features of management that reflect daily clinical practice. The observation that severe diabetics have urine that attracts sugar ants, crystallizes on leaves, or percolates slowly into the ground. The recognition that impotence often accompanies diabetes in men, and that large babies are produced through gestational diabetes in women. The fact that they also recognize gestational diabetes cannot be treated aggressively with some of the more potent preparations because this could lead to abortion. I could go on, the point is they know the disease, they consider it a disease of urbanization, and they are constantly working to find new combinations of plants that help them manage it.

As previously mentioned, and as is clear to many cultures Shaman has worked with, the resounding word that is associated with diabetes has the root word 'suk-' or 'sugar'. This will be noted on labels for healers preparations, on signs of village clinics, and many people have heard of this disease and, in many cases, have one or more family members affected by diabetes. Diabetes, or late-onset, NIDDM, disease is relatively new to the population and is thus found at higher levels because the disease is largely influenced by lifestyle choices. A collective effort to disease treatment and patient monitoring has greatly assisted in increasing public awareness in Uganda on new and changing health issues such as with diabetes. For instance, in the Nabayego Research Clinic, near Kampala, healers and Western-trained nurses are working together with diabetes patients. They monitor their urine for glucose levels and provide herbal remedies as treatment.

Note: Healers working steadily with disease treatment is also true of HIV/AIDS as Uganda has been flooded with case numbers over the last 15 years. For instance, it is estimated that some 60% of Southern Ugandans are HIV-positive. Many traditional healers have altered their practice, along with Western physicians, to treat HIV/AIDS patients in recent years. While some profess they have a cure, many healers have systematically provided beneficial treatments for patients and at a much lower cost than if they were to visit the city clinics, therefore, providing an invaluable service to the community, particularly in the villages where Western care is non-existent, too expensive for most, or found only in limited locations.

III. Origins of work with the Bugandan Traditional Healers Association (BTHA)

III.A. Invitation to working in Uganda

Beginning in early 1997, a relationship between Shaman and The Buganda Traditional Healers Association began as a result of a meeting between Shaman Advisor, Dr Maurice Iwu, and Ugandan physician Professor BR Kanyerezi at an

East African conference on traditional medicine. Professor Kanyerezi described how he worked with local healers specifically to treat HIV/AIDS-related symptoms. At the request of Dr Iwu, Professor Kanyerezi later contacted Steven King of Shaman Pharmaceuticals to discuss the possibility of collaboration with the healers of the Buganda region within Uganda. However, it was not stated at that time that any collaboration would ensue.

Upon further discussion, the Shaman Ethnobotany and Conservation Team decided that a physician–ethnobotanist team would travel to Uganda for preliminary discussions with Ugandan physicians, healers and botanists. As a result, two consecutive expeditions to Uganda were conducted by Shaman Pharmaceuticals field teams, led by Shaman ethnobotanist Susan Nelson-Harrison, over the course of three years. As described below, the stages of discussions and benefit sharing occurred throughout this period.

Prior to traveling to Uganda, S Nelson-Harrison researched the culture, environment, and political situation in Uganda. Research was conducted in order to provide herself and the accompanying physician of the team (first Dr Charles Limbach and secondly Dr Carey Jackson) with a cultural profile and survey of the ecogeographic region. This is standard practice for all research expeditions, in addition to training in the therapeutic area of research. Because Shaman was researching late-onset Diabetes at the time, a background search was also conducted as to the prevalence of diabetes within the population in Uganda. A variety of sources stated that the disease was spreading rapidly, especially within the past 10 years.

The first trip to Uganda commenced in May 1997 with botanist–physician team S Nelson-Harrison and Dr Charles Limbach traveling to the capital, Kampala, to meet first with the primary physician contact. Research was arranged to take place for a period of five weeks and was planned to be a series of discussions with healers' groups to explain the purpose of Shaman's research and the collaborations in other countries as examples of what might be possible in working in Uganda.

III.B. Preliminary discussions and prior informed consent

According to the CBD, prior informed consent (PIC) in article 15.5 is defined as:

> Access to genetic resources shall be subject to prior informed consent of the Contracting Party providing such resources, unless otherwise determined by the Party.[7]

During preliminary discussions in Uganda, considerable time was taken in explaining the purpose of drug discovery using the ethnobotanical approach and how the BTHA could participate in this type of collaboration with Shaman. In order to relay clearly the intentions and programs of the Shaman botanist–physician team, they were interviewed by numerous healers, government officials, and reporters during their first visit, for a period of two weeks, prior to any field work or further discussions about ethnobotanical research. PIC was discussed individually, and in later discussions, the Minister of Health of the Buganda Kingdom supported the BTHA collaboration with Shaman to conduct research, if they should desire, with Shaman. Still, PIC documents were reviewed by many healers before their own decisions were made.

Mr Andrew Galiwango, a local healer, teacher, and economist, accepted the position of in-country coordinator of the BTHA. Therefore, the Shaman team was able to explain their research program, and the healers could ask questions and have their concerns addressed through Mr Galiwango. However, this did not occur until many lengthy discussions and meetings ensued between government officials, Mr Galiwango, Dr Kanyerezi, and other healers in the BTHA. It was particularly helpful to work with Mr Galiwango as he is a healer as well and is fluent in Luganda and English. He also has worked with local physicians and can clearly relay medical terms that the healers and other parties could discuss in an open format.

While some issues, such as patents and working with pharmaceuticals, require detailed explanations, it was vital to have these ideas expressed in such a way so that everyone could come to a common understanding of these and other topics. Most people are familiar with the mistakes of large companies entering into new regions, eager to work with indigenous peoples, but not eager to slow down and take the time to explain their purpose and intent.

Many healers mentioned the following regarding herbal medicine practice and the difficulties they face in practising herbal medicine in Uganda. Some major points of concern and needed supplies were:

- a central building or meeting space for healers to obtain technical advice and assistance
- a central clinic for healers to obtain information and assistance with their practices in the villages
- bottle and tablet press supplies for herbal preparations
- preservatives for herbal preparations
- herbal gardens so healers could retain plant populations for their mixtures.

III.C. Legal framework in Uganda on regulation of access to natural resources and benefit sharing

Since Uganda signed the CBD in September, 1993, national policy and legal framework have not been firmly established regarding access to natural resources, regulation measures and benefit sharing. However, there are some national legal measures and statutes that are found in existing acts, or bills, that are relevant to the regulation and access of genetic resources in Uganda. A Comprehensive World Wildlife Fund (WWF)-sponsored report on Ugandan policies in relation to genetic resources and benefit sharing produced the following findings. Existing statutes include: The National Environment Statute 1995:

the only legislation which discusses the regulation of access to genetic resources in Section 45[2] states that National Environmental Management Authority. (NEMA) shall, in consultation with the lead agency, issue guidelines and prescribe measures specifying: appropriate arrangements for access to Uganda's genetic resources. The 1996 Uganda Wildlife Statue in Sections 23–27 prescribes some legal measures that regulate access to, and use of, resources within wildlife protected areas (national parks, game reserves, sanctuaries, and controled hunting areas). It implements the CITES laws as it prohibits protected species from being affected. Also, The Forest Act of 1962 requires a license from The Chief Conservator of Forests before any forest product is cut, taken, worked, or removed from the countrys' forest reserves....[15]

This applies only for forest reserve areas. Therefore, the need to revise this particular act to incorporate all forests is needed. The National Agricultural Research Organization. (NARO) in Uganda is also working with the FAO on drafting regulations to promote conservation, the 'safe exchange of plant genetic resources as well as exchange of related information and technologies', promote standards for collection, and the sharing of benefits derived from research.[15]

At the national level, ethnobotanical research has been carried out by Ugandan researchers, and screening has been accomplished at the Natural Chemotherapeutics Research Laboratory. (NCRL) in Kampala. However, these results are not published due to intellectual property rights issues. The Makerere University Botany and Forestry Departments have organized their own ethnobotany group and have held conferences, the first being in 1997 in order to assist in regulation efforts and work with students project concerns. Additional ethnobotanical research expeditions have been carried out by local researchers within Uganda, in collaboration with various NGOs, Makerere University, and private companies.[15]

The Constitution of Uganda provides for the protection of natural resources; this applies to non-cultivated trees, saplings, shrubs, cultivated produce, and hay or grass. The Uganda National Council for Science and Technology Statute (UNCST) of 1990 provides that no person should conduct, or engage in, any research or experimental activities without a valid research permit from UNCST. However, no specific provisions on regulation are noted in this statute. Also, the NARO Statute of 1992 is to work within the national agricultural research strategy, but it is not clear whether this covers medicinal plants.[15]

Apparently, there is no single agency responsible for granting permission to access to genetic resources in Uganda. Rather, there are six major agencies involved at different levels. For instance, NEMA is the principal agency that oversees the access to natural resources. In 1997, a standing Technical Committee on Biodiversity Conservation was established to inform the committee of biodiversity issues. The National Biodiversity Strategy and Action Plan has been in process, with Article 15 of the CBD being sent to the institutions involved, as a reminder of the CBD framework. The UWA is also an institution involved, which plays an important role in the regulation of access to genetic resources – mostly Wildlife Conservation Areas. (WCAs) – for the sustainable utilization of wildlife by, and for, the peoples in these protected areas. Other authorities include the National Seed Industry.[15]

In addition, the Forestry Department (particularly Nakawa) works within NEMA to issue plant exportation permits.[15]

All necessary permitting measures were taken by Shaman and Ugandan colleagues prior to any fieldwork, including discussions with the Ministry of the Buganda Kingdom, NEMA, and Makerere Botany Department as to policies that would require approval and permits. Permits were obtained from the Uganda Department of Agriculture (in accordance with the International Plant Convention 1951, Phytosanitary Convention for Africa 1971 and under Kawanda Inspection Services), but not until plant bulk collections were to be scheduled by the BTHA after Shaman's departure. These permits were applicable to the exporting of raw materials from screening of diabetes. All results of these screenings were returned to the healers. No product has been commercialized from these plants, and the data have been kept proprietary between Shaman and the healer who indicated each plant. All voucher specimens were deposited in duplicate at the National Herbarium at Makerere University, with the remaining voucher specimens transported to the Shaman Pharmaceuticals Herbarium for taxonomic verification.[15]

As recommended by the WWF, short-term access should be centralized by a government agency, such as NEMA. Licensing mechanisms for accessing genetic resources should be set in place with an advisory panel of local foresters, healers, and representatives from the Kingdom of Buganda, and the Ugandan government. Including local scientists, and nationals, in all phases of research should be constructed under the CBD. The promotion of a long-term strategy is clearly the key if adequate regulation can be balanced with sustainable development and use of Uganda's diversity of genetic resources.[15]

III.D. Agreement of principles/Memorandum of Understanding (MOU)

Article 15[4] of the Convention of Biological Diversity requires that access to genetic resources, where granted, should be on 'mutually agreed terms'. The convention's use of this phrase implies that those seeking access to genetic resources enter negotiations and try to reach an agreement with the provider of the resources.[8]

During pre-expedition research, it was discovered that only recently in Uganda, under the current President Museveni, the pre-war Kingdoms were in the process of being reinstated throughout Uganda. Only upon further understanding was it discovered that the kingdoms act mainly as grass-roots policy initiators and regulators of Ugandan Government policies. Furthermore, the Buganda Kingdom, where Shaman Pharmaceuticals was working, contains approximately 30% of the population of Uganda and is a very well-organized government body, with parliamentary headquarters at the old kingdom area of Bulange, just a short distance from downtown Kampala. However, this system instigated a challenging process of figuring out who to contact during the first Shaman expedition to Uganda. Only after additional research and questioning of local people did we find which avenues would be appropriate to take as our discussions with the BTHA took shape.

During the initial stages of negotiations between healers and Shaman Pharma-ceuticals, meetings were held between local physicians in Kampala prior to any contact with healers. These local Kampala-based physicians described the medical system in Uganda as starting to integrate both Western physicians and traditional healers, or herbalists, and at the same time, both parties were increasingly skeptical of one another. According to one physician, this was occurring because of skepticism on behalf of the healers who believed that the medical doctors were stealing their remedies for their clinical use. However, one doctor was working very well with some herbalists, who were working with his medical clinic on an ongoing basis. This was especially vital since the AIDS crisis became perilous. Both sides of the 'western to traditional bridge' were discussed in detail, as were the preconceived notions about researchers, such as ourselves, who had come to Uganda for so many years and promised so much, with little, if any, of those promises materializing.

The Agreement of Principles is a general document, which often has been supplemented by a specific contract of Memorandum of Understanding (MOU), which provides details about the specific program in each country. The detailed MOU has also been used to clarify the Agreement of Principles as it applies to an individual country. Lawyers from Uganda and the US reviewed documents initially brought by the Shaman research team in 1997. Questions and revisions were faxed back and forth to Shaman and Uganda during 1997–1998.

It often takes multiple revisions of these documents to state concisely how collaborations are formulated, such as that of Shaman–Uganda, in order to carry out joint research on topics that are of mutual interest, share expertise and facilities in research, collaborate and develop appropriate viable research methodologies. Since traditional medicine is the primary source of healthcare in Uganda, and most of the countries where Shaman works, it is important to outline how Shaman can assist in supporting traditional medicine.

III.E. Plant collections permits

During the preliminary discussions with healers, many issues were raised about the collection of medicinal plants and exporting of raw plant material from Uganda. Because the initial fieldwork in which Shaman would be involved was on private land, no permits were necessary to collect voucher specimens. However, The Nakawa Forestry Department, under the Ministry of Forestry, was to issue permits for any plant material that was collected and exported during Shaman's research. These permits were granted by the Ugandan government for all bulk plant shipments prior to any collection during the first year of collaboration with Shaman and Uganda when bulk collections took place. Each collection ranged from 10 to 30 kg of dried plant material per species.

It was decided that any vouchers to be collected were to be triplicate in number, with two left at the Makerere Herbarium – one set for the Makerere University National Herbarium and the other for the Nakawa Forestry Department. This is standard practice for all Shaman field collections. In addition, it was decided that most of the field equipment that was used in the field, such as plant presses, GPS system, and other collecting equipment, would be left at the Makerere University Herbarium for their own use.

IV. Benefit sharing

A total of approximately $68,485 has been provided to the Bugandan Healers and their communities over the past three years. This compensation has been provided during two research expeditions in the immediate form, and during the interim period as medium-term benefits.

IV.A. BTHA group discussions regarding benefit-sharing ideas

The important issues of the healers in Buganda were detailed in individual meetings, small herbal clinic meetings, large forum workshops, and meetings at the Buganda Palace. Clearly, the words 'mechanization' and 'preservation' of herbal medicine were foremost concerns to the herbal practices of the healers. While some healers mentioned the need for preservatives for many of their mixtures, others were more concerned with raw material supply for their own preparations. The difference here was between the more urbanized practitioners, as opposed to the village healers where perhaps supply might be more plentiful, but preserving a salve or liquid mixture might be much more troublesome.

Medicinal plant gardens were a main topic of discussion. Many healers were having more trouble finding a sustainable supply of their herbs. One healer mentioned his desire to farm medicinal plants and how he was learning which soil conditions worked best for cultivating some of his plants. Note: funds issued for reciprocity projects are given to communities where Shaman teams work during each expedition and are based on what each healers community decides, not what Shaman field teams deem appropriate.

IV.B. Direct compensation

On both expeditions, salaries were agreed upon prior to any work and were openly discussed with the healers, botanists, coordinators and drivers. Salaries were paid in cash, daily, or weekly to each person involved in the fieldwork, preparation of specimens, and to the healers themselves.

IV.C. Benefit sharing as integrated into two ethnobotanical research expeditions in 1997, 1998

The operating costs to maintain the collaboration with Uganda and in between visits by Shaman's team in 1997 and again 1998 for lodging, food, transportation, and communication for both Shaman personnel and Ugandan consultants are not considered as part of the compensation or benefit sharing with Uganda. Therefore, in the expeditions 'totals', these figures have been deleted in order to focus on the direct compensation and benefit sharing with the Bugandan Healers and their communities in addition to the field team we worked with there.

IV.D. Immediate-term benefits

Immediate-term benefits incorporate all compensation in the form of salaries and honorariums, in addition to projects funded by Shaman that are initiated during the expedition, not after the Shaman team leaves. This is an integral part of all Shaman expeditions. Because the BTHA had group meetings with the Shaman field team regarding project ideas at the onset of the BTHA–Shaman collaboration, the projects were decided upon with enough time to implement them during the expedition period.

Reciprocity is provided as benefit sharing at the community level, rather than to individual healers. Over the course of both expeditions in 1997 and 1998, seven projects were funded by Shaman to BTHA communities (see Appendices A and B) since five distinct districts were involved in ethnobotanical collections and/or field workshops. A total of $20,281 was provided by Shaman to the BTHA for immediate benefits, as described below, on the two consecutive expeditions. This was approximately 25% of each expedition total budget.

IV.E. Field work and benefit sharing summary: expedition 1: 1997

Seven healers worked with Shaman botanist/physician team (Ms Nelson-Harrison and Dr Limbach) in two districts – Kalangala and Mpigi. This incorporated four villages, all within the Buganda Kingdom. Paid salaries and honorariums were given to key physician contact, healers, translators, botanists and other field organizers. Two immediate-term reciprocity projects were decided upon by BTHA healers in group meetings with their communities. Medical care and supplies were distributed by Dr Limbach. Collection supplies and workshop supplies were left in Uganda with healers and The Makerere National Herbarium/Botany Department (see Appendix A for specifics).

IV.E.1. District 1: Mpigi
Funds were allocated to this district for three projects since we worked with two village community healers in the area. These projects were started when the Shaman team were there and were carried out through 1998 into 1999. Each project became a self-generating venture in that the benefits could be shared with the community and continued with little outside financial assistance. The first project decided upon was a broadcast Diabetes project, to be shared with the Kalangala District and others in the BTHA, which would include a weekly broadcast on Buganda Radio by traditional healers. The topics would be about traditional medicine, including the prevalence and treatment of diabetes, which the healers all concluded was an increasingly major illness. Funds were left with the BTHA for this project to get started after the Buganda Radio Manager of CBS, Bulange, was contacted, and an agreement was signed with the healers of Mpigi and Kalangala Districts. Additional

funds were left in a trust fund for the BTHA to be used to continue one Mpigi clinic construction and medicinal garden supplies for their district. This was soon created, and pictures were sent to Shaman about the progress of these projects. Lastly, funds were allocated for a road project as the healers mentioned that it was difficult to get water from the river to the outer villages as there was no clear path to carry the water. This road is in Kasanje Village in Mpigi and has helped many people obtain water supplies and also travel to the main road much easier.

IV.E.2. District 2: Kalangala
The healers in these islands decided upon a 'Shaman-Ssese' boat for healers to use for transportation of sick patients and to collect herbs on neighboring islands. This 33-foot wooden boat was purchased from a boat maker in Kalangala and a 15-hp Yamaha engine purchased in Kampala. The boat was later delivered to Baganda Kingdom Parliament, Bulange, and paint was purchased thereafter. Included were life jackets. The official launching of the boat took place in late 1997 when the Ministers of Health and Culture, along with District healers and the press, ventured to the launching site and took a trip to the Ssese Islands.

Medical and other supplies distributed included glucometers, test strips, blood-pressure cuffs, scale, medicine and other medical supplies, money for an orphaned child for one year, and surgery money for a patient with bacterial infection. Additional items left were workshop materials, binders, herbarium supplies, books, field equipment, office supplies, and additional life jackets for the boat.

The total in benefit sharing for this first expedition was $8033.20 (approximately 25% of total expedition budget. See Appendix A for specific costs and an overview).

IV.E.3. Post-expedition: project monitoring Ssese Island boat
This wooden boat, with motor, was officially launched in the fall of 1997 and has since provided transport to the mainland for healers, ill Ssese Island patients, fisherman, and other people who can pay for transport. It is also used to transport healers from the BTHA and Bugandan parliament officials to and from the islands to encourage the involvement of the healers of the Ssese Islands and have the Bugandan officials there. It was previous to 1960 that an official from the Kingdom of Buganda had visited Ssese until this time. The funds collected from those paying to use the boat are then put into the BTHA fund, where it is used to aid in the maintenance of the boat and other uses. Thus, the boat became a self-generating project as profits could be gained by transportation uses for the healers fund. Unfortunately, the boat engine sank in late-1999 and has not yet been replaced.

IV.E.4. Buganda radio program
This included a one-hour program every other week on Uganda CBS Radio and Bulange in order to discuss diabetes and traditional medicine in Uganda. This show ran for three months in Uganda. The healers decided not to continue the project with additional reciprocity funds because the funds could be used more efficiently elsewhere. Healers at Nabayego Research center noted that because of the radio program and the mention of their diabetes traditional medicine clinic, they received

an infiltration of patients. Both the ministers of Health and Culture were pleased with the airing of the program and wished it to continue, and asked if we could fund it again. However, it was explained that it was up to the healers and their communities to decide how they wish to use the funds.

IV.E.5. Medicinal plant project and Mpigi healers clinic
The Mpigi District initially decided upon this project, and three healers from this district received portions of the funds to use as they deemed best to start individual plant projects. The healers later agreed that a central location, for more healers in the area to use, would be best for the current reciprocity funds than individual projects. However, they had planted and intercropped some of their land with plants that were either becoming difficult to find or were in short supply. In general, the topic came up often with the healers of how to use funds to benefit more healers.

Later, the healers from Mukono, Luwero and Mpigi Districts were all interested in a centralized medicinal plant garden. The ideas for the new plant project were: start a few nurseries, maybe one in each district, and allow healers to contribute in a community meeting to decide which plants they wished to grow. The healers were also to decide which foresters and botanists would be involved since divulging which plants they use was a major concern of the healers. Lastly, organic mulching and animal fertilizers were decided upon, instead of chemicals. Assistance from the Makerere University Forestry Department, particularly from David Nelson-Nkuutu and Olivia Maganyi, were also of great aid in the healers' communities. The Clinic in Mpigi purchased cement and building materials and is currently in great demand as patients travel from far away as the main healer there is an expert mid-wife.

IV.E.6. Kasanje Road
A road was created, in the Mpigi District in order for the village to have access to the clean-water well built by an NGO in 1995. The funds came from plant collection funds from plant collections by the BTHA. The healers in this community would like to see the road eventually continue so that it can reach the next village over for water supplies.

IV.E.7. Medium-term benefit sharing
Because the interim between the research phase and product development can be many years, Shaman has developed a tiered system of benefit sharing with all cultures where they work. Mid-term benefit sharing incorporates technology transfer, community support in each country, plant collections profits, and other financial and training support. In Uganda, this support total was $45,278 over a three-year period (see Appendix B for specifics). The following are some of the ways in which Shaman fulfilled this medium-term support.

IV.E.8. Medical bills
An elderly healer that we worked with from the Mpigi District fell sick with tuberculosis. It was found, after X-rays, that she had a collapsed lung for some time. Still, she walked long distances and has recuperated after her treatment. She is

currently in good health and living in a remote area in the Mpigi District where patients travel long distances to see her. However, she still travels the same long distances to see patients, whom she says contact her through her dreams.

IV.E.9. Travel-technology transfer/conference attendance sponsorship
A Conference on Biodiversity at Makerere University was sponsored, in part, by Shaman so that students and faculty could discuss current biodiversity issues facing Africa. No Shaman personnel attended this conference: it was for students of the university alone. In addition, Shaman sponsored Professor Kanyerezi to attend a conference on African Herbal Medicine in East Africa in mid-1998. Lastly, Shaman travel funds were used by Mr Galiwango to travel to Shaman facilities from Uganda in order to visit Shaman and report to the healers in the BTHA his findings.

IV.E.10. Baby cots and medical supplies
As promised during the 1997 expedition, funds were spent to purchase baby cots for the Kalangala Medical center on the Ssese Islands. They were later delivered in 1998 to the clinic, as were life jackets and additional medical supplies.

IV.F. Field work and benefit sharing summary: expedition 2: 1998

Nine healers worked with Shaman botanist/physician team (Ms Nelson-Harrison and Dr Limbach) in four districts – Mpigi, Mukono, luwero, and Mubende. This incorporated six villages, all in Buganda. Paid salaries and honorariums were given to key physician contacts, healers, translators, botanists and other field organizers. Four immediate-term reciprocity projects were decided upon by BTHA healers in group meetings with their communities. Dr Carey Jackson distributed medical care and supplies. Collection supplies and workshop supplies were left in Uganda with healers and The Makerere National Herbarium/Botany Department. Approximately 28% of the total expedition budget was in support of immediate benefit sharing, as can be seen below (for specifics, see Appendix C).

IV.F.1. District 1: Mpigi district medicinal plant gardens and clinics
Mpigi decided to continue to use funds for a medicinal plant garden project. One healer requested the assistance of the forestry department to build a centralized nursery and to expand the project so that the whole community could have input. In addition, clinical structure support for the healers' clinics (bricks and herbal preparation supplies) was also derived from these funds.

IV.F.2. District 2: Luwero: Nabayego Research Center
Partial funds were used for cement and sheet metal for the Nabayego Research Center. The clinic also requested roofing materials, which they will likely deduct from this reciprocity money. The remaining funds were used to strengthen the facilities in the herbalists clinic.

IV.F.3. District 3: Mubende transportation/clinic
Since transport is a major problem in this district (it is a three- to five-hour journey to Kampala, and the roads are very poor), there was talk of having a healers' fund for transport to BTHA meetings and also to purchase some bicycles for the village. Later, partial funds were used to help build a central clinic in the town of Mityana within the Mubende District.

IV.F.4. District 4: Mukono district, Sezibwa Falls, Bunyagira School Project
This is a primary school where the focus is to balance a 'Western' education with cultural tradition, where they teach the youngsters about Buganda music and traditional medicine. There are currently four teachers for 120 primary school children, ages 3–12 or 13. Each teacher has only one textbook. They have chalk but need all other supplies, especially syllabus, textbooks and charts. Only half of the classroom building was finished when we first arrived, so many of the students had to study in one small room.
    Shaman supplied funds initially for 14 iron sheets to cover an additional classroom and 12 wooden benches. Additional writing books and the roof were made from the funds Shaman gave them, and half the benches were made. The school is registered by the government, so they should receive funding soon since Uganda, just in 1997, is supplying 'free' primary-school education. Currently, parents pay the teachers. The school is in its fourth year of operation. This was clearly an excellent project that the healers and community supported, so Shaman gladly assisted.

IV.F.5. Bunyigira, Kyagwe Women's/Students' Medicinal Plant Project
The parents of the children of Bunyagira were interested in collectively donating land for a medicinal garden for the children to learn about growing and collecting the local plants that healers use in their area. The women gathered to show us the land, and so some of the reciprocity money will go towards starting the garden.
    Both projects in the Bunyagira area have received individual funding and continue to do so to this day by Dr Carey Jackson, attending physician on the second expedition, and an individual Shaman investor, Mr Robert Marrow, of Ohio.

IV.F.6. Additional funds: Makerere University Forestry Department
Funds were donated by Shaman to the head of the department to begin a nursery for medicinal plants on school property. The site was observed behind the Forestry building. The garden was to be initiated and cared for by the students.
    Workshop Exchange and Clinic Discussions as Immediate Benefit sharing (technology transfer) has proved to be a valuable exchange during many Shaman expeditions in such countries as Tanzania, Guinea, Belize, and Cameroon. The format of the workshops varies and is open to students, healers, university students, in-country botanists, and other community and government members in order for field work to develop not just on a structured protocol from other country exchanges, but as a collaboration built on the desires and needs of that particular community. In Uganda, a two-day Diabetes Workshop Exchange at Makerere

University and a clinic discussion on Diabetes at the Nabayego Clinic were held during the Shaman team's second visit to Uganda.

The Diabetes Exchange Workshop was sponsored by Shaman and Makerere University. Local healers, botany and forestry students and professors, local physicians were in attendance. Sixty-five people attended the first day of the workshop, which involved discussions about Diabetes in Uganda and how it is interpreted and treated between healers and physicians. Also, Shaman's ethnobotanical research methodology was discussed, and questions were raised about Shaman's activities in Uganda to date. This methodology was offered as a way for students to conduct their own fieldwork and not necessarily that of Shaman. A few students, physicians, and university professors commented that this was the first workshop of its kind at the university. The overall impression was that it was a step in the right direction in building bridges with the healers and physicians. Also, it was noted that it was a workshop that allowed students and healers to work together through educating one another and formulate their own dialog about traditional medicine in an open format.

The second day of the workshop was a full day in the field in the Mpigi District. The participants were forestry, botany and medical professionals and Makerere University students from the workshop exchange the day before, in addition to five healers from various districts who were interested in attending. A prominent healer, Mr Kato, the chairman of the BTHA, was the leader of the fieldwork. The day included a walk into the forest to collect plants that he uses in his herbal preparations. Students took notes on the collection methods and, with the consent of Kato, the names of the plants. Kato also took everyone into his ceremonious house, where he showed them how he uses various spiritual indications as to the patients health concerns and solutions. The day ended with the plants being pressed for the students to take to their herbarium, with the consent of the healer, and some photos being taken of the students, teachers, healers, and other participants.

Photos and certificates of workshop completion were later sent back to the participants and healers through Mr Galiwango, the coordinator of the BTHA, and the Makerere University Botany and Forestry Departments, respectively.

IV.G. Additional medium-term benefit sharing: post-expedition: technology transfer/ transportation

Transportation has been noted as a key issue for healers in the BTHA as they are from various districts within the Buganda region. Funds to the sum of $6000 were sent by Shaman to the BTHA to purchase a vehicle in 1998. The BTHA now has a truck in order to transport healers to group meetings, for plant collections, and other transportation needs such as going out to the Ssese Islands port landing. The roads in Uganda are getting better as infrastructure has been a key point for the Uganda Government for the past five years. However, the danger at many borders due to political unrest still creates barriers to much travel in those areas.

Additional medium-term reciprocity has been initiated with the BTHA in working to strengthen existing projects from immediate benefit sharing, such as the Nabayego

Clinic building. These funds have been made possible through plant collections, where a portion of these funds goes into the BTHA trust fund to be channeled into different projects in the districts within the Buganda Kingdom. Also, each district within the BTHA has notably gained benefits from plant collection funds. For instance, Mpigi now has a larger clinic for patients, and Mubende has a medical office and patient waiting area (see Appendix B medium-term support specifics).

IV.H. Long-term benefit sharing

Should Shaman have a product that is selling on the market and show promising product sales, long-term benefit sharing will be started with Uganda, and the over 70 other cultural groups Shaman has worked with throughout the last 10 years of research and development. In 1997, the Healing Forest Conservancy, the non-profit organization that Shaman founded at the onset of Shaman Pharmaceuticals, and Shaman representatives initiated a pilot project in Nigeria to follow how a trust fund mechanism can work to the benefit of the healers and their communities once long-term benefit sharing is in practice by Shaman. Funds were given to this trust fund and are currently being distributed within the Nigerian healers' community where Shaman has worked over that past 10 years.

IV.I. Plant collections and front-end screening results for NIDDM research in Uganda from both expeditions

One hundred and six distinct species were collected (in voucher form only, not to be tested) on two consecutive expeditions by Shaman and BTHA research teams (some duplicate collections of the same species not included in this number). The assistance of the Makerere University Botany Department and Department of Forestry, particularly from Olivia Maganyi and David Nelson Nkuutu, was vital to the species identification of these collections. Plant vouchers were collected in triplicate so that each collection could be deposited at the Makerere University Herbarium in addition to each voucher being left for NEMA or a Ugandan agency of choice by Makerere Botanists. After the Shaman research team departed, duplicate voucher specimens were transported to the Shaman Herbarium, and field data were reviewed, in addition to chemistry, biology, and ethnomedical literature at the genus level of each targeted species.

Due to the fact that the Shaman Diabetes Program tested 100 plants per year from various tropical regions throughout the world, this highly selective process would allow for only a few Ugandan plants to be screened. Thus, after the literature and data 'elimination' or selection process, 13 species were requested by Shaman to have bulk plant material collected by the botanist and healer in Uganda who indicated plants for diabetes. These are considered recollections from the original expeditions. All collections were also cross-checked with previous Shaman collections at the Shaman Herbarium by Ms Nelson-Harrison and sent to the Missouri Botanical Garden for additional taxonomic opinion to species. Therefore, 13 species were recollected in bulk, even though the 106 distinct species are used for diabetes and related symptoms by BTHA healers.

Of the 13 bulk collections, five species were found to be active in the Shaman NIDDM program. Activity is based on the lowering of glucose, triglycerides, and study of animal body weight and food consumption. Of these five active plants, one plant proceeded through to the fractionation status, but no patent was ever filed, as no compound was isolated at the time of the termination of Shaman's diabetes screening program in February 1999.

IV.J. Returning data: front-end screening results key

A common complaint among Ugandan healers was their negative experiences with Western companies, foreign students, and NGOs. Often times, the promise was as simple as returning footage or a few photographs to the healers. In addition, data was rarely sent back either published or not. The returning of screening data, plant lists, and/or taxonomy revisions, can act as a common bridge of understanding.

However, there must be a way in which data are returned so that the healers can understand, and their traditional practices can benefit from this knowledge. Article 16.3 of the CBD clearly states that the access to and transfer of, technology using genetic resources to countries providing the genetic resources. This transfer of information, including sharing data, is a vital part of the research equation that has sadly been left out of many foreign research programs, particularly in the US, as the CBD was never officially signed by the US.

All results from the Shaman–BTHA Collaborative Diabetes Project were returned to healers through the coordinator of the BTHA. Shaman routinely returns these data on all screened plants in order to relay the testing results in a clear and concise manner. However, because testing is performed in a specific model, Shaman does not advocate that any plant tested can or will not work in subsequent screens, or that the positive or negative result does not indicate that the botanical preparation in Uganda, or elsewhere, when used by traditional healers is not effective.[2]

Data were returned to Uganda on two occasions, according to the year of screening. 1997 results were returned in mid-1998, and 1998–early 1999 results were returned in mid-1999. All results were accompanied with an explanation of what each screening result meant. For instance, if the glucose lowering was positive for a particular extract, this was noted, along with any possible adverse affects noted in the animals tested (for instance, food consumption low, and/or mortality of some mice). Mr Galiwango, the coordinator of the BTHA, then brought results to individual healers and translated the information in Luganda.

IV.K. Follow-up to fieldwork: future possibilities in working in Uganda

Note: Shaman's drug discovery department was discontinued in late-1998 to redirect efforts into the new division Shaman Botanicals. Still, in the future, there might be a possibility of resuming discovery work, but at this time, it is uncertain whether or not such an arrangement can be solidified.

The districts within the BTHA were never more than a four-hour drive from the main city, Kampala, thus greatly assisting in the data and plant collection logistics as more remote areas would make recollections more difficult and costly. It is also recognized that limiting research to within the BTHA could have caused some factions within healers' groups outside the BTHA, that were previously non-existent.

Some minor field setbacks during the second expedition included arriving to the Ssese Islands to find that much of the island had received an abundance of rain and turned to swamp. Because the majority of the Ssese healers lived on neighboring islands, the trip was unproductive as many people were either unable to meet or too ill to travel. Also, there were a few days that our fieldwork had to be limited or changed since some of the field team had malaria or some other health complication.

In late 1998, Shaman Pharmaceuticals ceased Discovery fieldwork in Diabetes in order to focus on the new division Shaman Botanicals. None the less, the communication between Shaman and the BTHA has remained constant in explaining this change and continuing to support BTHA reciprocity projects. The marketing of the first product by Shaman Botanicals, if successful, will allow the long-term benefit sharing funding to be established in the form of a trust fund for the BTHA and other communities of healers Shaman has worked with. This trust fund will apply to all indigenous communities regardless of the fact that the product is being developed from a Latin American plant. With the assistance of the Healing Forest Conservancy, the non-profit Shaman founded 10 years ago, the implementation of this return of funds to Uganda, and elsewhere, will have the protocol of a pilot project to return benefits to Nigerian healers.[12] However, at this time, it is uncertain whether or not Shaman will be profitable. Thus, long-term benefit sharing might not occur.

The BTHA has since obtained additional sources of funding from individuals and NGOs for traditional healers to attend conferences on plant medicine and to strengthen their herbal practices. One project in particular is the Bunyagira School, where individual donations totaling $700 have been sent in 1999 to help the children's school in that village. Future possibilities for collaboration with the BTHA will depend on the success of Shaman's current product and future research expeditions that are not yet planned at this time. In the meantime, the relationship between Shaman and the BTHA will continue, if only to assist the BTHA in obtaining assistance from other companies or NGOs, at the request of the BTHA.

V. Conclusion

Since the last visit by Shaman to Uganda in July, 1998, the secretary and coordinator of the BTHA, Mr Andrew Galiwango, arranged meetings with the BTHA and the Ministers of Health and Culture at the Buganda Parliament. During these meetings, many issues about Uganda's collaboration with Shaman were openly discussed. A direct contact link between Shaman and the healers, government officials, and other interested parties has been carried through since primary visits, which is essential for the information exchange to continue and to be mutually beneficial. The coordination of these activities could not have happened without the constant work of Mr Galiwango and Kato, the director of the BTHA. It is invaluable to have such coordination in the source country so that communication can be clear, and recommendations can be made moving forward.

The Minister of Health in the Buganda Kingdom 'commended the Shaman contribution to the development of Buganda'. He said that Shaman is the 'first international company of its kind in Uganda which operates at the grassroots level and has not only helped the traditional herbalists, in modernizing their work, but also boosted the recognition of native doctors, which was non-existent'.[14]

As far as laws that will affect research into the 21st century, the Ugandan government is working closely with collaborating scientists and non-government organizations to monitor the use of biological resources, both in national parks and in different kingdoms where the land is owned by the separate kings. The regulatory agencies in Uganda continue to be NEMA and the Ugandan Government Forestry Department. All foreign researchers at this time are to obtain a permit to conduct research prior to entry into Uganda, which was not strictly regulated until late 1999.

During a meeting of healers at the Bugandan Kingdom Palace, one healer from the BTHA said 'There will be no secrecy in meetings. Everything must be established in the open. It is the way we work here in Uganda. It is the only way to agree'. Collaborative research between Western companies and traditional healers in source countries must incorporate community needs and make sure that they are adequately met in order for any collaboration to be mutually beneficial and grow into a sustainable future. As we enter a new century, we hope that as industry continues to engage in partnerships with source countries, a balance rather than a sieve can be put in place for partnerships to be based on guidelines that encourage participation of local peoples in the drug discovery, conservation, and research process.

Acknowledgments

We wish to express our gratitude to the people and governments of the Buganda Kingdom and Uganda, who worked together with Shaman Pharmaceuticals on the BTHA–Shaman collaboration over the past four years. In particular, we appreciate the hospitality and relationships with the communities in the Buganda Kingdom where we conducted research: Mpigi, Mubende, Mukono, Luwero, and the Ssese Islands in the Kalangala District. In addition, the assistance and field efforts of the Makerere University Botany and Forestry Departments were remarkable, especially the botanical expertise of David Nelson Nkuutu and Olivia Maganyi-both in the field and at the Makerere Herbarium. Dr Bukenya-Ziraba, director of the Botany Department, was particularly helpful in establishing our connection to the National Herbarium at Makerere, organizing the Diabetes Conference, and introducing us to Olivia and Nkuutu. We are also thankful to the head of the Forestry Department, Dr Kabogozza, for allowing us to use the department vehicle for the workshop in Mpigi and also for allowing his students to utilize Shaman reciprocity funds to build a university medicinal plant garden. We appreciate the guidance of the Honorable

Ministers of Health and Culture of Buganda and use of facilities in the Buganda Kingdom Palace at Mengo for healers' discussions, meetings, and for the establishment of the radio program on Diabetes.

The introduction to working in Uganda could not have occurred without the mutual interest and friendship built by Dr Maurice Iwu, Shaman advisor, and Dr BR Kanyerezi of Kampala, Uganda. Their guidance was essential to this collaboration being formed. Mr Andrew Galiwango, and Sirimani Kato were the guiding forces that allowed this collaboration to flourish and continue. Thank you to Missouri Botanical Garden for their continuous support in species identification and to Capp at Herbarium Supply Company. David Nelson Nkuutu, Olivia Maganyi, Dr BR Kanyerezi, Dr Thomas Carlson, Carina Romero, Beto Borges, Lisa Conte, Katy Moran, Dr Mike Balick, Maurice Iwu, Julie Chinnock, and others provided valuable comments on this manuscript.

Appendix A Immediate compensation/benefit sharing for Ugandan expedition 1997

Immediate-term compensation: $3680
Immediate-term reciprocity: $4940
Medical supplies, botany supplies: $1889
Total benefits 1997: $10,509
Percentage of expedition budget: 26.5%

Appendix B Medium-term benefit sharing in Uganda

Vehicle $6000 + medical bills: $300, conference sponsorship $1228 and Shaman transfer $3200 = $10,728
Plant collections: $34,450
Field logistics for Shaman Botanicals plant surveying: $500
Conference on biodiversity sponsorship, Makerere University: $600
Total: $45,728 medium-term benefits to BTHA, Uganda over a 3-year period

Appendix C Compensation/benefit sharing for Shaman expedition 1998

Immediate-term compensation: $5510
Immediate-term reciprocity: $6738
Supplies to clinic in Ssese Island, workshop supplies, medical supplies, school donations: $2053
Total benefits 1998: $12,248
Total expedition budget: 27.2%

References

1. Black MB, Elisabetsky E, Laird S, editors. (1996) Medicinal resources of the tropical forest: biodiversity and its importance for human health. New York: Colombia University Press.
2. Brush S, Stabinsky D, editors, King SR, Carlson TJ, Moran K. (1996) Biological diversity, indigenous knowledge, drug discovery, and intellectual property rights. Valuing local knowledge: indigenous people and intellectual property rights. Washington, DC: Island Press.

3. Carlson TS, Iwu MM, King SR, Obialor C, Ozioko A. (1997) Medicinal plant research in Nigeria: an approach for compliance with the convention on biological diversity. Diversity 13: 29–33.
4. Chinnock JA, Balick MJ, Camberos S (in press). Traditional healers and modern science – bridging the gap: Belize, a case study. In: Paul A, Wigston D, Pepers D, editors. Building bridges with traditional knowledge. New York: New York Botanical Garden Press.
5. Farnsworth NR, Akerele O, Binge IAS, Soejarto DD, Guo Z-G. (1985) Medicinal plants in therapy. Bull WHO 63: 965–981.
6. Finlay H, Crowther G. (1997) East Africa: lonely planet, 4th edition. USA: Lonely Planet Publications.
7. Iwu MM. (1986) African ethnomedicine. Enugu, Nigeria: Snap Press, p. 8.
8. Iwu MM. (1996) Implementing the biodiversity treaty: how to make international co-operative agreements work. TibTech 3:14.
9. Johnston B. (1999) WHO acknowledges African healers. Herbalgram 43:16.
10. King SR, Carlson T. (1995) Biomedicine, biotechnology, and biodiversity: the Western hemisphere experience. Interciencia 3:134–139.
11. Greaves T (editor), King SR. (1994) Establishing reciprocity: biodiversity conservation and new models for cooperation between forest dwelling people and the pharmaceutical industry. In: Intellectual property rights for indigenous peoples: a sourcebook. Oklahoma City, OK: The Society for Applied Anthropology.
12. Moran K. (1998) Convention on biological diversity: case studies on benefit sharing arrangements. Mechanisms for benefit sharing: Nigerian case study for the Convention on Biological Diversity. In: Plotkin M, editor. Bratislava: Convention on Biological Diversity.
13. Moran K. (1992) In: Plotkin M, Favolare L, editors. Ethnobiology and U.S. policy. Sustainable harvest and marketing of rainforest products. Washington, DC: Island Press.
14. Njuba Times (Tuesday, January 10, 1998) p. 3, Buganda hails U.S. projects. Kampala, Uganda.
15. Ntambirweki J, Tamale E. (1996) Existing and planned measures on regulation of access to genetic resources and benfit sharing in Uganda. In: Measures to control access and promote benefit sharing, World Wildlife Fund report, Conference Proceedings. WWF.
16. Nzita R, Mbaga. (1995) Niwampa Peoples and Cultures of Uganda, Fountain Publishers Ltd., Kampala, Uganda.
17. Oubre AY, Carlson TJ, King SR, Reaven GM. (1997) For debate: from plant to patient: an ethnomedical approach to the identification of new drugs for the treatment of NIDDM. Diabetologia 40:615–617.
18. Panafrican News Agency December 18, 1999 WHO urges African Countries To Develop Traditional Medicine Distributed via Africa News Online (www.africanews.org).
19. Panafrican News Agency March 6, 2000, FAO Says African Forests Seriously Endangered, Africa News Online (www.africanews.org) The report, entitled 'The Challenges To Sustainable Forestry Development In Africa' is available on the following internet site: http://www.fao.org/fao regional conferences 2000.
20. Strauss S. (1997) Uganda's leader crafts an African success story. In: The San Francisco Chronicle, California.
21. Tempesta MS, King S. (1994) Tropical plants as a source of new pharmaceuticals. In: Pharmaceutical manufacturing international. London: Sterling, pp. 47–50.
22. ten Kate K, Laird S. (1999) The commercial use of biodiversity: access to genetic resources and benefit-sharing, European Communities. London: Earthscan Publications, pp. 5–7.
23. Wamboga-Mugirya E, Muwanga-Bayego H. (1998) Diabetes named 5th top killer. In: The Monitor. Kampala, Uganda.

Iwu and Wootton (eds.), Ethnomedicine and Drug Discovery
© 2002 Elsevier Science B.V. All rights reserved.

CHAPTER 25

Ethnobotanical approach to pharmaceutical drug discovery: strengths and limitations

MAURICE M IWU

Abstract

Higher plants are still regarded as potential sources of new medicinal compounds. About half of all known drugs are derived from natural products and their semi-synthetic derivatives. There are many approaches available for the selection of plants for drug discovery; however, the ethnobotanical approach to pharmaceutical lead drug discovery may significantly enhance the probability of identifying a potential drug molecule from medicinal plants. The method involves the integration of many disciplines including anthropology, botany, ecology, pharmacy, linguistics, medicine and ethnography. The International Center for Ethnomedicine and Drug Development (InterCEDD) has developed a simple scheme for the rapid evaluation of plant samples used in traditional medicine and their analysis for the presence of bioactive molecules.

Keywords: *ethnobotany, lead discovery, dereplication, InterCEDD protocol*

I. Introduction

I.A. People, plants and molecules

Plants provide a good source of starting material for the discovery of biologically active molecules that could be developed into new medicines. Plant products are also important as sources of intermediates for the synthesis of very important medicinal agents. Enormous costs are saved by the use of plant materials as starting materials in the manufacture of complex molecules than when they are prepared completely from synthetic substances. The preparation of steroidal hormones from plant sapogenins, for example, involves simple transformation reactions, whereas the complete synthesis of such compounds is achieved through many intermediate stages (Woodward's cholesterol synthesis involves 42 stages) and at prohibitive cost. Another advantage in the hemisynthesis of complex molecules from phytochemicals is the availability of the desired conformation in the starting material, which in turn guarantees the stereospecificity required for some pharmacological activity and therapeutic respose.

Naturally occurring compounds and their derivatives constitute about one-half of all drugs in current use. They have also provided the molecular template or intellectual stimulus for the synthesis of about half of all synthetically produced medicinal compounds.[1] A statistical analysis of compounds isolated from natural products and those derived by total synthesis employed in drug development has shown that a mere 90,000 known naturally occurring compounds contributed about 40% of the total possible new drug molecules, whereas the several millions of synthetic molecules accounted for the remaining 60%.[2] Prof. Tyler has attributed this remarkable difference in productivity to the fact that only a limited number of different molecules are involved in, or have a beneficial effect upon, life processes, and that nature has performed a pre-selection of molecules that influence specific metabolic roles in all living things.[3] It is noteworthy that despite the large investment by the pharmaceutical industry in modern drug-discovery technologies, such as combinatorial chemistry and robotic-based high-throughput screening, natural products are still one of the major sources of new structural entities for drug development.

The transition between a new chemical entity to a clinically accepted drug involves an integration of the activities of many disciplines. Only a small number of pharmacologically active compounds ever make it to advanced drug development, and an even smaller number make it from development to the market. It is estimated that of each 5000 chemical entities synthesized, only 250 will reach animal testing, five will reach the level of clinical testing in healthy volunteers or patients, and only one will ultimately make it to the market place.[4] It is estimated that of all drugs that reach the first phase of clinical testing, approximately 10% will reach the market as a drug entity.[5] The statistics look even bleaker if one considers the fact that most of the new drugs that make it to the market place are not exactly new chemotypes. Dimarsi et al.[6] have reported that between 1976 and 1990, 269 new chemical entities were approved by the Food and Drug Administration in the United States. Of these entities, 131 (49%) were found to have little or no therapeutic advantage over existing products, 94 (35%) were described as having a modest advantage over existing products, and 41 (15%) showed a significant improvement over current drugs. Only three drugs, representing 1% of all the approved drugs, were targeted to fill gaps in areas of greatest therapeutic need, such as AIDS.

There are several approaches used in the search for pharmacologically active compounds described in the literature.[7,8] These can be grouped into five, namely (a) the random approach, which involves the collection of all plants found in a study area, (b) the phytochemical targeting, which entails the collection of all members of a plant family known to be rich in certain class(es) of compounds, (c) the chemotaxonomic approach, (d) the ecological approach, in which the relationship between the plant and its ecosystem is used as a means of selection, and (e) the ethnobotanical-directed sampling approach, based on traditional medicinal use(s) of a plant (including ethnomedicine). These different methods will be briefly described, and the ethnobotanical approach will be discussed in greater detail.

I.A.1. Random approach

This involves collection of plants in a given locality to generate a large number of plant materials. The underlying philosophy is that having a large pool of plant material will provide a great chemical diversity, which increases the chance of obtaining a biologically active molecule. With recent developments in bioassay technologies and extraction methods, it is indeed possible to screen a very large pool of extracts in a relatively short time. It is best suited for targeted screens in which a clearly defined disease or molecular mechanism is screened against a battery of purified extracts. The random approach provides too large a sample size for in-vivo evaluations. There are two examples of drug-discovery programs in which this approach has been used. The first is the anti-cancer program of the US National Cancer Institute (NCI), and the other is the general screening for biological activity by the Indian Central Drug Research Institute. (CDRI). The NCI program has screened more than 35,000 species for anticancer activity. The program is credited with the discovery of taxol and camptothecin, among several other compounds undergoing development.[9–11] The NCI has also started screening for anti-HIV activity and at present has identified several molecules at different stages of development (see the review by Gordon Cragg in this volume). The CDRI program tested more than 2000 plants against several biological activities, such as antimicrobial, anti-infertility, antihypercholestremic, anti-inflammatory, cardiovascular, diuretic, central nervous system activity, and antitumor.[12–13] It has been observed that although a large number of known and novel bioactive compounds were isolated from biological active plants, to date, no new drugs for human use could be traced to that program.[14] In general terms, there are two possible approaches that have been used to guide the collection of plant materials in order to ensure that a large diversity of plant materials is included:

• Approach 1: taxonomic breadth – which will attempt to sample as many families and genera as possible. In this approach, it is possible to prepare a list of plants in a given location and then collect as many representative species as possible to reflect the biological diversity in the area.
• Approach 2: replicate samples within a genus or family in order to evaluate the taxonomic effect on plant chemistry diversity. A typical project using this approach could, for example, choose to capture all genera with less than four species and all families with four genera before collecting individual species from dominant families or genera.

I.A.2. Phytochemical approach

The phytochemical approach is not a true plant-selection method since the plants to be screened are collected before the phytochemical tests are conducted to determine the presence of certain types of compounds. The follow-up bioassays are then restricted only to plants that contain the target chemical class or group. It is useful as a field method to pre-screen for the presence of chemical groups associated with certain types of activities, for example, anthraquinoles, cardiotonic glycosides, alkaloids, saponins, flavonoids, etc. It can also be valuable when there is a need to obtain diverse structures for structure–activity relationships.

I.A.3. Chemotaxonomic approach
This method is similar to the phytochemical approach but with a notable difference in the consideration of plant chemotaxonomy as a pre-selection tool. Plants are collected based on the known presence of certain chemotaxonomic markers in a given family or genus. A fascinating modification of this method is to combine it with knowledge of biosynthesis of the plant constituents and to identify those plant phytochemicals that are key 'event blockers' and use them as molecular probes in animal systems. By this method, it is possible to identify compounds capable of enhancing or stopping metabolic pathways in plants and relate these to specific enzyme systems in animal systems.

I.A.4. Ecological approach
The survival of any organism depends on its ability to adapt to its ecosystem. Plants have evolved sophisticated mechanisms for co-existing with their neighbors and adapting to physical environmental stresses. It is believed that some of the essential plant constituents have been produced by the plants either for their own physiological processes or as a defense against predators. Careful observations of plant–animal interaction within a given ecosystem can serve as indicator for possible activity. For example, antimicrobial agents have been isolated from plants that showed resistance to attack by fungi and other microbes, and a possible insecticidal activity could be discerned from observing the feeding habits of insects in the wild.

I.B. Ethnobotanical approach

Ethnobotany is the study of interrelations between humans and plants; however, current use of the term implies the study of indigenous or traditional knowledge of plants. It involves the indigenous knowledge of plant classification, cultivation, and use as food, medicine and shelter. Although most of the early ethnobotanists studied plant used in cultures other than their own, the term ethnobotany does not necessarily mean the study of how 'other' people use plants. It is also not restricted to the study of medicinal plants by indigenous cultures. The use of ethnobotany in plant selection entails a careful recording of the relationship between indigenous communities and plants. It is a very complex undertaking that often requires collaboration of experts drawn from various disciplines such as anthropology, botany, ecology, pharmacy, linguistics, medicine and ethnography. Ethnobotany has now emerged as a discipline by itself that studies all types of interrelations between people and plants. As Ford[15] noted: 'ethnobotany lacks a unifying theory but it does have a common discourse'. The central theme is the recognition of the reciprocal and dynamic nature of the relationship between humans and plants.[16] There are excellent publications available on the general introduction to the protocols and ethical issues concerning ethnobotanical work.[16,20–22]

I.C. Ethnobotany and drug development

The sequence for development of pharmaceuticals usually begins with the identification of active lead molecules, detailed biological assays, and formulation of dosage forms in that order, and followed by several phases of clinical studies designed to establish safety, efficacy and pharmacokinetics profile of the new drug. Possible interaction with food and other medications may be discerned from the clinical trials. In the development of phytotherapeutic agents a top-bottom approach is usually advocated. This consists of first conducting a clinical evaluation of the treatment modalities and therapy as administered by traditional doctor or as used by the community as folk medicine. This should be followed by acute and chronic toxicity studies in animals. Studies should when applicable include cytotoxicity studies. It is only if the substance has an acceptable safety index would it be necessary to conduct detailed pharmacological/biochemical studies. It has to be noted, however, that in therapeutics, it is not the toxicity of the agents per se that determines their suitability for use as medicinal substances but the width of the margin between the toxic dose and the therapeutic dose, or what is known as the therapeutic index. Because of their long history of human use, plant materials collected through the ethnobotanical approach often display higher therapeutic index than their synthetic counterparts.

Examples of drugs derived directly from leads from traditional medicine or produced from simple modification of such leads are shown in Table 1. The advantage of this method in selecting plants for biological screening is demonstrated by the fact that fewer drugs have discovered from natural product not used in traditional medicine. Table 2 lists drugs that did not originate from their ethnomedical use. In a comparative analysis of plants collected either by random or ethno-directed methods from different districts of the Sinai region of Egypt,[23] it was found that plant sampling based on an ethnobotanical approach produced a greater number of plants showing antimicrobial activity. Results from a study conducted by our research group (as part of the International Cooperative Biodiversity Group) on plants with antimalarial, antifungal, and antiviral activities indicate that there is consistently a higher level of activity for plants used in traditional medicine than those selected randomly.[24]

The ethnobotanical approach has it obvious limitations (see Schuster in this volume), which are mainly due to the fact that some of the diseases may be unfamiliar to the indigenous population, and therefore, no effective intervention or therapy has been developed.

I.D. De-replication of extracts for drug discovery

One of the most difficult problems associated with screening of plant extracts for biological activity is the identification of substances with a truly unique biological activity from other substances that may be positive in a given bioassay but are not desirable during a lead drug-discovery program. A key objective in pharmaceutical

Table 1
Drugs derived from ethnomedical leads[a]

Drug	Action or clinical use	Plant source
Adoniside	Cardiotonic	*Adonis vemalis* L.
Aescin	Anti-inflammatory	*Aesculus hippocastanum* L.
Aesculetin	Antidysentery	*Fraxinus rhynchophylla* Hance
Agrimophol	Antihelmintic	*Agrimonia eupatoria* L.
Ajmaline	Anti-arrhythmic	*Rauvolfia vomitoria* Kurz
Ajmalicine	Circulatory disorders	*Rauvolfia serpentina* (L.) Benth ex.
Allyl isothiocyanate	Rubefacient	*Brassica nigra* (L.) Koch
Andrographolide	Bacillary dysentery	*Andrographis paniculata* Nees
Anisodamine	Anticholinergic	*Anisodus tanguticus* (Maxim.) Pascher
Anisodine	Anticholinergic	*Anisodus tanguticus* (Maxim.) Pascher
Arecoline	Antihelmintic	*Areca catechu* L.
Asiaticoside canescens	Vulnerary	*Centella asiatica* (L.) Urban
Aspirin	Analgesic anti-inflammatory	*Filipendula ulmaria*
Atropine	Anticholinergic	*Atropa belladonna* L.
Berberine	Bacillary dysentery	*Berberis vulgaris* L.
Benzoin	Oral disinfectant	*Styrax tonkinensis*
Bergenin	Antitussive	*Ardisia japonica* Bl.
Bromelain	Anti-inflammatory; proteolytic agent	*Ananas cosmosus* (L.) Merrill
Caffeine	CNS stimulant	*Camelilia sinensis* (L.) Kuntze
(+)-Catechin	Hemostatic	*Potentilla fragaroides* L.
Chymopapain	Proteolytic; mucolytic	*Carica papaya* L.
Cocaine	Local anesthetic	*Eyhthroxylum coca* Lamk.
Codeine	Analgesic; antitussive	*Papaver somniferum* L.
Colchicine	Antitumor agent; antigout	*Colchicum autumnale* L.
Convallotoxin	Cardiotonic	*Convallaria majalis* L.
Curcumin	Choleretic	*Curcuma longa* L.
Cynarin	Choleretic	*Cynara scolymus* L.
Danthron	Laxative	*Cassia* spp.
Deserpidine	Antihypertensive; tranquilizer	L. *Rauvolfia vomitor*
Deslanoside	Cardiotonic	*Digitalis lanata* Ehrh.
Digitalin	Cardiotonic	*Digitalis purpurea* L.
Digoxin	Cardiotonic	*Digitalis lanata* Ehrh.
Emetine Richard	Amebicide; emetic	*Cephaelis ipecacuanha* (Brotero) A.
Ephedrine	Sympathomimetic	*Ephedra sinica* Staph.
Etoposide	Antitumor agent	*Podophyllum peltatum* L.
Gitalin	Cardiotonic	*Digitalis purpurea* L.
Glaucaroubin	Amebicide	*Simarouba glauca* DC
Glycyrrhizin	Sweetener	*Glychrrhizia glabra* L.
Gossypol	Male contraceptive	*Gossypium* spp.
Hemsleyadin	Bacillary dysentary	*Helmsleya amabilis* Diels
Hydrastine	Hemostatic; astringent	*Hydrastis canadensis* L.
Hyoscamine	Antisholinergic	*Hyoscamus niger* L.
Kainic acid	Ascaricide	*Digenea simplex* (Wulf.) Agardh
Kawain	Tranquilizer	*Piper methysicum* Forst. f.
Khelin	Brohchodilator	*Ammi visnaga* (L.) Lamk.
Lanatosides A, B, C	Cardiotonic	*Digitalis lanata* Ehrh.

Table 1 (*continued*)

Drug	Action or clinical use	Plant source
Lobeline	Smoking deterrent; respiratory stimulant	*Lobelia inflata* L.
Monocrotaline	Antitumor agent	*Crotolaria sessiliflora* L.
Morphine	Analgesic	*Papaver somniferum* L.
Neoandrographolide	Bacillary dysentery	*Andrographis paniculata* Nees
Noscapine	Antitussive	*Papaver somniferum* L.
Ouabain	Cardiotonic	*Strophanthus gratus* Baill.
Papain	Proteolytic; mucolytic	*Carica papaya* L.
Phyllodulcin	Sweetener	*Hydrangea macrophylla* (Thunb.) DC
Physostigmine	Cholinesterase inhibitor	*Physostigma venenosum* Balf.
Picrotoxin	Analeptic	*Anamirta cossulus* (L.) W. &A.
Pilocarpine	Parasympathomimetic	*Pilocarpus jaborandi* Holmes
Podophyllotoxin	Condylomata acuminata	*Podophyllum peltatum* L.
Protoveratrines A & B	Antihypertensive	*Veratrum album* L.
Pseudoephedrine	Sympathomimetic	*Ephedra sinica* Staph.
Pseudoephedrine, nor-	Sympathomimetic	*Ephedra sinica* Staph.
Psoralen	Management of vertigo	*Psoralea coryiifolia*
Quinine Trimen	Antimalarial	*Cinchona ledgeriana* Moens ex.
Quinidine Trimen	Anti-arrhythmic	*Cinchona ledgeriana* Moens ex.
Quisqualic acid	Antihelmintic	*Quisqualis indica* L.
Rescinnamine Kurz	Antihypertensive; tranquilizer	*Rauvolfia serpentina* (L.) Benth ex.
Reserpine Kurz	Antihypertensive; tranquilizer	*Rauvolfia serpentina* (L.) Benth ex.
Rhomitoxin	Antihypertensive	*Rhododendron molle* G. Don
Rorifone	Antitussive	*Rorippa indica* L. Hochr.
Rotenone	Piscicide	*Lonchocarpus nicou* (Aubl.) DC.
Rotundine	Analgesic; sedative	*Stephania sinica* Diels
Salicin	Analgesic	*Salix alba* L.
Santonin	Ascaricide	*Artemisia maritima* L.
Scillarin A	Cardiotonic	*Urginea maritima* (L.) Baker
Scopolamine	Sedative	*Datura metel* L.
Sennosides A & B	Laxative	*Cassia* spp.
Silymarin	Antihepatoxic	*Silybum marianum* (L.) Gaertn.
Stevioside	Sweetener	*Stevia rebaudiana* Bertoni
Strychnine	Antihepatotoxic	*Strychnos nux-vomica* L.
Teniposide	Antitumor agent	*Podophyllum peltatum* L.
Tetrahydropalmatine Schltal.	Analgesic; sedative	*Corydalis ambigua* (Pallas) Cham. &
Theobromine	Diuretic; bronchodilator	*Theobroma cacao* L.
Theophylline	Diuretic; bronchodilator	*Camellia sinensis* (L.) Kuntze
Trichosanthin	Abortifacient	*Thymus vulgaris* L.
Toxiferine	Relaxant in surgery	*Strychnos guianensis*
Tubocurarine	Skeletal muscle relaxant	*Chondodendron tomentosum* R. & P.
Valepotriates	Sedative	*Valeriana officinalis* L.
Vincamine	Cerebral stimulant	*Vinca minor* L.
Xanthotoxin	Leukoderma; vitiligo	*Ammi majus* L.
Yohimbine Pierre	Aphrodisiac	*Pausinystalia yohimbe* (K. Schum.)
Yuanhuacine	Abortifacient	*Daphne genkwa* Seib. & Zucc.

[a] Data adapted from Fabricant and Farnsworth,[14] Cox, P.[19]

Table 2
Plant-derived drugs and their sources not developed on the basis of ethnomedical information[a]

Drug	Plant sources
Allantoin	Several plants
Anabasine	*Anabasis aphylla* L.
Benzyl benzoate	Several plants
Borneol	Several plants
Camphor	*Cinnamonum camphora* (L.) J.S. Presl
Camptothecin	*Camptotheca acuminata* Decne.
Cissampeline	*Cissampelos pareira* L.
Colchicaine amide	*Colchicum autumnale* L.
Demecolcine	*Colchicum autumnale* L.
L-Dopa	*Mucuna deeringiana.* (Bort) Merr.
Galanthamine	*Lycoris squamigera* Maxim.
Glaucine	*Glaucium flavum* Crantz
Glaziovine	*Ocotea glazovii* Mex
Hesperidin	*Citrus spp.*
Huperzine A	*Huperzia serrata* (Thunb. Ex Murray) Trevis.
Menthol	*Mentha spp.*
Methyl salicylate	*Gaultheria procumbens* L.
Nicotine	*Nicotiana tabacum* L.
Nordihydroguaiaretic acid	*Larrea divaricata* Cav.
Pachycarpine	*Sophora pachycarpa* Schrenk ex C.A. Meyer
Palmatine	*Coptis japonica* Makino
Papaverine	*Papaver somniferum* L.
Pinitol	Several plants
Rutin	*Citrus spp.*
Sanguinarine	*Sanguinaria candensis* L.
Sparteine	*Cytisus scoparius* (L.) Link
Taxol	*Taxus brevifolia* Nutt.
Tetrahydrocannabinol	*Cannabis sativa* L.
Tetrandrine	*Stephania tetrandra* S. Moore
Thymol	*Thymus vulgaris* L.
Vasicine (peganine)	*Adhatoda vasica* Nees
Vinblastine	*Catharanthus roseus* (L.) G. Don
Vincristine	*Catharanthus roseus* (L.) B. Don

[a] Data adapted from Fabricant and Farnsworth.[14]

lead discovery is the enhancement of the efficiency and speed of identifying novel biological active compounds from a pool of molecules in plant extracts. Dereplication is a term used in drug discovery to describe efforts to identify and characterize, with speed and efficiency,[17] the following types of leads:

1. Novel chemical entities that are specifically active in the target bioassay under consideration.
2. Commonly reoccurring classes of 'nuisance compounds' that provide a non-specific response in a large number of bioassays. Such compounds typically give a positive response to a broad range of in vitro bioassays but with negative results in animal studies. Examples include tannins in terrestrial plants, alkylated

pyridinium polymers from marine sponges and polyphloroglucinols in brown algae.[18]
3. Known secondary metabolites that may or may not occur widely, but which provide a specific response to the target bioassay and are no longer required.
4. Compounds that had been previously identified in a previous assay with specific activity but are no longer required.

The availability of sophisticated phytochemical analytical tools has greatly simplified the dereplication process by making it possible to couple the separation ability of modern chromatographic techniques with spectrophotometers in order to defocus nuisance samples from the discovery process. A simple technique that has found wide application is the chemical removal of tannins from extracts before subjecting them to bioassays. Extracts are also profiled chemically before bioassays in order to identify the presence of unwanted substances. Thin-layer chromatography and HPLC are particularly useful for this purpose. It is now routine in high-throughput screening, for example, to identify only certain regions of the chromatogram for bioassay based on prior chemical profiling of the fractions. Chemotaxonomic information can also be a very valuable tool in dereplication by enabling the investigator to remove from the screen at a very early stage of the process those plant species that contain compounds that are usually positive in a given target bioassay. The US National Cancer Institute provides a good example with its dereplication of members of the plant family Clusiaceae (Guttiferae), which yielded the two anti-HIV compound classes, the calanolide and the guttiferones. The NCI has now incorporated the dereplication of these two compound classes in their preliminary evaluation and prioritization of screening of leads from that plant family.[17]

I.E. A method for application of ethnomedical information for drug discovery

Developments in our understanding of ethnobotany and the natural history of various indigenous cultures have made it possible to enhance significantly the probability of identifying naturally occurring compound for drug development. Described below is a simple scheme developed by the International Center for Drug Development (InterCEDD) for the rapid evaluation of plant samples used in traditional medicine and their investigation for the presence of bioactive molecules. The process, known as the InterCEDD protocol, allows for both the development of phytomedicines and the discovery of pharmaceutical lead compounds.

The essential features of the method are outlined below (Figure 1). A working group is convened prior to any ethnographic fieldwork to evaluate available information and to prepare a regional study on the epidemiology, traditional medicine, culture and ecology of the people and their environment. This pre-collection study also includes a thorough review of information on plants known to be used in the area for the preparation of traditional remedies. This involves using existing databases and information gathered from the consultant herbalists and local health officials.

Table 3
Summary of the African ICBG drug-development leads using the InterCEDD protocol

Diseases	Extracts tested	Activity (%)	Leads[a]
Malaria	500	343 (70%)	20
Leishmania	130	52 (39%)	6
Cytotoxicity	20	16 (80%)	5
Viral	30	16 (80%)	2
Trypanosomiasis	27	13 (48%)	3
Trichomonas	25	10 (40%)	7
Opportunistic infection			
Cryptosporidium	22	7 (31%)	2
Toxoplasmosis	22	6 (27%)	2

[a] Isolated and characterized molecular leads.

An advisory team consisting of botanists, medical doctors (Western-trained) and a pharmacognosist then briefs the working group using case presentations of diseases identified in the epidemiology report. The case descriptions of specific diseases are presented to traditional healers or shaman. Photographs of diseases with readily visible symptoms are also utilized. The working group interviews the informants in a carefully pre-determined manner. In seeking information, emphasis is on the identifiable symptoms rather than diseases. The approach relies upon recognition of common signs and symptoms. Terms such as 'malaria' or 'cancer' are not utilized, even when the informant is educated or familiar with the terms. As much as possible, only individuals with a good knowledge of the area and who speak the local language are used.

I.E.1. Ethnomedical evaluation
The ethnomedical information obtained from the above team is analyzed, as indicated in the scheme depicted in Figure 1. For the ICBG, the following pre-screens supplement the Interceed protocol.

I.E.2. Prescreens and preliminary bioassays
About 1000 plant samples are subjected to preliminary bioassays each year. About 200 of the most active extracts (or fractions/isolates) are tested against each disease category. Out of this number, the 25 most active extracts are selected for in vivo studies, tertiary screens and further development. Strict selectivity is required to ensure that only extracts with the greatest promise of containing therapeutically useful compounds are investigated. Three pre-screens are used for the initial selection. They are rapid, reliable and inexpensive and can screen a relatively large number of plant extracts at field laboratories. Highly automated and specific receptor binding then augments these screens and enzyme-based assays in the US laboratories.

The pre-screens are:

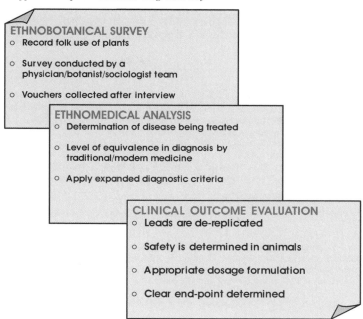

Fig. 1. The InterCEED protocol.

- Brine shrimp (*Artemia salina*) lethality bioassay: a simple general bioassay to predict possible activity of crude extracts and isolates.
- Potato disc bioassay: a test for detecting compounds with possible anticancer activity. It is based on the inhibition of crown gall, a neoplasm of plants induced by the bacterium *Agrobacterium tumefaciens*, in suitably prepared potato discs.

I.F. Antifungal bioautographic assay and microbial evaluation

This tests extracts against two fungi, two Gram-positive and two Gram-negative bacteria. Active extracts are selected for bioassay-directed fractionation, and active constituents can be identified rapidly in two steps:

- Secondary assays: The results of the pre-screen are evaluated and extracts directed to the appropriate secondary biological evaluations.
- Bioactivity-directed fractionation: The results from the primary and secondary screens direct the fractionation of active extracts. The active fraction will be fractionated further to a highly active fraction, which will be separated chromatographically to obtain the active compounds.
- Identification of active compounds: Isolates with biological activity are subjected to spectral analysis (UV, IR, proton NMR, carbon-13 NMR and mass spectra)

and the determination of melting points, optical rotation, etc. Spectroscopic equipment will also be used in the US laboratories.

References

1. Baerheim Svendsen A. (1999) Biogene Arzneistoffe-heutte noch oder heute wieder? In: Czygan F-C, editor. Biogene Arzneistoffe Braunschweig: Freidr. Vieweg and Sohns, pp. 27–39.
2. Mueller H. (2000) Natural products and their importance in drug discovery. In: Luijendijk T, de Graf P, Remmelzwaal A, Verporte R, editors. Years of natural products research: past, present and future. Leiden: University of Leiden: L05.
3. Tyler VE. (2001) The future of botanical drugs and the rainforest. J Herb Pharmacother 1:5–12.
4. Office of Technological Assessment. (1993) Pharmaceutical R&D: costs, risk and rewards. Washington DC: OTA.
5. Drayer JI, Burns JP. (1995) From discovery to the market: the development of pharmaceuticals. In: Wolff ME, editor. Burger's medicinal chemistry and drug discovery, 5th edition. New York: Wiley, Ch. 9, p. 257.
6. DiMasi JA, Hansen RW, Grabowski HG, Lasagna L. (1991) J Health Econ 10:107–142.
7. Farnsworth NR, Bingel AS. (1977) Problems and prospects of discovering new drugs from plants by pharmacological screening. In: Wagner H, Wolf P, editors. New natural products and plant drugs with pharmacological, biological or therapeutical activity. Berlin: Springer, pp. 1–22.
8. CIBA Foundation. (1994) Ethnobotany and the search for new drugs, Symposium 185. New York: Wiley.
9. Suffness M, Douros J. (1982) Current status of the NCI Plant and Animal Products Program. J Nat Prod 45:1–44.
10. Cragg GM, Boyd MR, Cardellina JH, Newman DJ, Snadder KM, McCloud TG. (1994) Ethnobotany and drug discovery: the experience of the US National Cancer Institute. CIBA Found Symp 185:178–190.
11. Cragg GM, Newman DJ, Snadder KMF. (1997) Natural products in drug discovery and development. J Nat Prod 60:52–60.
12. Dhar M, Dhar MM, Dhawan BN, Mehrotra B, Ray C. (1968) Screening of Indian plants for biological activity 1. Indian J Exp Biol 6:232–247.
13. Dhawan BN, Dubey MF, Mehrotra B, Rastogi RP, Tandon S. (1980) Screening of Indian plants for biological activity 1. Indian J Exp Biol 18:594–606.
14. Fabricant DS, Farnsworth NR. (2001) The value of plants used in traditional medicine for drug discovery. Environ Health Perspect 109 (Suppl 1):69–76.
15. Ford RI. (1978) The nature and status of ethnobotany. Anthropological Papers No. 67. Introduction. Ann Arbor, MI: University of Michigan, pp. 29–32.
16. Alexiades MN, Sheldon JW, editors. (1992) Selected Guidelines for ethnobotanical research: a field manual.
17. Cardellina JH, Fuller RW, Gamble WR, Westergaard C, Boswell J, Currens M, Boyd M. (1999) Evolving strategies for the selection, dereplication and prioritization of antitumor and HIV-inhibitory natural products. In: Bohlin L, Bruhn JG, editors. Bioassay methods in natural product research and drug development, proceeding of the Phytochemical Society of Europe. Dordrecht: Kluwer Academic, p. 201
18. Cardellina JHI, Munro MHG, Fuller RW, Manfredi KP, McKee TC, Tischler M, Bokesch HR, Gustafson KR, Beutler JA, Boyd MR. (1993) A chemical screening strategy for the dereplication and prioritization of HIV-inhibitory aqueous natural products. J Nat Prod 56:1123–1129.
19. Cox P. (1994) The ethnobotanical approach to drug discovery: strengths and limitation. In: Prance GT, Chadwick DJ, Marsh J, editors. Ethnobotany and the search for new drugs, Ciba Foundation Symp 185. New York: Wiley, pp. 25–36.
20. Cotton CM. (1996) Ethnobotany: principles and application. Chichester, UK: Wiley.
21. Martin G. (1995) Ethnobotany, a methods manual. London: Chapman & Hall, p. 268.
22. Cox PA, Balick MJ. (1994) The ethnobotanical approach to drug discovery. Sci Am 270:60–65.
23. Khafagi IK, Dewedar A. (2000) The efficiency of random versus ethno-directed research in the evaluation of Sinai medicinal plants for bioactive compounds. J Ethnopharmacol 71:365–376.
24. Schuster BG, Jackson JE, Obijiofor CN, Okunji CO, Milhous W, Losos E, Ayafor JF, Iwu MM. (1999) Drug development and conservation of biodiversity in West and Central Africa: a model for collaboration with indigenous people. Pharm Biol 37(Suppl):1–16.

Contributors

Rosita Arvigo – Ix Chel Tropical Research Foundation. Address: Ix Chel Tropical Research Foundation, General Delivery, Cayo District, San Ignacio, Belize, Central America.

Michael J Balick – Vice President and Chair of Research and Training Director and Philecology Curator of the New York Botanical Garden Institute of Economic Botany. Address: The New York Botanical Garden Institute of Economic Botany, Bronx, NY 10458–5126, USA.

David G. Campbell – Department of Biology at Grinnell College, PO Box 805, Grinnell, IA 60112–0806, USA.

Thomas JS Carlson – Associate Adjunct Professor, Department of Integrative Biology, University of California, Berkeley, and Director of the Health, Ecology, Biodiversity, and Ethnobiology (HEBE) Center at the Berkeley Natural History Museums, University of California. Address: 1001 Valley Life Sciences Building # 2465, University of California, Berkeley, CA 94720–2465, USA.

Richard A Cech – Herbalist, Horizon Herbs. Address: Horizon Herbs, PO Box 69, Williams, OR 97544, USA.

Gordon Cragg – Chief of the Natural Products Branch, Developmental Therapeutics Program Division of Cancer Treatment and Diagnosis, of the National Cancer Institute. Address: National Cancer Institute, Fairview Center, Room 206, PO Box B, Frederick, MD 21702–1201, USA.

Angela Duncan Diop – Chief Operating Officer of Axxon Biopharm Inc. Address Axxon Biopharm Inc. 11303 Amherst Avenue, Suite 2, Silver Spring, MD 20902, USA.

S Mbua Ngale Efange – Chemist in the Department of Radiology, Department of Medicinal Chemistry, Department of Neurosurgery and in the Graduate Program in Neuroscience, at the University of Minnesota. Address: University of Minnesota, Minneapolis, MN, 55455, USA.

Elaine Elisabetsky – Laboratório de Etnofarmacologia, ICBS, Universidade Federal do Rio Grande do Sul. Address: Universidade Federal do Rio Grande do Sul, CP 5072, 90041–970, Porto Alegre, RS, Brazil.

Andrew Galiwango – Coordinator of the Buganda Traditional Healers Association, Uganda. Address: PO Box 6353, Kampala, Uganda.

Nigel Gericke – African Natural Health in South Africa. Address: African Natural Health, PO Box 937, Sun Valley, Cape Town 7985, South Africa.

Michael A Gollin – Partner, VENABLE, Attorneys at Law. Address: VENABLE 1201 New York Avenue, N.W., Suite 1000, Washington, DC 20005–3917, USA.

Joerg Gruenwald – President at Phytopharm Consulting, Institute for Phytopharmaceuticals. Address: Waldseewed 6, 13467 Berlin, Germany.

Marianne Guerin-McManus – Director of the Conservation International Finance Program. Address: Conservation International, 1919 M St. NW, Suite 500, Washington, DC 20036, USA.

Alan Harvey – University of Strathclyde, Institute for Drug Research. Address: The University of Strathclyde, S. Institute for Drug Research, 27 Taylor St., Glasgow, G4 0NR, UK.

Maurice M Iwu – Executive Director of Bioresources Development and Conservation Programme, Nsukka Nigeria and Senior Research Associate at the Division of Experimental Therapeutics of Walter Reed Army Institute of Research. Address: 11303 Amherst Avenue, Suite 2, Silver Spring, MD 20902, USA.

Carey Jackson – Shaman Pharmaceuticals, Inc. Address: 213 E. Grand Avenue, South San Francisco, CA 94080–4812, USA.

BR Kanyerezi – Professor Emeritus, Makere University, Medical School, Kampala, Uganda.

Steven R King – Shaman Pharmaceuticals, Inc. Address: 213 E. Grand Avenue, South San Francisco, CA 94080–4812, USA.

Sirimani Kisingi Kato – Director of the Buganda Traditional Healers Association, Uganda. Address: PO Box 6353, Kampala, Uganda.

Sarah A Laird – Rainforest Alliance Address: Rainforest Alliance, 65 Bleeker Street, New York, NY 10011, USA.

Robert Lettington – Legal and Policy Consultant, International Congress of Insect Physiology and Ecology, (ICIPE). Address: ICIPE, PO Box 30772, Nyayo Stadium, Nairobi, Kenya.

Charles Limbach – Shaman Pharmaceuticals, Inc. Address: 213 E. Grand Avenue, South San Francisco, CA 94080–4812, USA.

Lisa Meserole – Associate Director of Bioresources Development and Conservation Program and a Physician for One Sky Medicine.

Katy Moran – Executive Director of the Healing Forest Conservancy. Address: Healing Forest Conservancy 3521 St. NW Washington, DC 20007, USA.

David J Newman – Chemist in the Natural Products Branch at the National Cancer Institute. Address: National Cancer Institute, Fairview Center, Room 206, PO Box B, Frederick, MD 21702–1201, USA.

Kent C Nnadozie – Bioresources Development and Conservation Programme. Address: 1 Tinuade Street, Off Allen Ave. Ikeja Lagos, Nigeria.

Jody Noé – Botanical Medicine Academy in Vermont and the Association of Naturopathic Physicians. Address: 138 Elliot, Brattleboro, VT 05301, USA.

Chioma Obijiofor – Bioresources Development and Conservation Program. Address: BDCP 11303 Amherst Avenue, Suite 2, Silver Spring, MD 20902, USA.

Joseph I Okogun – National Institute for Pharmaceutical Research and Development. Address: NIPRD, P. M. B. 21, Abuja, Nigeria.

Chris O Okunji – Bioresources Development and Conservation Programme, Nsukka Nigeria and Senior Research Associate at the Division of Experimental Therapeutics of Walter Reed Army Institute of Research, Washington, DC. Address: BDCP 11303 Amherst Avenue, Suite 2, Silver Spring, MD 20902, USA.

Anthony Onugu – Economist at Bioresources Development and Conservation Program. Address: BDCP 11303 Amherst Avenue, Suite 2, Silver Spring, MD 20902, USA.

Gabriel E Osuide – National Agency for Food and Drug Administration and Control. Address: NAFDAC, lot 1057, Ikeja Cresent, Off Oyo Road, Area 2 Section 1, Garki, Abuja, Nigeria.

Leopoldo Romero – Ix Chel Tropical Research Foundation. Address: Ix Chel Tropical Research Foundation, General Delivery, Cayo District, San Ignacio, Belize, Central America.

Brian G Schuster – Walter Reed Army Institute of Research. Address: WRAIR, 503 Robert Grant Rd, Silver Spring, MD 20910, USA.

Gregory Shropshire – Ix Chel Tropical Research Foundation. Address: Ix Chel Tropical Research Foundation, General Delivery, Cayo District, San Ignacio, Belize, Central America.

V Srini Srinivasan – Director of the Dietary Supplements Division of the US Pharmacopoeial Convention Inc. Address: US Pharmacopoeial Convention Inc., Dietary Supplements Division, 12601 Twinbrook Parkway, Rockville, MD 20852.

Susan Tarka Nelson-Harrison – Shaman Pharmaceuticals, Inc. Address: 213 E. Grand Avenue, South San Francisco, CA 94080–4812, USA.

Jay Walker – Department of Botany, University of Wisconsin-Madison. Address: Department of Botany, University of Wisconsin-Madison, 132 Birgge Hall, 430 Lincoln Drive, Madison, WI 53706–1381, USA.

Jackie Wootton – President of the Alternative Medicine Foundation, Executive Director of the Journal of Alternative and Complementary Medicine. Address: Alternative Medicine Foundation, Inc., 5411 West Cedar Lane, Suite 205A, Bethesda, MD 20814

Subject Index